Fields Institute Monographs

VOLUME 30

The Fields Institute is a centre for research in the mathematical sciences, located in Toronto, Canada. The Institutes mission is to advance global mathematical activity in the areas of research, education and innovation. The Fields Institute is supported by the Ontario Ministry of Training, Colleges and Universities, the Natural Sciences and Engineering Research Council of Canada, and seven Principal Sponsoring Universities in Ontario (Carleton, McMaster, Ottawa, Toronto, Waterloo, Western and York), as well as by a growing list of Affiliate Universities in Canada, the U.S. and Europe, and several commercial and industrial partners.

For further volumes:
http://www.springer.com/series/10502

Fields Institute Monographs

VOLUME 30

The Fields Institute for Research in the Mathematical Sciences

Jan Pachl

Uniform Spaces
and Measures

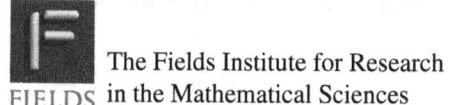

The Fields Institute for Research
in the Mathematical Sciences

Springer

Jan Pachl
The Fields Institute for Research
   in Mathematical Sciences
Toronto, ON, Canada

ISSN 1069-5273   ISSN 2194-3079 (electronic)
ISBN 978-1-4899-9258-1   ISBN 978-1-4614-5058-0 (eBook)
DOI 10.1007/978-1-4614-5058-0
Springer New York Heidelberg Dordrecht London

Printed on acid-free paper

Springer is part of Springer Science+Business Media (www.springer.com)

Preface

When measures are considered, by means of integration, as functionals on spaces of functions, certain continuity properties of measures and of sets of measures often play a prominent role. This is the case in topological measure theory (smoothness of measures in duality with continuous functions), abstract harmonic analysis (conditions for well-behaved convolution of measures on topological groups) and probability theory (uniform tightness and its relationship with weak convergence of probability distributions on complete metric spaces).

The main thesis of this treatise is that a number of continuity properties commonly imposed on measures are instances of a general property defined in the language of uniform spaces and that several fundamental results about measures remain valid in such a setting. The general property singles out a class of functionals called *uniform measures* on the space of bounded uniformly continuous functions on a uniform space.

Although a uniform measure is a functional on a function space, not a genuine measure (a countably additive set function on a σ-algebra), in some respects uniform measures behave like measures, and for some purposes they may be used as a substitute for measures. Moreover, in many problems, the underlying space carries a natural uniform structure, which leads to questions about functionals on spaces of uniformly continuous functions; uniform measures often feature in answers to such questions.

The basic theory of uniform measures was developed by a number of researchers in the 1960s and 1970s, but an interested reader would need considerable patience to track down scattered sources, some unpublished, written using a variety of definitions and differing notations. Recently the need for an accessible exposition became more apparent in view of new results in abstract harmonic analysis. In this monograph I offer a unified treatment of the theory of uniform measures, with a view towards such applications and others that I expect still to come.

This is primarily a reference for the theory of uniform measures and related functionals on spaces of uniformly continuous functions. It is also suitable for graduate or advanced undergraduate courses on selected topics in topology and functional analysis.

Part I is a self-contained development of the necessary results about uniform spaces. Part II is a systematic presentation of the basic theory of uniform measures, concluding with applications in the study of convolution algebras. Part III complements the basic theory with results in several related areas.

Uniform measures may be, and have been, defined in several equivalent ways. Although it adds to their usefulness, it also means that anyone attempting to cover all their major facets in a linear text is faced with a number of choices about starting points and the order of presentation. In selecting and organizing the material, my main objective has been to assemble a foundation for applying the theory in functional analysis, in a way that is likely to appeal to those interested in such applications. I have omitted or deferred some developments that would distract from the main objective. Part III includes several such developments that are intrinsically interesting and supply a broader context for the theory in Part II.

Despite a distance in time and space, this book owes much to the late Zdeněk Frolík and to the supportive environment for young mathematicians that he created in Prague in the 1970s. He made major contributions to the theory described here, and uniform measures were often discussed in his seminars. I obtained my first results about uniform measures with his advice and encouragement.

I am grateful to Henri Buchwalter, David Fremlin, Ramon van Handel and Miloš Zahradník for allowing me to use their unpublished ideas and materials. I thank Anthony Hager, Petr Holický, Matthias Neufang and Juris Steprāns for their comments and valuable suggestions, Carl Riehm for shepherding the text through the publication process, and my wife Cynthia for her patience and support.

The content incorporates ideas and techniques that I learnt from many mathematicians over the years. I appreciate their implicit contributions, even if I cannot name them all here.

The monograph was written while I was a visitor at the Fields Institute in Toronto. Contacts with colleagues at the institute and access to its facilities helped me a great deal in assembling the material that follows.

Toronto, Canada   Jan Pachl

Contents

Prerequisites

The theory developed in this book builds on a number of standard results in topology, functional analysis and measure theory. The prerequisites are summarized in this chapter, along with notation conventions. Definitions of common terms and most proofs are omitted. Detailed explanations and proofs may be found in the reference works cited in each section.

The traditional QED symbol □ marks the end of a proof, or of a lemma or theorem when the proof is omitted. The square dot ■ marks the end of a definition, example or exercise. The expression $A := B$ means that A is defined to be B. The expression "iff" stands for "if and only if" in definitions.

P.1 Sets and Mappings

Reference: Jech [102].

To have a concrete setting for the notions of cardinal and ordinal numbers, I assume the ZFC set theory (Zermelo–Fraenkel with Choice). However, any other mainstream set theory with the axiom of choice would do just as well. The definitions and results in this treatise use only elementary properties of sets and do not require understanding of the intricacies of axiomatic set theories.

The cardinality of a set S is $|S|$. The cardinal successor of a cardinal κ is κ^+. The smallest infinite cardinal is $\aleph_0$. Infinite cardinals $\aleph_\alpha$ are indexed by ordinals α. The cofinality of a cardinal κ is $\mathrm{cf}(\kappa)$.

The reader should be familiar with Zorn's Lemma and other consequences of the Hausdorff Maximal Principle, as described, for example, in [109, 0.25].

A *mapping* φ from a set A to a set B, written as $\varphi : A \to B$, always means a total mapping; that is, $\varphi(x)$ is defined for every $x \in A$. For a fixed set S and $A \subseteq S$, the *characteristic function of A in S* is $\mathrm{I}_A : S \to \{0, 1\}$. When A and B are sets, B^A denotes the set of all mappings from A to B.

J. Pachl, *Uniform Spaces and Measures*, Fields Institute Monographs 30, DOI 10.1007/978-1-4614-5058-0_1,

I use a simple version of the general lambda-calculus notation to mark domain restriction for multivariate mappings. If φ is a mapping of several variables $x, x', \ldots$, then $\backslash_x \varphi$ stands for the mapping $x \mapsto \varphi(x, x', \ldots)$. For example, if φ maps $X \times Y$ to Z, then the mappings $\backslash_x \varphi(x,y) \colon X \to Z$ and $\backslash_y \varphi(x,y) \colon Y \to Z$ are defined by $\backslash_x \varphi(x,y)(x_0) := \varphi(x_0, y)$, $\backslash_y \varphi(x,y)(y_0) := \varphi(x, y_0)$ for $x_0 \in X$, $y_0 \in Y$. Similarly for mappings from the product of three or more sets.

P.2 Topological Spaces and Groups

References: Császár [31], Engelking [47], Kelley [109].

I assume the reader knows the basics of general topology as covered in the references. Several results needed later are stated in this section. Many other concepts and results from the theory of metric and topological spaces are used throughout the treatise without explicit mention. I do not assume knowledge of uniform spaces. Necessary parts of uniform space theory are developed in Part I.

By definition, all topological and vector spaces are non-empty. On occasion, it will be convenient to allow non-Hausdorff topological spaces. However, all completely regular and all compact topological spaces are Hausdorff. When A is a subset of a topological space, $\overline{A}$ is the closure of A and $\mathrm{int}A$ is the interior of A.

In dealing with topology, I frequently use (directed) nets and Moore–Smith convergence [109, Ch. 2]. A *net* is a collection $\{x_\gamma\}_{\gamma \in \Gamma}$ of elements x_γ indexed by an upwards-directed partially ordered set Γ. Most of the time I omit Γ and write simply $\{x_\gamma\}_\gamma$. When $P(\gamma)$ is a property that depends on $\gamma \in \Gamma$, the statement "$P(\gamma)$ holds for almost all γ" means that there exists $\gamma_0 \in \Gamma$ such that $P(\gamma)$ holds for all $\gamma \geq \gamma_0$.

A *sequence* is a net $\{x_j\}_{j \in \omega}$ indexed by the totally ordered set $\omega := \{0, 1, 2, \ldots\}$.

A net $\{x_\gamma\}_{\gamma \in \Gamma}$ in a set S is *universal in S* iff for every set $A \subseteq S$ either $x_\gamma \in A$ for almost all γ or $x_\gamma \in S \setminus A$ for almost all γ.

Lemma P.1. *Let S be a non-empty set. Every net of elements of S has a subnet that is universal in S.*

Proof. [109, 2.J]. □

Let $\mathbb{N}$ denote the set ω with the discrete topology (and later on also with the discrete uniformity).

The ordered field of real numbers is denoted by $\mathbb{R}$, and $\mathbb{R}^+ := \{r \in \mathbb{R} \mid r \geq 0\}$. Besides the algebraic operations and the order that make it an ordered field, $\mathbb{R}$ carries also the usual metric (absolute value of the difference) and the topology defined by the metric. The symbol $\mathbb{R}$ stands for the set of real numbers or the set with one or more of these structures. The meaning will be clear from the context and will not cause any confusion.

When S is a set, $\mathbb{R}^S$ is the set of all real-valued functions on S. Unless stated otherwise, $\mathbb{R}^S$ is considered with the algebraic operations and partial order defined pointwise at each point of S. When $\mathscr{F} \subseteq \mathbb{R}^S$, the expression $\mathscr{F} \nearrow h$ means that $\mathscr{F}$

is upwards directed and $\sup\{f(x) \mid f \in \mathscr{F}\} = h(x)$ for every $x \in S$. The expression $\mathscr{F} \searrow h$ means that $\mathscr{F}$ is downwards directed and $\inf\{f(x) \mid f \in \mathscr{F}\} = h(x)$ for every $x \in S$. Similarly for $f_\gamma \nearrow h$ and $f_\gamma \searrow h$ when $\{f_\gamma\}_\gamma$ is a net of functions.

If $\mathfrak{S}$ is a set of subsets of S, $\{A_\gamma\}_\gamma$ is a net of subsets of S and $A \subseteq S$, then $\mathfrak{S} \nearrow A$, $\mathfrak{S} \searrow A$, $A_\gamma \nearrow A$ and $A_\gamma \searrow A$ stand for the same expressions with sets replaced by their characteristic functions. Thus for example, $\mathfrak{S} \nearrow A$ means that $\mathfrak{S}$ is upwards directed by inclusion and $\bigcup \mathfrak{S} = A$.

The space of all continuous real-valued functions on a topological space T is $\mathsf{C}(T)$, and $\mathsf{C}_b(T)$ is the space of uniformly bounded functions in $\mathsf{C}(T)$. When T is a completely regular topological space, βT is its Čech–Stone compactification and T is identified with a subspace of βT.

A set $\mathscr{W}$ of subsets of a topological space T is a *base of the topology* of T if every member of $\mathscr{W}$ is an open set and every non-empty open set in T is a union of some subset of $\mathscr{W}$. The *weight* of a topological space T is the least cardinal κ for which T has a base of cardinality κ.

Definition P.2. A *pseudometric* on a set S is a function $\Delta: S \times S \to \mathbb{R}^+$ such that

(M1) $\Delta(x,x) = 0$ for all $x \in S$.

(M2) $\Delta(x,y) = \Delta(y,x)$ for all $x,y \in S$.

(M3) $\Delta(x,y) + \Delta(y,z) \geq \Delta(x,z)$ for all $x,y,z \in S$.

A *metric* on S is a pseudometric such that

(M4) $\Delta(x,y) > 0$ for all $x,y \in S$ such that $x \neq y$.   ∎

Every pseudometric on a set S defines a topology on S. The topology is Hausdorff if and only if the pseudometric is a metric.

When Δ is a pseudometric on S, $x \in X$ and $A,B \subseteq S$, $A \neq \emptyset \neq B$, define

$$\Delta(A,B) := \inf\{\Delta(a,b) \mid a \in A, b \in B\}$$

$$\Delta(x,A) := \Delta(A,x) := \Delta(\{x\},A).$$

The Δ-*diameter* of a non-empty set $A \subseteq S$ is a number in $\mathbb{R}^+$ or ∞, defined by

$$\Delta\text{-diam}(A) := \sup\{\Delta(a,b) \mid a,b \in A\}.$$

Every pseudometric Δ on a non-empty set S defines the *associated metric space* S/Δ with metric $\Delta^\bullet$ and the canonical surjection $\chi_\Delta: S \to S/\Delta$, as follows. The equivalence relation $\overset{\bullet}{\sim}$ on S is defined by $x \overset{\bullet}{\sim} y$ iff $\Delta(x,y) = 0$, for $x,y \in S$. The points of the metric space S/Δ are the equivalence classes of $\overset{\bullet}{\sim}$, and the metric $\Delta^\bullet$ is given by $\Delta^\bullet(x^\bullet, y^\bullet) := \Delta(x,y)$ where $x^\bullet, y^\bullet \in S/\Delta$, $x,y \in S$, and $x^\bullet$ and $y^\bullet$ are the equivalence classes of x and y, respectively. The mapping χ_Δ sends every $x \in S$ to its equivalence class.

A subset A of a Hausdorff topological space T is

- *relatively compact in T* iff the closure $\overline{A}$ of A in T is compact;

- *relatively sequentially compact in* T iff every sequence in A has a subsequence that converges in T;
- *relatively countably compact in* T iff every sequence in A has a subnet that converges in T.

A Hausdorff space T is *sequentially compact* iff it is relatively sequentially compact in itself, and T is *countably compact* iff it is relatively countably compact in itself.

A Hausdorff topological space T is *paracompact* iff every open cover of T has an open locally finite refinement. Equivalently, iff every open cover has an open σ-discrete refinement [109, 5.28]. Every metrizable topological space is paracompact.

Let S be a non-empty set, $\mathscr{W}$ a set of subsets of S and $\mathscr{F} \subseteq \mathbb{R}^S$. Then $\sigma(\mathscr{W})$ is the smallest σ-algebra Σ on S such that $\Sigma \supseteq \mathscr{W}$, and $\sigma(\mathscr{F})$ is the smallest σ-algebra on S for which all functions in $\mathscr{F}$ are measurable.

When T is a topological space, the *Borel σ-algebra* $\mathsf{Bo}(T)$ on T is the smallest σ-algebra of subsets of T that includes all open sets. The sets in $\mathsf{Bo}(T)$ are called *Borel sets*. A mapping φ from T to a topological space T' is *Borel measurable* iff $\varphi^{-1}(V) \in \mathsf{Bo}(T)$ for every open set $V \subseteq T'$. Clearly $\sigma(\mathsf{C}(T)) \subseteq \mathsf{Bo}(T)$ for every topological space T. If T is metrizable, then $\sigma(\mathsf{C}(T)) = \mathsf{Bo}(T)$.

For topological groups, I follow the standard terminology as defined, for example, by Császár [31]. All topological groups are assumed to be Hausdorff.

The binary operation in groups, and more generally in semigroups, is written as $(x,y) \mapsto xy$ or $(x,y) \mapsto x \cdot y$. When G is a group, its identity element is e_G, the inverse of $x \in G$ is x^{-1}, and if $A \subseteq G$, then $A^{-1} := \{x^{-1} \mid x \in A\}$.

A pseudometric Δ on a semigroup S is *right-invariant* if $\Delta(x,y) = \Delta(xz,yz)$ for all $x,y,z \in S$.

Theorem P.3. *Let G be a topological group. For every neighbourhood V of e_G there is a right-invariant pseudometric Δ on G such that the function $x \mapsto \Delta(e_G,x)$ is continuous on G and $\Delta(e_G,x) \geq 1$ for every $x \in G \setminus V$.*

Proof. The statement follows from [97, 8.2]. □

Category theory is not a prerequisite for the material presented in this treatise. However, the reader acquainted with categorical notions will recognize many well-behaved categories and functors. In fact, C, $\mathsf{C_b}$ and β as well as other operations defined further on are functors on familiar categories and as such act not only on objects but also on morphisms.

Moreover, a number of "forgetful" functors appear throughout, such as the obvious functor from topological groups to topological spaces or from topological spaces to sets. In most cases, when no confusion results, I use the same symbol for an object and for the same object with some structure "forgotten". For example, if G is a topological group, then G denotes also the induced topological space in expressions such as $\mathsf{C_b}(G)$ and βG. Similarly, when X is a topological space or a topological group, the symbol X denotes also the set of points of X in expressions such as $x \in X$ and $Y \subseteq X$. This convention causes no confusion, as long as we are careful with expressions that include equality; for example, if G and G' are two topological groups, then $G = G'$ means that G and G' are equal as topological groups, not merely as sets or groups or topological spaces.

P.3   Topological Vector Spaces

References: Fabian et al. [48], Schaefer [164].

All vector spaces considered in this treatise are over the field $\mathbb{R}$ of real numbers, and accordingly a *linear functional* on a vector space E is a linear mapping from E to $\mathbb{R}$. It is a simple exercise to extend the results proved here to vector spaces over the field of complex numbers.

All locally convex topological vector spaces are assumed to be Hausdorff. When E is a locally convex space, E^* is its topological dual.

When α is a seminorm on a vector space E, the *pseudometric of α* is the function $(x,y) \mapsto \alpha(x-y)$, $x,y \in E$. The topology of any locally convex space is defined by a set of seminorms [164, II.4] or, equivalently, by the corresponding set of pseudometrics. The pseudometric of α is a metric if and only if α is a norm.

When f is a real-valued function on a set S, define

$$\|f\|_A := \sup \Big\{ |f(x)| \mid x \in A \Big\}$$

for $\emptyset \neq A \subseteq S$ and $\|f\|_\emptyset := 0$. Thus $\|f\|_A$ is a non-negative real number or ∞.

Lemma P.4. *Let E be a vector space and let $\mathscr{F}$ be a finite set of linear functionals on E. If g is a linear functional on E such that $\bigcap \{ f^{-1}(0) \mid f \in \mathscr{F} \} \subseteq g^{-1}(0)$, then g is a linear combination of the functionals in $\mathscr{F}$.*

Proof. [164, IV.1.1]. □

Theorem P.5. *Let A be closed convex non-empty subset of a locally convex space E. For every $x_0 \in E \setminus A$ there exists $f \in E^*$ such that $\sup_{x \in A} f(x) < f(x_0)$.*

Proof. [164, II.9.2]. □

Corollary P.6. *Let E be a locally convex space. A vector subspace E_0 of E is dense in E if and only if every functional in E^* that is identically 0 on E_0 is identically 0 on E.* □

Let $\mathfrak{S}$ be a set of subsets of a non-empty set S that is upwards directed by inclusion $\subseteq$. For any $\mathscr{F} \subseteq \mathbb{R}^S$, the sets $\{ f \in \mathscr{F} \mid \|f - f_0\|_A < \varepsilon \}$, where $\varepsilon > 0$, $f_0 \in \mathscr{F}$, and $A \in \mathfrak{S}$, form a base for a topology on $\mathscr{F}$, called the $\mathfrak{S}$-*topology* (or the *topology of uniform convergence on the sets in* $\mathfrak{S}$). The next theorem gives a sufficient condition for the $\mathfrak{S}$-topology to be locally convex.

Theorem P.7. *Let T be a topological space and let $\mathfrak{S}$ be a set of subsets of T that is upwards directed and whose union is dense in T. If F is a vector subspace of $\mathbb{R}^T$ whose elements are continuous on T and bounded on each set in $\mathfrak{S}$, then F with the $\mathfrak{S}$-topology is a locally convex space.*

Proof. [164, III.3.2]. $\square$

Several instances of the $\mathfrak{S}$-topology have their own names. When $\mathfrak{S}$ is the set of all finite subsets of S, the $\mathfrak{S}$-topology is called the *S-pointwise topology*. When E and F are two vector spaces in duality, the *F-weak topology* on E is the F-pointwise topology obtained by identifying E with a subspace of $\mathbb{R}^F$. When T is a topological space and $\mathfrak{S}$ is the set of all compact subsets of T, the $\mathfrak{S}$-topology is called the *compact–open topology*.

When E and F are two vector spaces in duality and E is given the F-weak topology, the dual of E is F [164, IV.1.2]. In fact, the F-weak topology is the coarsest locally convex topology on E for which its dual is F. It follows from the next theorem that there exists also the finest locally convex topology on E with dual F.

Theorem P.8 (Mackey–Arens). *Let E and F be two vector spaces in duality, and identify each element of F with the linear functional it defines on E. Let $\mathfrak{S}$ be a set of absolutely convex E-weakly compact subsets of F such that $\mathfrak{S} \nearrow F$ and $rA \in \mathfrak{S}$ for any $A \in \mathfrak{S}$, $r \in \mathbb{R}^+$. When E is endowed with the $\mathfrak{S}$-topology, the dual of E is F.*

Proof. [164, IV.3.2]. $\square$

A net $\{x_\gamma\}_\gamma$ in a locally convex space E is called *Cauchy* iff for every neighbourhood V of 0 in E there exists $y \in E$ such that $x_\gamma \in V + y$ for almost all γ. The space E is *complete* if every Cauchy net in E converges in E. This definition agrees with the more general notion of a complete uniform space, defined in Sect. 1.1. The space E is *sequentially complete* if every Cauchy sequence in E converges in E.

Theorem P.9 (Hahn–Banach). *Let E be a vector space and E_0 a vector subspace of E. Let α be a seminorm on E and f_0 a linear functional on E_0, and assume that $|f_0(x)| \le \alpha(x)$ for all $x \in E_0$. Then there exists a linear functional f on E that extends f_0 and such that $|f(x)| \le \alpha(x)$ for all $x \in E$.*

Proof. [164, II.3.2]. $\square$

Corollary P.10. *If α is a seminorm on a vector space E, then*

$$\alpha(x) = \sup\{f(x) \mid f \text{ is a linear functional on } E \text{ and } |f(y)| \le \alpha(y) \text{ for all } y \in E\}$$

for every $x \in E$. $\square$

A subset B of a locally convex space E is *bounded* iff for every neighbourhood V of 0 there exists $r \in \mathbb{R}$ such that $B \subseteq rV$. In the next theorem, the condition $\sup_{b \in B}|b(x)| < \infty$ for every $x \in E$ means that B is bounded in the E-weak topology.

Theorem P.11 (Banach–Steinhaus). *Let E be a Banach space with norm $\|\cdot\|$. If $B \subseteq E^*$ and $\sup_{b \in B} |b(x)| < \infty$ for every $x \in E$, then $\sup_{b \in B} \|b\| < \infty$.*

Proof. [48, 3.15]. □

Theorem P.12. *Let E be a complete locally convex space and f a linear functional on E^*. If the restriction of f to every equicontinuous subset of E^* is E-weakly continuous, then f is E-weakly continuous on E^*.*

Proof. This follows from Corollary 2 in [164, IV.6.2]. □

Theorem P.13. *Let E be a locally convex space, and let B be a bounded subset of E whose closed convex hull is complete. These properties of B are equivalent:*

(i) B is relatively E^*-weakly compact in E.
(ii) For every equicontinuous sequence $\{f_i\}_i$ in E^* and for every sequence $\{x_j\}_j$ in B, if the two double limits $\lim_i \lim_j f_i(x_j)$ and $\lim_j \lim_i f_i(x_j)$ exist, then they are equal.

Proof. [164, IV.11.2]. □

Theorem P.14 (Ascoli). *Let T be a compact space, and let $\mathscr{F}$ be a $\|\cdot\|_T$ bounded subset of $C_b(T)$. The set $\mathscr{F}$ is relatively compact in $C_b(T)$ with the $\|\cdot\|_T$ topology if and only if $\mathscr{F}$ is equicontinuous on T.*

Proof. [100, III.37]; [109, 7.17]. □

For any non-empty set S, the space $\ell_\infty(S)$ of all bounded real-valued functions on S with the vector operations defined pointwise and the sup norm $\|\cdot\|_S$ is a Banach space. The space $\ell_1(S)$ of the functions $d \in \mathbb{R}^S$ for which $\|d\|_1 := \sum_{x \in S} |d(x)|$ is finite, with the vector operations defined pointwise and the $\|\cdot\|_1$ norm, is also a Banach space, and $\ell_1(S)^* = \ell_\infty(S)$. Clearly every function in $\ell_1(S)$ vanishes outside of a countable subset of S. Write $\ell_1 := \ell_1(\omega)$ and $\ell_\infty := \ell_\infty(\omega)$. Since ℓ_∞ is the dual of ℓ_1, the ℓ_∞-weak topology is simply the weak topology of the Banach space ℓ_1.

Theorem P.15. *1. The space ℓ_1 is ℓ_∞-weakly sequentially complete.*
2. Every ℓ_∞-weakly convergent sequence in ℓ_1 converges in the $\|\cdot\|_1$ norm.
3. Every relatively ℓ_∞-weakly countably compact set in ℓ_1 is relatively $\|\cdot\|_1$ norm compact in ℓ_1.
4. The $\|\cdot\|_1$ unit ball in ℓ_1 is not metrizable in the ℓ_∞-weak topology.

Proof. Parts 1 and 2 are in [48, 5.19]. Part 3 follows from Part 2 and [48, 4.47]. Part 4 follows from [48, 3.28]. □

When Σ is a σ-algebra of subsets of a non-empty set S, $\ell_\infty(S, \Sigma)$ denotes the space of bounded real-valued Σ-measurable functions on S.

Definition P.16. Let Δ be a pseudometric on a non-empty set S and $f \colon S \to \mathbb{R}$. The function f is *1-Lipschitz for* Δ, or Δ-*1-Lipschitz*, iff $|f(x) - f(y)| \leq \Delta(x, y)$ for all $x, y \in S$. For any function $h \colon S \to \mathbb{R}^+$, define

$$\mathsf{BLip}(\Delta, h) := \{f \in \mathbb{R}^S \mid |f| \leq h \text{ and } f \text{ is } 1\text{-Lipschitz for } \Delta\},$$

$$\mathsf{Lip}(\Delta, h) := \bigcup_{j \in \omega} j\,\mathsf{BLip}(\Delta, h).$$

Write $\mathsf{BLip_b}(\Delta) := \mathsf{BLip}(\Delta, 1)$ and $\mathsf{Lip_b}(\Delta) := \mathsf{Lip}(\Delta, 1)$. ∎

The set $\mathsf{BLip}(\Delta, h)$ is compact in the S-pointwise topology. Every $\mathsf{Lip}(\Delta, h)$ is a Banach space with the norm that assigns

$$\inf\Big\{r \in \mathbb{R}^+ \mid |f| \leq rh \text{ and } |f(x) - f(y)| \leq r\Delta(x,y) \text{ for all } x, y \in S\Big\}$$

to $f \in \mathsf{Lip}(\Delta, h)$, and $\mathsf{BLip}(\Delta, h)$ is its closed unit ball. Weaver [178] surveys the theory of such *Lipschitz spaces*.

Lemma P.17. *Let Δ be a pseudometric on a non-empty set S, and let $h: S \to \mathbb{R}^+$ be a Δ-1-Lipschitz function. Let $\chi_\Delta: S \to S/\Delta$ be the canonical surjection onto the associated metric space S/Δ with metric $\Delta^\bullet$. Then there is a unique $g: S/\Delta \to \mathbb{R}^+$ such that $h = g \circ \chi_\Delta$. The function g is $\Delta^\bullet$-1-Lipschitz and*

$$\mathsf{BLip}(\Delta, h) = \{f \circ \chi_\Delta \mid f \in \mathsf{BLip}(\Delta^\bullet, g)\},$$

$$\mathsf{Lip}(\Delta, h) = \{f \circ \chi_\Delta \mid f \in \mathsf{Lip}(\Delta^\bullet, g)\},$$

$$\mathsf{BLip_b}(\Delta) = \{f \circ \chi_\Delta \mid f \in \mathsf{BLip_b}(\Delta^\bullet)\},$$

$$\mathsf{Lip_b}(\Delta) = \{f \circ \chi_\Delta \mid f \in \mathsf{Lip_b}(\Delta^\bullet)\}.$$

Proof. If $x, y \in S$, $x \overset{\bullet}{\sim} y$, then $|h(x) - h(y)| \leq \Delta(x,y) = 0$; hence there is a function $g: S/\Delta \to \mathbb{R}^+$ such that $h = g \circ \chi_\Delta$. The function g is unique because χ_Δ is surjective, and it is $\Delta^\bullet$-1-Lipschitz because

$$|g(x^\bullet) - g(y^\bullet)| = |h(x) - h(y)| \leq \Delta(x,y) = \Delta^\bullet(x^\bullet, y^\bullet).$$

The same argument proves the second statement in the lemma. □

P.4 Riesz Spaces

Reference: Fremlin [59, Ch.35].

When F is a partially ordered vector space, $H^+ := \{h \in H \mid h \geq 0\}$ for any $H \subseteq F$. The *positive cone* of F is F^+.

Definition P.18. Let F with the partial order $\leq$ be a Riesz space (=vector lattice).

1. For $f \in F$, write $f^+ := f \vee 0$, $f^- := (-f) \vee 0$ and $|f| := f \vee -f$.

2. A vector subspace E of F is *solid in F* iff $g \in E$ whenever $g \in F$ and $|g| \leq |f|$ for some $f \in E$.
3. A vector subspace E of F is a *band in F* iff E is solid in F and $\sup H \in E$ whenever $H \subseteq E$ is a non-empty upwards directed set such that $\sup H$ exists in F.
4. The *dual partial order* $\leq$ on the space of real-valued linear functionals on F is defined by

$$\mathfrak{m} \leq \mathfrak{n} \text{ iff } \mathfrak{m}(f) \leq \mathfrak{n}(f) \text{ for every } f \in F^+,$$

   for functionals $\mathfrak{m}$ and $\mathfrak{n}$.
5. Let $F^\sim$ denote the *order-bounded dual* of F with the dual partial order.   ∎

Lemma P.19. *If F is a Riesz space, then $F^\sim$ is a Riesz space. If $\mathfrak{m}, \mathfrak{n} \in F^\sim$ and $f \in F^+$, then*

$$\mathfrak{m}^+(f) = \sup\{\mathfrak{m}(g) \mid g \in F \text{ and } 0 \leq g \leq f\},$$

$$|\mathfrak{m}|(f) = \sup\{\mathfrak{m}(g) \mid g \in F \text{ and } |g| \leq f\},$$

$$\mathfrak{m} \vee \mathfrak{n} = \mathfrak{m} + (\mathfrak{n} - \mathfrak{m})^+.$$

Proof. The first two identities are in 356B of [59]. The last identity holds in every Riesz space [59, 352D].   □

Lemma P.20. *Let F be a Riesz space, $h \in F^+$, $\mathfrak{m} \in F^\sim$ and $\varepsilon > 0$. Then there exists $g \in F^+$ such that $\mathfrak{m}^+(f) < \mathfrak{m}(g \wedge f) + \varepsilon$ whenever $f \in F^+$, $f \leq h$.*

Proof. By Lemma P.19, there is $g \in F^+$, $g \leq h$, such that $\mathfrak{m}^+(h) < \mathfrak{m}(g) + \varepsilon$. Take any $f \in F^+$ such that $f \leq h$. Then $0 \leq g - g \wedge f \leq h - f$; therefore

$$\mathfrak{m}(g) - \mathfrak{m}(g \wedge f) = \mathfrak{m}(g - g \wedge f) \leq \mathfrak{m}^+(g - g \wedge f) \leq \mathfrak{m}^+(h) - \mathfrak{m}^+(f)$$

and $\mathfrak{m}^+(f) \leq \mathfrak{m}(g \wedge f) + \mathfrak{m}^+(h) - \mathfrak{m}(g) < \mathfrak{m}(g \wedge f) + \varepsilon$.   □

Let S be a non-empty set. The space $\mathbb{R}^S$ with the vector operations and partial order defined pointwise at every $x \in S$ is a Riesz space. The space $\ell_\infty(S)$ is a Riesz subspace of $\mathbb{R}^S$; moreover, $\ell_\infty(S)$ is a Banach lattice with the norm $\|\cdot\|_S$.

The space $\ell_1(S)$ with the vector operations and partial order defined pointwise at every $x \in S$ is also a Riesz space, and it is a Banach lattice with the norm $\|\cdot\|_1$.

Theorem P.21. *Let S be a non-empty set and let F be a Riesz subspace of the Riesz space $\ell_\infty(S)$ containing constant functions.*

1. *The order-bounded dual $F^\sim$ of F is also its normed-space dual F^*. With the dual norm $\|\cdot\|$, $F^\sim$ is a Banach lattice.*
2. $\|\mathfrak{m}\| = |\mathfrak{m}|(1)$ *for every $\mathfrak{m} \in F^\sim$.*
3. *Any $\|\cdot\|$-closed subspace of $F^\sim$ that is solid in $F^\sim$ is a band in $F^\sim$.*

Proof. Parts 1 and 2 follow from 356N(a,b) in [59]. Part 3 follows from 356N(a), 354N and 354E(a) in [59].   □

Lemma P.22. *Let S be a non-empty set and let F be a Riesz subspace of the Riesz space $\ell_\infty(S)$ containing constant functions. If $\mathfrak{m} \in F^\sim$ and $\{\mathfrak{m}_\gamma\}_\gamma$ is a net in $F^\sim$ such that $\lim_\gamma \mathfrak{m}_\gamma(f) = \mathfrak{m}(f)$ for every $f \in F$ and $\lim_\gamma \|\mathfrak{m}_\gamma\| \leq \|\mathfrak{m}\|$, then*

$$\lim_\gamma \mathfrak{m}_\gamma^+(f) = \mathfrak{m}^+(f)$$

$$\lim_\gamma \mathfrak{m}_\gamma^-(f) = \mathfrak{m}^-(f)$$

for every $f \in F$.

Proof. The proof is adapted from [19, 3.3]. Since the closed $\|\cdot\|$ balls in $F^\sim$ are F-weakly compact, it is enough to prove the conclusion when the nets $\{\mathfrak{m}_\gamma^+\}_\gamma$ and $\{\mathfrak{m}_\gamma^-\}_\gamma$ are F-weakly convergent. Let $\mathfrak{n}, \mathfrak{n}' \in F^\sim$ be such that $\lim_\gamma \mathfrak{m}_\gamma^+(f) = \mathfrak{n}(f)$ and $\lim_\gamma \mathfrak{m}_\gamma^-(f) = \mathfrak{n}'(f)$ for every $f \in F$.

Clearly $\mathfrak{m} = \mathfrak{m}^+ - \mathfrak{m}^- = \mathfrak{n} - \mathfrak{n}'$. By Lemma P.19, if $f \in F^+$, then

$$\mathfrak{m}^+(f) = \sup\left\{\lim_\gamma \mathfrak{m}_\gamma(g) \mid g \in F \text{ and } 0 \leq g \leq f\right\} \leq \lim_\gamma \mathfrak{m}_\gamma^+(f) = \mathfrak{n}(f)$$

which means that also $\mathfrak{m}^-(f) \leq \mathfrak{n}'(f)$. Then

$$\mathfrak{n}(1) + \mathfrak{n}'(1) = \lim_\gamma |\mathfrak{m}_\gamma|(1) = \lim_\gamma \|\mathfrak{m}_\gamma\| \leq \|\mathfrak{m}\| = \mathfrak{m}^+(1) + \mathfrak{m}^-(1),$$

hence $\mathfrak{m}^+(1) = \mathfrak{n}(1)$ and $\mathfrak{m}^-(1) = \mathfrak{n}'(1)$.

Take any $f \in F$ such that $0 \leq f \leq 1$. Then

$$\mathfrak{n}(f) = \mathfrak{n}(1) - \mathfrak{n}(1-f) \leq \mathfrak{m}^+(1) - \mathfrak{m}^+(1-f) = \mathfrak{m}^+(f) \leq \mathfrak{n}(f),$$

so that $\mathfrak{n}(f) = \mathfrak{m}^+(f)$ for all such f and hence for all $f \in F$.   □

P.5   Measures

References: Fremlin [60].

The basic reference for the results in this section is Volume 4 of Fremlin's Measure Theory [60], but I use a slightly modified notation and terminology. The main difference is that I use the term *measure* for Fremlin's *countably additive functional* on a σ-algebra (also known as a *finite signed measure*).

A *measure* is a σ-additive real-valued function on a σ-algebra Σ. For any measure $\mu \colon \Sigma \to \mathbb{R}$, its *positive part* μ^+, *negative part* μ^- and *total variation* $|\mu|$ are defined by

$$\mu^+(A) := \sup\{\mu(B) \mid B \in \Sigma \text{ and } B \subseteq A\}$$

$$\mu^-(A) := \mu^+(A) - \mu(A)$$

$$|\mu|(A) := \mu^+(A) + \mu^-(A)$$

for $A \in \Sigma$. The functions μ^+, μ^- and $|\mu|$ are non-negative measures on Σ. The space of measures on a σ-algebra of subsets of S is a Banach space with the norm $\|\mu\|_{\text{TV}} := |\mu|(S)$.

The next theorem is used in only one place further on, in an exercise solution.

Theorem P.23 (Nikodým). *If M is a set of measures on a σ-algebra Σ and $\sup\limits_{\mu \in M} \mu(E) < \infty$ for every $E \in \Sigma$, then M is $\|\cdot\|_{\text{TV}}$ bounded.*

Proof. [45, IV.9.8]. □

When μ is a measure on a σ-algebra Σ of subsets of S and $f \colon S \to \mathbb{R}$ is a Σ-measurable function, say that f is *μ-integrable* iff it is $|\mu|$-integrable in the sense of [57, 122M], and write $\int f \, d\mu := \int f \, d\mu^+ - \int f \, d\mu^-$. If f is μ-integrable, then $\int f \, d\mu = \lim_{j \in \omega} \int (-j) \vee f \wedge j \, d\mu$. For $\mu \geq 0$ and $f \geq 0$, we get that f is μ-integrable if and only if the finite $\lim_{j \in \omega} \int f \wedge j \, d\mu$ exists, and then $\int f \, d\mu = \lim_{j \in \omega} \int f \wedge j \, d\mu$.

Let μ be a measure on a σ-algebra Σ of subsets of S, Σ' a σ-algebra of subsets of S', and $\varphi \colon S \to S'$ a measurable mapping. The *image of μ under φ*, denoted by $\varphi(\mu)$, is the measure on Σ' defined by $\varphi(\mu)(A) := \mu(\varphi^{-1}(A))$, $A \in \Sigma'$. A function $f \colon S' \to \mathbb{R}$ is $\varphi(\mu)$-integrable if and only if $f \circ \varphi$ is μ-integrable, and in that case $\int f \circ \varphi \, d\mu = \int f \, d\varphi(\mu)$.

Theorem P.24 (Daniell representation). *Let S be a non-empty set and F a Riesz subspace of $\ell_\infty(S)$ that contains constant functions. The following two properties of a linear functional $\mathfrak{m} \colon F \to \mathbb{R}$ are equivalent:*

(i) There is a measure $\mu \colon \sigma(F) \to \mathbb{R}$ such that $\mathfrak{m}(f) = \int f \, d\mu$ for every $f \in F$.
(ii) If $\{f_j\}_j$ is a sequence in F such that $f_j \searrow 0$, then $\lim_j \mathfrak{m}(f_j) = 0$.

If (i) holds, then there is a unique such measure μ, and moreover, $\mathfrak{m}^+(f) = \int f \, d\mu^+$, $\mathfrak{m}^-(f) = \int f \, d\mu^-$ and $|\mathfrak{m}|(f) = \int f \, d|\mu|$ for $f \in F$, and $\|\mathfrak{m}\| = \|\mu\|_{\text{TV}}$.

Proof. Combine 437C and 437X(a) in [60]. □

Theorem P.25. *Let $\mathscr{W}$ be a non-empty set of subsets of a non-empty set S such that:*

(a) If $A_0, A_1 \in \mathscr{W}$, then $A_0 \cup A_1 \in \mathscr{W}$.
(b) If $A_j \in \mathscr{W}$ for $j \in \omega$, then $\bigcap_j A_j \in \mathscr{W}$.
(c) If $A \in \mathscr{W}$, then $S \setminus A$ is a countable union of sets in $\mathscr{W}$.

Then

$$|\mu|(A) = \sup\left\{ |\mu|(B) \mid B \subseteq A \text{ and } B \in \mathscr{W} \right\}$$

for every measure $\mu \colon \sigma(\mathscr{W}) \to \mathbb{R}$ and every $A \in \sigma(\mathscr{W})$.

Proof. Apply [60, 412C] with $\mathscr{K} = \mathscr{W}$ and $\mathscr{A} = \mathscr{W} \cup \{S \setminus A \mid A \in \mathscr{W}\}$. (See also the proof of [60, 412D].)   □

Let T be a topological space. A τ-*additive Borel measure* on T is a measure $\mu: \mathrm{Bo}(T) \to \mathbb{R}$ with the following property: If $\mathscr{W}$ is a non-empty upwards-directed set of open sets in T, then $\sup_{V \in \mathscr{W}} |\mu|(V) = |\mu|(\bigcup \mathscr{W})$.

Clearly a measure $\mu: \mathrm{Bo}(T) \to \mathbb{R}$ is τ-additive if and only if μ^+ and μ^- are.

Theorem P.26. *Let T be a completely regular space and μ a τ-additive Borel measure on T, $\mu \geq 0$. Let g be a non-negative real-valued function on T, and $\{f_\gamma\}_\gamma$ a net in $C_b(T)$ such that $f_\gamma \searrow g$. Then g is Borel measurable and*

$$\lim_\gamma \int f_\gamma \, d\mu = \int g \, d\mu.$$

Proof. [60, 414B].   □

A closed subset Z of T is a *support* of a measure $\mu: \mathrm{Bo}(T) \to \mathbb{R}$ iff $|\mu|(T \setminus Z) = 0$ and $|\mu|(V \cap Z) > 0$ for every open subset V of T such that $V \cap Z \neq \emptyset$. If a measure has a support, then it is unique.

Theorem P.27. *Let T be a completely regular space.*

1. *Every τ-additive Borel measure on T has a support.*
2. *If a metrizable set $Z \subseteq T$ is the support of a measure on $\mathrm{Bo}(T)$, then Z is separable.*

Proof. Part 1 is proved in [60, 411N]. To prove part 2, it is enough to observe that, since all measures considered here are finite, the support of any measure satisfies the countable chain condition.   □

Theorem P.28. *Let T be a completely regular topological space and F a Riesz subspace of $\ell_\infty(T)$ that contains constant functions. Assume that the topology of T is the coarsest topology for which every $f \in F$ is continuous. The following two properties of a linear functional $\mathrm{m}: F \to \mathbb{R}$ are equivalent:*

(i) *There is a τ-additive Borel measure μ on T such that $\mathrm{m}(f) = \int f \, d\mu$ for every $f \in F$.*

(ii) *If $\{f_\gamma\}_\gamma$ is a net of functions $f_\gamma \in F$ such that $f_\gamma \searrow 0$, then $\lim_\gamma \mathrm{m}(f_\gamma) = 0$.*

If (i) holds, then there is a unique such τ-additive Borel measure μ, and moreover, $\mathrm{m}^+(f) = \int f \, d\mu^+$, $\mathrm{m}^-(f) = \int f \, d\mu^-$ and $|\mathrm{m}|(f) = \int f \, d|\mu|$ for every $f \in F$, and $\|\mathrm{m}\| = \|\mu\|_{\mathrm{TV}}$.

Proof. Combine 437H and 437X(a) in [60].   □

A *tight Borel measure* on a topological space T is a measure $\mu \colon \mathrm{Bo}(T) \to \mathbb{R}$ such that

$$|\mu|(A) = \sup\Big\{|\mu|(K) \mid K \subseteq A \text{ and } K \text{ is closed and compact}\Big\}$$

for every $A \in \mathrm{Bo}(T)$.

Tight Borel measures are also known as regular, compact regular, or Radon measures. Clearly a measure $\mu \colon \mathrm{Bo}(T) \to \mathbb{R}$ is tight if and only if μ^+ and μ^- are, and every tight measure is τ-additive.

A set M of tight Borel measures on T is said to be *uniformly tight* iff the set $\{|\mu|(T) \mid \mu \in M\} \subseteq \mathbb{R}$ is bounded and for every $\varepsilon > 0$ there exists a closed compact set $K \subseteq T$ such that $|\mu|(T \setminus K) < \varepsilon$ for all $\mu \in M$.

Theorem P.29. *Let T be a completely regular topological space and F a Riesz subspace of $\ell_\infty(T)$ that contains constant functions. Assume that the topology of T is the coarsest topology for which every $f \in F$ is continuous. The following two properties of a set $\mathfrak{A}$ of linear functionals on F are equivalent:*

(i) *There is a uniformly tight set M of tight Borel measures on T such that $\mathfrak{A}$ is the set of functionals of the form $f \in F \mapsto \int f \, d\mu$, where $\mu \in M$.*

(ii) *The set of restrictions of the functionals in $\mathfrak{A}$ to the $\|\cdot\|_T$-unit ball in F is equicontinuous in the compact–open topology.*

Proof. To prove (i)$\Rightarrow$(ii), take any $\varepsilon > 0$. There is a compact set $K \subseteq T$ such that $|\mu|(T \setminus K) < \varepsilon$ for every $\mu \in M$. By (i), every $\mathrm{m} \in \mathfrak{A}$ is represented by some $\mu \in M$, and if $f, g \in F$ are such that $\|f\|_T \le 1$, $\|g\|_T \le 1$ and $\|f - g\|_K < \varepsilon$, then

$$|\mathrm{m}(f) - \mathrm{m}(g)| = \left| \int f \, d\mu - \int g \, d\mu \right|$$

$$\le \int_K |f - g| \, d\mu + \int_{T \setminus K} |f - g| \, d\mu < \varepsilon |\mu|(T) + 2\varepsilon.$$

Since the values $|\mu|(T)$ are bounded for $\mu \in M$, this proves (ii).

To prove the converse, assume (ii). By Theorem P.28, for every $\mathrm{m} \in \mathfrak{A}$ there is a τ-additive Borel measure μ on T such that $\mathrm{m}(f) = \int f \, d\mu$ and $|\mathrm{m}|(f) = \int f \, d|\mu|$ for $f \in F$. It remains to be proved that the set M of such measures μ is uniformly tight. The equicontinuity of $\mathfrak{A}$ implies that the values $|\mathrm{m}|(1)$, $\mathrm{m} \in \mathfrak{A}$, are bounded; hence $|\mu|(T)$, $\mu \in M$, are bounded.

Since $\mathfrak{A}$ satisfies (ii), so does the set of $|\mathrm{m}|$ where $\mathrm{m} \in \mathfrak{A}$. There is a compact set $K \subseteq T$ such that if $\mathrm{m} \in \mathfrak{A}$, $f \in F$, $\|f\|_T \le 1$ and $\|f\|_K = 0$, then $|\mathrm{m}|(f) < \varepsilon$. The topology of T is the coarsest one for which every $f \in F$ is continuous. Thus for every $x \in T \setminus K$ there is $f \in F^+$ such that $\|f\|_K = 0$ and $f(x) = 1$. It follows that there is a net of functions $f_\gamma \in F^+$ such that $\|f_\gamma\|_K = 0$ for every γ, $f_\gamma \le f_\beta$ for $\gamma \le \beta$, and $\lim_\gamma f_\gamma(x) = 1$ for every $x \in T \setminus K$. Let $V_\gamma := \{x \in T \mid f_\gamma(x) > \frac{1}{2}\}$. For every $\mu \in M$, we have

$$|\mu|(V_\gamma) \le 2 \int f_\gamma \mathrm{d}|\mu| = 2|\mathfrak{m}|(f_\gamma) < 2\varepsilon,$$

and $|\mu|(T \setminus K) = \lim_\gamma |\mu|(V_\gamma) \le 2\varepsilon$ because $|\mu|$ is τ-additive. $\qquad\square$

Theorem P.30. *Let T be a completely regular topological space and F a Riesz subspace of $\ell_\infty(T)$ that contains constant functions. Assume that the topology of T is the coarsest topology for which every $f \in F$ is continuous. The following two properties of a linear functional $\mathfrak{m}: F \to \mathbb{R}$ are equivalent:*

(i) There is a tight Borel measure μ on T such that $\mathfrak{m}(f) = \int f \mathrm{d}\mu$ for every $f \in F$.
(ii) The restriction of $\mathfrak{m}$ to the $\|\cdot\|_T$-unit ball in F is continuous in the compact–open topology.

If (i) holds, then there is a unique such tight Borel measure μ, and moreover, $\mathfrak{m}^+(f) = \int f \mathrm{d}\mu^+$, $\mathfrak{m}^-(f) = \int f \mathrm{d}\mu^-$ and $|\mathfrak{m}|(f) = \int f \mathrm{d}|\mu|$ for every $f \in F$, and $\|\mathfrak{m}\| = \|\mu\|_{\mathrm{TV}}$.

Proof. The equivalence (i) $\Leftrightarrow$ (ii) is a special case of Theorem P.29. The remaining statements follow from Theorem P.28. $\qquad\square$

Theorem P.31 (Riesz Representation). *Let T be a compact space and let $\mathfrak{m}$ be a $\|\cdot\|_T$-continuous functional on $C_b(T)$. Then there is a unique tight Borel measure μ on T such that $\mathfrak{m}(f) = \int f \mathrm{d}\mu$ for all $f \in C_b(T)$.*

Proof. Apply Theorem P.30 with $F = C_b(T)$. $\qquad\square$

The following definition is adapted from [60, 438A].

Definition P.32. Let κ be a cardinal and S a set with $|S| = \kappa$. The cardinal κ is *measure-free* iff there is no measure $\mu \ge 0$ defined on the σ-algebra of all subsets of S such that $\mu(S) = 1$ and $\mu(\{x\}) = 0$ for every $x \in S$. $\qquad\blacksquare$

The cardinals that are not measure-free are sometimes called *real-valued measurable* in the literature. However, that term has a different meaning in standard set-theory terminology; the connection is explained in [61, 543B]. The assumption that all cardinals are measure-free is consistent with the ZFC set theory.

Theorem P.33. *1. The cardinal $\aleph_0$ is measure-free.*
2. If κ is a measure-free cardinal, then κ^+ is measure-free.
3. The cardinal $2^{\aleph_0}$ is measure-free if and only if the Lebesgue measure on the interval $[0,1]$ cannot be extended to a measure defined on all subsets of $[0,1]$.
4. If $\kappa \ge 2^{\aleph_0}$ is a measure-free cardinal, then 2^κ is measure-free.

Proof. [60, 438C]. $\qquad\square$

Theorem P.34. *Let μ be a measure on a σ-algebra Σ of subsets of S, and let T be a metrizable topological space of measure-free weight. If a mapping $\varphi: S \to T$ is measurable with respect to Σ and $\mathrm{Bo}(T)$, then there is a closed separable set $T_0 \subseteq T$ such that $|\mu|(S \setminus \varphi^{-1}(T_0)) = 0$.*

Proof. [60, 438D]. $\qquad\square$

P.6 Prerequisites for Part III

This section gathers several additional prerequisites that are used only in isolated places in Part III.

The next theorem simplifies a proof in Sect. 10.5. Let T be a Hausdorff topological space. A point x in T is a *P-point* iff the intersection of every countable set of neighbourhoods of x is also a neighbourhood of x. The space T is a *P-space* iff every point of T is a P-point.

Theorem P.35. *Let T be a Hausdorff P-space. Then every compact subset of T is finite.*

Proof. [81, 4K]. $\qquad\qquad\qquad\qquad\qquad\qquad\qquad\qquad\qquad\qquad\qquad\qquad\qquad\square$

Several standard concepts of probability theory are needed in Chap. 11, as covered, for example, by Dudley [44] and Fremlin [58]. A *probability measure* is a non-negative measure of total mass 1. The probability of an event Θ is denoted $\mathrm{Prob}(\Theta)$. Random variables with values in a topological space are assumed to be Borel measurable; that is, if ζ is a T-valued random variable, then $[\zeta \in A]$ is an event for every set $A \in \mathrm{Bo}(T)$. The *(probability) distribution of* ζ is the probability measure μ on the σ-algebra $\mathrm{Bo}(T)$ defined by $\mu(A) = \mathrm{Prob}(\zeta \in A)$ for $A \in \mathrm{Bo}(T)$. The *expected value* (the *mean*) of a real-valued random variable ζ is denoted $\mathrm{E}(\zeta)$.

Theorem P.36. *Let $\{\zeta_j\}_j$ be a sequence of real-valued random variables.*

1. *If* $\lim_j \mathrm{E}(|\zeta_j|) = 0$, *then there is a subsequence of* $\{\zeta_j\}_j$ *that converges to 0 almost surely.*
2. *If* $|\zeta_j| \leq 1$ *for all j and the sequence* $\{\zeta_j\}_j$ *converges to 0 almost surely, then* $\lim_j \mathrm{E}(|\zeta_j|) = 0$.

Proof. These are special cases of 4.3.5 and 9.2.1 in [44] or 245C and 245K in [58]. $\qquad\qquad\qquad\qquad\qquad\qquad\qquad\qquad\qquad\qquad\qquad\qquad\qquad\qquad\qquad\square$

The remaining three theorems are needed in Chap. 12. A real-valued function on a topological space is of *Baire class 1* iff it is a pointwise limit of a sequence of continuous functions.

Theorem P.37. *Let T be a compact space and f a real-valued function on T. If f is of Baire class 1, then there is a dense subset A of T such that f is continuous at every point of A.*

Proof. The set of points of continuity of f is comeagre [116, 31.X] , and every comeagre set in a compact space is dense [31, 9.2]. $\qquad\qquad\qquad\qquad\qquad\qquad\square$

A subset A of a topological space T has the *Baire property in T* iff there is an open set $V \subseteq T$ such that the sets $A \setminus V$ and $V \setminus A$ are meagre in T. A subset A of a compact space K is *universally BP in K* iff for every compact space K' and every continuous mapping $\varphi \colon K' \to K$ the set $\varphi^{-1}(A)$ has the Baire property in K' (cf. 1A and 1E in [62]).

A subset A of a topological space T is a *CBP set* iff the set $A \cap K$ is universally BP in K for every compact $K \subseteq T$. A function $f : T \to \mathbb{R}$ is *CBP measurable* iff for every open set $V \subseteq \mathbb{R}$ the set $f^{-1}(V)$ is CBP in T. Evidently every Borel subset of T is a CBP set, and if $f : T \to \mathbb{R}$ is CBP measurable, then so is its restriction to any subspace of T.

Theorem P.38. *Let K_0 and K be compact spaces and $\varphi_0 : K_0 \to K$ a continuous surjective mapping. Let $A \subseteq K$ be such that the set $\varphi_0^{-1}(A)$ is universally BP in K_0. Then A is universally BP in K.*

Proof. The proof is due to Fremlin. A slightly different version appears in [62, 1E].

Take any continuous mapping $\varphi_1 : K_1 \to K$ from a compact space K_1. For $i = 0, 1$ let $\pi_i : K_0 \times K_1 \to K_i$ be the canonical projections. Then

$$L := \{ (x_0, x_1) \in K_0 \times K_1 \mid \varphi_0(x_0) = \varphi_1(x_1) \}$$

is a compact subset of $K_0 \times K_1$, and $\pi_1(L) = K_1$ because $\varphi_0(K_0) = K$. By 4A2Gi in [60], there is a closed set $L' \subseteq L$ such that the restriction of π_1 to L' is irreducible and $\pi_1(L') = K_1$.

Now use the fact that if a set has the Baire property in a compact space, then so does its image under any irreducible continuous surjection [144, L.2][166, 25.2.3]. Since $\varphi_0^{-1}(A)$ is universally BP, the set $L' \cap \pi_0^{-1}(\varphi_0^{-1}(A))$ has the Baire property in L', and the set

$$\pi_1(L' \cap \pi_0^{-1}(\varphi_0^{-1}(A))) = \varphi_1^{-1}(A)$$

has the Baire property in K_1. $\square$

Corollary P.39. *Let K_0 and K be compact spaces, $\varphi_0 : K_0 \to K$ a continuous surjective mapping and $f : K \to \mathbb{R}$ a function such that $f \circ \varphi_0$ is CBP measurable. Then f is CBP measurable.* $\square$

The "quotient" property in the corollary has an analogue for the functions of Baire class 1:

Lemma P.40. *Let K_0 and K be compact spaces, $\varphi_0 : K_0 \to K$ a continuous surjective mapping, and $f : K \to \mathbb{R}$ a function such that $f \circ \varphi_0$ is of Baire class 1. Then f is of Baire class 1.*

Proof. This follows as a simple special case from 5.9.13 and 6.1.1 in [160]. $\square$

When S is a non-empty set, say that a function $\psi : \{0, 1\}^S \to \mathbb{R}$ is *finitely additive* iff $\psi(I_A) + \psi(I_B) = \psi(I_{A \cup B})$ whenever $A, B \subseteq S$ are disjoint.

Theorem P.41. *Let S be a non-empty set and $\psi : \{0, 1\}^S \to \mathbb{R}$ a finitely additive function. If ψ is CBP measurable with respect to the product topology on $\{0, 1\}^S$, then $\sum_{x \in S} |\psi(I_{\{x\}})| < \infty$ and $\psi(I_A) = \sum_{x \in A} \psi(I_{\{x\}})$ for every $A \subseteq S$.*

Proof. This is Lemma 2.1 in [28]. $\square$

Part I
Uniform Spaces

This part has a dual purpose. For the reader who knows the theory of metric but not of uniform spaces, I develop the latter as a natural generalization of the former, with complete proofs of the results that will be needed later on. For the reader who already knows uniform spaces, Chaps. 1, 2 and 4 are a review of the necessary definitions and theorems, presented in the language of pseudometrics (rather than the more traditional languages of entourages or uniform covers). Chapter 3 introduces several new notions extending those traditionally studied for semitopological semigroups, and includes a general result about ambitable topological groups.

In selecting the topics presented here I am guided by the needs of subsequent chapters. The scope is necessarily limited. Other treatments of uniform spaces, with additional concepts and results, may be found in the references listed in Sect. 1.5.

Chapter 1
Uniformities and Topologies

In this chapter I define uniform structures and uniformly continuous mappings in the language of pseudometrics. I derive their basic properties and their relationship to topologies.

1.1 Uniform Structures and Mappings

There are several equivalent ways to define uniform spaces. I find the definition based on pseudometrics most convenient, because it simplifies the construction of uniformly continuous functions.

When Δ_0 and Δ_1 are two pseudometrics on a non-empty set S, consider the pointwise maximum $\Delta_0 \vee \Delta_1$ of Δ_0 and Δ_1:

$$(\Delta_0 \vee \Delta_1)(x,y) := \Delta_0(x,y) \vee \Delta_1(x,y) \text{ for } x,y \in S.$$

Clearly $\Delta_0 \vee \Delta_1$ is also a pseudometric.

When $\mathscr{P}$ is a set of pseudometrics on a set S and Δ is a pseudometric on S, write $\Delta \ll \mathscr{P}$ iff

$$\forall \varepsilon > 0 \, \exists \Delta_\varepsilon \in \mathscr{P} \, \exists \theta > 0 \, \forall x,y \in S \, [\Delta_\varepsilon(x,y) < \theta \Rightarrow \Delta(x,y) < \varepsilon].$$

Definition 1.1. A *uniform structure* (or a *uniformity*, for short) on a non-empty set S is a set $\mathscr{U}$ of pseudometrics on S with these properties:

(U1)   If $\Delta_0, \Delta_1 \in \mathscr{U}$, then $\Delta_0 \vee \Delta_1$.
(U2)   If Δ is a pseudometric on S such that $\Delta \ll \mathscr{U}$, then $\Delta \in \mathscr{U}$.
(U3)   If $x,y \in S$, $x \neq y$, then there is $\Delta \in \mathscr{U}$ such that $\Delta(x,y) > 0$.

A *uniform space* X is a non-empty set S (*set of points of* X) together with a uniformity on S. Let $\mathsf{UP}(X)$ denote the uniformity of X (the set of pseudometrics), and let $\mathsf{UP_b}(X)$ denote the set of bounded pseudometrics in $\mathsf{UP}(X)$.   ∎

J. Pachl, *Uniform Spaces and Measures*, Fields Institute Monographs 30,
DOI 10.1007/978-1-4614-5058-0_2,
© Springer Science+Business Media New York 2013

Some authors define uniform spaces using only conditions equivalent to (U1) and (U2); in their terminology, uniform spaces as defined here would be called *Hausdorff* (or *separated*) uniform space.

From now on, the set of points of any uniform space X is written simply as X in expressions such as $|X|$, $x \in X$ and $Y \subseteq X$, and "the set X" means the set of points of X. Note however that extra care is needed for expressions that include equality. When X and Y are uniform spaces, $X = Y$ means that X and Y are equal as uniform spaces, not merely as sets or some other objects with some structure "forgotten".

Bounded pseudometrics are in some respects better behaved than general (unbounded) ones. Fortunately it is often possible to replace a general pseudometric by a bounded one. If Δ is a pseudometric, then $1 \wedge \Delta$ is also a pseudometric, and $\Delta \in \mathsf{UP}(X)$ if and only if $1 \wedge \Delta \in \mathsf{UP}(X)$.

Every uniform space defines a topology on its set of points, as follows: If $x \in X$ and Δ is a pseudometric on X, write

$$\odot[x, \Delta] := \{y \in X \mid \Delta(x,y) < 1\}.$$

Theorem 1.2. *If X is a uniform space, then the collection of sets $\odot[x, \Delta]$, where $\Delta \in \mathsf{UP}(X)$ and $x \in X$, is a base of a Hausdorff topology on the set of points of X.*

Proof. To prove that the collection is a base of a topology, take any $x_0, x_1 \in X$, $\Delta_0, \Delta_1 \in \mathsf{UP}(X)$ and $x \in \odot[x_0, \Delta_0] \cap \odot[x_1, \Delta_1]$. Then

$$\odot[x, \Delta/r] \subseteq \odot[x_0, \Delta_0] \cap \odot[x_1, \Delta_1]$$

where $r := \min(1 - \Delta(x, x_0), 1 - \Delta(x, x_1)) > 0$ and $\Delta := \Delta_0 \vee \Delta_1$. The pseudometric Δ/r belongs to $\mathsf{UP}(X)$ because $\Delta/r \ll \{\Delta\}$ and $\Delta \in \mathsf{UP}(X)$.

In view of property (U3) in Definition 1.1, the resulting topology is Hausdorff. $\qquad\square$

For any uniform space X, the topology defined in Theorem 1.2 is called simply *the topology* of X. Thus every uniform space X is also a topological space, which gives meaning to open and closed subsets of X, continuous mappings and homeomorphisms to and from X, and expressions such as $\mathsf{C}(X)$ and $\mathsf{C_b}(X)$. By Corollary 1.23, the topology of every uniform space is completely regular.

The following description of the closure operation in the topology of X is often useful:

Theorem 1.3. *For every uniform space X, $x \in X$ and $A \subseteq X$, the following are equivalent:*

  (i) *$x \in \overline{A}$ (x is in the closure of A).*
 (ii) *$\Delta(x, A) = 0$ for every $\Delta \in \mathsf{UP}(X)$.*
(iii) *$\Delta(x, A) < 1$ for every $\Delta \in \mathsf{UP}(X)$.*
 (iv) *$\Delta(x, A) = 0$ for every $\Delta \in \mathsf{UP_b}(X)$.*
  (v) *$\Delta(x, A) < 1$ for every $\Delta \in \mathsf{UP_b}(X)$.*

Proof. Evidently (ii)⇒(iii)⇒(v) and (ii)⇒(iv)⇒(v), and (i)⇔(iii) from the definition of the topology of X.

For every $\varDelta \in \mathsf{UP}(X)$ and $r > 0$, the pseudometric $1 \wedge r\varDelta$ is in $\mathsf{UP_b}(X)$. That proves (v)⇒(ii). $\qquad\square$

For a mapping $\varphi\colon X \to Y$ and a pseudometric $\varDelta$ on Y, the pseudometric $\overleftarrow{\varphi}\varDelta$ on X is defined by $\overleftarrow{\varphi}\varDelta(x,y) := \varDelta(\varphi(x),\varphi(y))$ for $x,y\in X$. For a set $\varPhi$ of mappings from X to Y and a bounded pseudometric $\varDelta$ on Y, the pseudometric $\overleftarrow{\varPhi}\varDelta$ on X is defined by

$$\overleftarrow{\varPhi}\varDelta(x,y) := \sup\{\,\overleftarrow{\varphi}\varDelta(x,y) \mid \varphi\in\varPhi\,\}, \; x,y\in X.$$

Definition 1.4. When X and Y are uniform spaces, a mapping $\varphi\colon X \to Y$ is *uniformly continuous* iff $\overleftarrow{\varphi}\varDelta \in \mathsf{UP}(X)$ for every $\varDelta \in \mathsf{UP}(Y)$. A mapping $\varphi\colon X \to Y$ is a *uniform isomorphism* iff it is bijective and both φ and its inverse are uniformly continuous.

A set $\varPhi$ of mappings from X to Y is *uniformly equicontinuous* iff $\overleftarrow{\varPhi}\varDelta \in \mathsf{UP}(X)$ for every $\varDelta \in \mathsf{UP_b}(Y)$. $\qquad\blacksquare$

Clearly $\varphi\colon X \to Y$ is uniformly continuous if and only if $\overleftarrow{\varphi}\varDelta \in \mathsf{UP}(X)$ for every bounded pseudometric $\varDelta\in\mathsf{UP}(Y)$. Thus φ is uniformly continuous if and only if the singleton set $\{\varphi\}$ is uniformly equicontinuous.

Theorem 1.5. *Let X and Y be two uniform spaces. Every uniformly continuous mapping from X to Y is continuous.*

Proof. Let $\varphi\colon X \to Y$ be a uniformly continuous mapping. To prove that φ is continuous, it is enough to show that the set $\varphi^{-1}(\odot[y,\varDelta])$ is open in X for every $\varDelta \in \mathsf{UP}(Y)$ and $y\in Y$.

Take any $x\in\varphi^{-1}(\odot[y,\varDelta])$. Then $\varphi(x)\in\odot[y,\varDelta]$; hence there exists $r>0$ such that $\odot[\varphi(x),\varDelta/r] \subseteq \odot[y,\varDelta]$. The pseudometric $\overleftarrow{\varphi}\varDelta$ is in $\mathsf{UP}(X)$ because φ is uniformly continuous, and

$$\varphi(\odot[x,\overleftarrow{\varphi}\varDelta/r]) \subseteq \odot[\varphi(x),\varDelta/r] \subseteq \odot[y,\varDelta],$$

hence $\odot[x,\overleftarrow{\varphi}\varDelta/r] \subseteq \varphi^{-1}(\odot[y,\varDelta])$. Thus $\varphi^{-1}(\odot[y,\varDelta])$ is an open set. $\qquad\square$

When X is a uniform space and T is a topological space, say that X or the uniformity of X *compatible with T* or with the topology of T iff X and T have the same set of points and the topology of X is the topology of T.

Typically many uniformities on a set are compatible with the same topology, and in that sense the passage from a uniform space to its topological space "forgets" some structure. However, no structure is forgotten when the space is compact:

Theorem 1.6. *Every continuous mapping from a compact uniform space to any uniform space is uniformly continuous.*

Proof. Let X be a compact uniform space, and $\varphi\colon X \to Y$ a continuous mapping from X to a uniform space Y.

Take any $\Delta \in \mathsf{UP}(Y)$ and $\varepsilon > 0$. I claim there is $\Delta_\varepsilon \in \mathsf{UP}(X)$ such that

$$\forall x, y \in X \ [\, \Delta_\varepsilon(x,y) < 1/2 \ \Rightarrow \ \overleftarrow{\varphi}\Delta(x,y) < \varepsilon\,].$$

Since φ is continuous, the set $\varphi^{-1}(\odot[\varphi(x), 2\Delta/\varepsilon])$ is open in X for every $x \in X$. Hence there are pseudometrics $\Delta_x \in \mathsf{UP}(X)$, $x \in X$, such that

$$\odot[x, \Delta_x] \subseteq \varphi^{-1}(\odot[\varphi(x), 2\Delta/\varepsilon])$$

for every $x \in X$. By compactness, there is a finite set $D \subseteq X$ such that the sets $\odot[z, 2\Delta_z]$, $z \in D$, cover X.

The pseudometric $\Delta_\varepsilon := \max_{z \in D} \Delta_z$ is in $\mathsf{UP}(X)$. If $x, y \in X$, $\Delta_\varepsilon(x,y) < 1/2$, find $z \in D$ such that $\Delta_z(z,x) < 1/2$. Then

$$\Delta_z(z,y) \leq \Delta_z(z,x) + \Delta_\varepsilon(x,y) < 1\,,$$

which means that $x, y \in \odot[z, \Delta_z] \subseteq \varphi^{-1}(\odot[\varphi(z), 2\Delta/\varepsilon])$, and

$$\overleftarrow{\varphi}\Delta(x,y) = \Delta(\varphi(x), \varphi(y)) \leq \Delta(\varphi(x), \varphi(z)) + \Delta(\varphi(z), \varphi(y)) < \varepsilon.$$

That proves the claim, and it follows that $\overleftarrow{\varphi}\Delta \in \mathsf{UP}(X)$.   □

Corollary 1.7. *Every homeomorphism of compact uniform spaces is a uniform isomorphism.*   □

The set of uniform structures on a fixed set of points is partially ordered by inclusion. When X and Y are two uniform spaces on the same set of points, say that the uniformity of X is *finer than* the uniformity of Y and that the uniformity of Y is *coarser than* the uniformity of X iff $\mathsf{UP}(X) \supseteq \mathsf{UP}(Y)$; that is, iff the identity mapping from X to Y is uniformly continuous. Say that X is finer or coarser than Y iff the same holds for the uniformities of X and Y.

The *discrete uniformity* on a set S is the set of all pseudometrics on S. It is the finest of all uniformities on S. A uniform space is *uniformly discrete* iff its uniformity is the discrete uniformity.

1.2 Cardinal Reflections, Compactness and Completeness

When Δ is a pseudometric on a set S and $Y \subseteq S$, say that Y is Δ-*uniformly discrete* iff there exists $\varepsilon > 0$ such that $\Delta(x,y) \geq \varepsilon$ for all $x, y \in Y$, $x \neq y$. When X is a uniform space and $Y \subseteq X$, say that Y is *uniformly discrete in* X iff it is Δ-uniformly discrete for some $\Delta \in \mathsf{UP}(X)$. Evidently a uniform space is uniformly discrete if and only if it is uniformly discrete in itself.

Theorem 1.8. *Let κ be an infinite cardinal and S a non-empty set. The following two properties of a pseudometric Δ on S are equivalent:*

(i) If $Y \subseteq S$ is Δ-uniformly discrete in S, then $|Y| < \kappa$.
(ii) For every $\varepsilon > 0$, there exists $Y \subseteq S$ such that $|Y| < \kappa$ and $\Delta(x,Y) < \varepsilon$ for every $x \in S$.

Proof. To prove (i)$\Rightarrow$(ii), suppose that (i) holds, and take any $\varepsilon > 0$. Consider the family of all sets $Y \subseteq S$ such that $\Delta(x,y) \geq \varepsilon$ whenever $x,y \in Y$, $x \neq y$. By Zorn's Lemma, this family contains a maximal set Y_0. We have $\Delta(x,Y_0) < \varepsilon$ for every $x \in S$ because Y_0 is maximal, and $|Y_0| < \kappa$ by (i).

To prove (ii)$\Rightarrow$(i), suppose that (ii) holds, and take any Δ-uniformly discrete set $Y \subseteq S$. There is some $\varepsilon > 0$ such that $\Delta(x,y) \geq \varepsilon$ for all $x,y \in Y$, $x \neq y$. By (ii), there exists $Y' \subset S$ such that $|Y'| < \kappa$ and $\Delta(x,Y') < \varepsilon/2$ for every $x \in S$. Thus there is a mapping $\varphi \colon Y \to Y'$ such that $\Delta(x, \varphi(x)) < \varepsilon/2$ for every $x \in Y$, and φ is injective because $\Delta(x,y) \geq \varepsilon$ for $x,y \in Y$, $x \neq y$. It follows that $|Y| \leq |Y'| < \kappa$. $\qquad\square$

Corollary 1.9. *Let κ be an infinite cardinal. The following three properties of a uniform space X are equivalent:*

(i) If $Y \subseteq X$ is uniformly discrete in X, then $|Y| < \kappa$.
(ii) For every $\Delta \in \mathsf{UP}(X)$ and $\varepsilon > 0$, there exists $Y \subseteq X$ such that $|Y| < \kappa$ and $\Delta(x,Y) < \varepsilon$ for every $x \in X$.
(iii) For every $\Delta \in \mathsf{UP}(X)$, there exists $Y \subseteq X$ such that $|Y| < \kappa$ and $\Delta(x,Y) < 1$ for every $x \in X$. $\qquad\square$

Lemma 1.10. *Let κ be an infinite cardinal, let $\{\Delta_0, \Delta_1, \ldots, \Delta_j\}$ be a non-empty finite set of pseudometrics on a non-empty set S and let Δ be a pseudometric on S such that $\Delta \ll \Delta_0 \vee \Delta_1 \vee \cdots \vee \Delta_j$. If the pseudometrics Δ_i, $i = 0, 1, \ldots, j$, have property (ii) in Theorem 1.8, then so does Δ.*

Proof. Write $\Delta' := \Delta_0 \vee \Delta_1 \vee \cdots \vee \Delta_j$. To prove that Δ has property (ii) in Theorem 1.8, take any $\varepsilon > 0$. There is $\theta > 0$ such that if $\Delta'(x,y) < \theta$, then $\Delta(x,y) < \varepsilon$. There are sets $Y_i \subseteq S$, $i = 0, 1, \ldots, j$, such that $|Y_i| < \kappa$ and $\Delta'(x,Y_i) < \theta$ for every $x \in S$. The set $Y := \bigcup_i Y_i$ satisfies $|Y| < \kappa$ and $\Delta'(x,Y) < \theta$, and hence also $\Delta(x,Y) < \varepsilon$, for every $x \in X$. $\qquad\square$

Lemma 1.11. *Let X be any uniform space and κ an infinite cardinal. Then*

$$\mathscr{U} := \{ \Delta \in \mathsf{UP}(X) \mid \text{if } Y \subseteq X \text{ is } \Delta\text{-uniformly discrete in } X \text{ then } |Y| < \kappa \}$$

is a uniform structure that is compatible with the topology of X.

Proof. By Lemma 1.10, $\mathscr{U}$ has properties (U1) and (U2) in Definition 1.1.

If $\Delta \in \mathsf{UP}(X)$ and $x_0 \in X$, then the pseudometric Δ' defined by

$$\Delta'(x,y) := 1 \wedge |\Delta(x,x_0) - \Delta(y,x_0)|$$

for $x,y \in X$ is in $\mathscr{U}$ (in fact, every Δ'-uniformly discrete set in X is finite), and $\bigodot[x_0, \Delta/\varepsilon] = \bigodot[x_0, \Delta'/\varepsilon]$ whenever $0 < \varepsilon < 1$. It follows that $\mathscr{U}$ has property (U3) because $\mathsf{UP}(X)$ does, and the topology of $\mathscr{U}$ is the same as that of $\mathsf{UP}(X)$. $\qquad\square$

Definition 1.12. Let X be any uniform space and α any ordinal. The α-*th cardinal reflection* of X, denoted $\mathsf{p}_\alpha X$, is the uniform space with the same points as X and the uniformity $\mathscr{U}$ defined in Lemma 1.11 with $\kappa = \aleph_\alpha$.   ■

By Lemma 1.11, X and $\mathsf{p}_\alpha X$ have the same topology. Clearly a uniform space X satisfies the conditions in Corollary 1.9 with $\kappa = \aleph_\alpha$ if and only if $X = \mathsf{p}_\alpha X$. (As was noted above, the equality $X = \mathsf{p}_\alpha X$ means that X and $\mathsf{p}_\alpha X$ are equal as uniform spaces, not merely as sets or topological spaces.)

Among the cardinal reflections, p_0 and p_1 will be of most interest in the sequel. The reflection p_0 captures the concept of a precompact (sometimes called totally bounded) space.

Definition 1.13. A uniform space X is *precompact* iff it satisfies the conditions in Corollary 1.9 with $\kappa = \aleph_0$. The 0-th cardinal reflection $\mathsf{p}_0 X$ is called the *precompact reflection* of X and written simply as $\mathsf{p}X$.   ■

Compact and precompact spaces are closely related, as the next theorem shows. First note that every convergent net in a uniform space is Cauchy, in the sense of the following definition.

Definition 1.14. A net $\{x_\gamma\}_\gamma$ in a uniform space X is *Cauchy* iff for every $\Delta \in \mathrm{UP}(X)$ the estimate $\Delta(x_\beta, x_\gamma) < 1$ holds for almost all β, γ. The space X is *complete* iff every Cauchy net in X converges in X.   ■

Theorem 1.15. *A uniform space is compact if and only if it is precompact and complete.*

Proof. Let X be a uniform space whose topology is compact. To prove X is precompact, take any $\Delta \in \mathrm{UP}(X)$. The collection of open sets $\odot[x, \Delta]$, $x \in X$, covers X. Being compact, X is covered by a finite subcollection. Thus X has property (iii) in Corollary 1.9 with $\kappa = \aleph_0$.

Next observe that if a Cauchy net has a cluster point x, then it converges to x. Since X is compact, every net in X has a cluster point, and therefore every Cauchy net in X converges.

To prove the converse, let X be precompact and complete. From property (iii) in Corollary 1.9 with $\kappa = \aleph_0$, we get that every universal net in X is Cauchy. Thus every universal net in X converges, which means that X is compact.   □

1.3   Metric Spaces and Real Functions

When X is a metric space and Δ is its metric,

$$\mathscr{U}(\Delta) := \{\Delta' \mid \Delta' \text{ is a pseudometric on } X \text{ and } \Delta' \ll \{\Delta\}\}$$

is a uniformity on the set X. In fact, it is the uniformity induced by Δ, as defined further on in Sect. 2.1. Say that a uniform space is *metrizable by* Δ iff its uniformity is $\mathscr{U}(\Delta)$, and *metrizable* iff it is metrizable by some metric Δ.

Many metrics Δ on a given set may determine the same uniform structure $\mathscr{U}(\Delta)$, so that the passage from a metric space to the uniform space "forgets" some structure. Nevertheless, the passage preserves uniform concepts. Indeed, let X_i with metrics Δ_i, $i = 0, 1$, be two metric spaces and let φ be a mapping from X_0 to X_1. Then φ is uniformly continuous with respect to the uniform structures $\mathscr{U}(\Delta_0)$ and $\mathscr{U}(\Delta_1)$ if and only if

$$\forall \varepsilon > 0 \, \exists \theta > 0 \, \forall x, y \in X_0 \; [\, \Delta_0(x,y) < \theta \; \Rightarrow \; \Delta_1(\varphi(x), \varphi(y)) < \varepsilon \,].$$

Thus uniform continuity with respect to $\mathscr{U}(\Delta_0)$ and $\mathscr{U}(\Delta_1)$ is equivalent to the usual notion of uniform continuity for metric spaces. Clearly the topology of a metric Δ is the same as the topology of the uniformity $\mathscr{U}(\Delta)$. Similarly, by the following lemma, the completeness as usually defined for a metric space is the same as the completeness of the corresponding uniform space.

Lemma 1.16. *A metrizable uniform space X is complete if and only if every Cauchy sequence in X converges.*

Proof. The condition is obviously necessary. To prove its necessity, let X be metrizable by a metric Δ, assume that every Cauchy sequence in X converges and let $\{x_\gamma\}_{\gamma \in \Gamma}$ be a Cauchy net in X. There are $\gamma(j) \in \Gamma$ for $j \in \omega$ such that $\gamma(0) \leq \gamma(1) \leq \cdots$ and $\Delta(x_\beta, x_\gamma) < 1/(j+1)$ for all $\beta, \gamma \geq \gamma(j)$. The sequence $\{x_{\gamma(j)}\}_j$ is Cauchy, and its limit is also the limit of the net $\{x_\gamma\}_\gamma$. $\square$

These observations show that there is no danger of confusion or ambiguity when we adopt the view that every metric space *is* a uniform space. The meaning of expressions such as $\mathsf{UP}(X)$, $\mathsf{p}_\alpha X$ and $Y \Subset X$ (defined further on) for a metric space X is obtained by treating X as a uniform space.

$\mathbb{R}$ is a metric space with the usual metric $\Delta_{\mathbb{R}}(x,y) := |x - y|$, $x, y \in \mathbb{R}$. In the sequel, the uniformity induced by $\Delta_{\mathbb{R}}$ is called simply *the uniformity* of $\mathbb{R}$, and $\mathbb{R}$ is always considered as a uniform space with this uniformity.

For a uniform space X, let $\mathsf{U}(X)$ denote the space of uniformly continuous real-valued functions on X, and let $\mathsf{U_b}(X)$ denote the space of bounded functions in $\mathsf{U}(X)$.

Lemma 1.17. *Let X be a uniform space and let f be a real-valued function on X. If $\{f_\gamma\}_\gamma$ is a net of functions $f_\gamma \in \mathsf{U}(X)$ such that $\lim_\gamma \|f_\gamma - f\|_X = 0$, then $f \in \mathsf{U}(X)$.*

Proof. I shall prove that $\overleftarrow{f} \, \Delta_{\mathbb{R}} \in \mathsf{UP}(X)$. Take any $\varepsilon > 0$. There exists γ such that $\|f_\gamma - f\|_X < \varepsilon$. Since $f_\gamma \in \mathsf{U}(X)$, there are $\Delta' \in \mathsf{UP}(X)$ and $\theta > 0$ such that

$$\forall x, y \in X \; [\, \Delta'(x,y) < \theta \; \Rightarrow \; |f_\gamma(x) - f_\gamma(y)| < \varepsilon \,].$$

For $x, y \in X$, we have

$$\overleftarrow{f} \Delta_{\mathbb{R}}(x,y) = |f(x) - f(y)| \leq |f(x) - f_\gamma(x)| + |f_\gamma(x) - f_\gamma(y)| + |f_\gamma(y) - f(y)| < 3\varepsilon,$$

which shows that $\overleftarrow{f} \Delta_{\mathbb{R}} \ll \mathsf{UP}(X)$ and therefore $\overleftarrow{f} \Delta_{\mathbb{R}} \in \mathsf{UP}(X)$.   □

Theorem 1.18. *Let X be any uniform space.*

1. $\mathsf{U}(X)$ *is a Riesz subspace of* $\mathbb{R}^X$.
2. *If* $f \in \mathsf{U}(X)$ *and* $\inf\{|f(x)| \mid x \in X\} > 0$, *then* $1/f \in \mathsf{U_b}(X)$.
3. $\mathsf{U_b}(X)$ *is a Riesz subspace of* $\mathbb{R}^X$ *and a Banach lattice with the norm* $\|\cdot\|_X$.
4. *If* $f_0, f_1 \in \mathsf{U_b}(X)$, *then* $f_0 f_1 \in \mathsf{U_b}(X)$.

Proof. 1. Take any $f_0, f_1 \in \mathsf{U}(X)$, and let $f := f_0 \vee f_1$, $g := f_0 + f_1$. Then

$$\overleftarrow{f} \Delta_{\mathbb{R}} \leq \overleftarrow{f_0} \Delta_{\mathbb{R}} \vee \overleftarrow{f_1} \Delta_{\mathbb{R}},$$
$$\overleftarrow{g} \Delta_{\mathbb{R}} \leq \overleftarrow{f_0} \Delta_{\mathbb{R}} + \overleftarrow{f_1} \Delta_{\mathbb{R}} \leq 2(\overleftarrow{f_0} \Delta_{\mathbb{R}} \vee \overleftarrow{f_1} \Delta_{\mathbb{R}}),$$

hence $f, g \in \mathsf{U}(X)$. A similar but simpler argument shows that $rf_0 \in \mathsf{U}(X)$ for every $r \in \mathbb{R}$.

2. Take any $f \in \mathsf{U}(X)$ such that $r := \inf\{|f(x)| \mid x \in X\} > 0$. Then $g := 1/f$ is in $\mathsf{U}(X)$ because

$$\overleftarrow{g} \Delta_{\mathbb{R}} \leq (1/r^2)\overleftarrow{f} \Delta_{\mathbb{R}}$$

and clearly $\|g\|_X \leq 1/r$.

3. Since $\mathsf{U_b}(X) = \mathsf{U}(X) \cap \ell_\infty(X)$ and $\ell_\infty(X)$ is a Riesz subspace of $\mathbb{R}^X$, $\mathsf{U_b}(X)$ is a Riesz subspace of $\mathbb{R}^X$ by Part 1. By Lemma 1.17, $\mathsf{U_b}(X)$ is complete in the norm $\|\cdot\|_X$.

4. Take any $f_0, f_1 \in \mathsf{U_b}(X)$, and let $f := f_0 f_1$. Then

$$\overleftarrow{f} \Delta_{\mathbb{R}} \leq \|f_1\|_X \overleftarrow{f_0} \Delta_{\mathbb{R}} + \|f_0\|_X \overleftarrow{f_1} \Delta_{\mathbb{R}} \leq (\|f_0\|_X + \|f_1\|_X)(\overleftarrow{f_0} \Delta_{\mathbb{R}} \vee \overleftarrow{f_1} \Delta_{\mathbb{R}})$$

which shows that $f \in \mathsf{U_b}(X)$.   □

Definition 1.19. Let X be any uniform space. A set of real-valued functions on X is said to be $\mathsf{UE}(X)$, or simply UE when X is understood, iff it is uniformly equicontinuous and bounded at every point of X. A set of real-valued functions on X is said to be $\mathsf{UEB}(X)$, or simply UEB when X is understood, iff it is uniformly equicontinuous and $\|\cdot\|_X$ bounded.   ■

Lemma 1.20. *Let X be a uniform space.*

1. *A set* $\mathscr{F} \subseteq \mathbb{R}^X$ *is* $\mathsf{UE}(X)$ *if and only if there are a pseudometric* $\Delta \in \mathsf{UP}(X)$ *and a Δ-1-Lipschitz function* $h\colon X \to \mathbb{R}^+$ *such that* $\mathscr{F} \subseteq \mathsf{BLip}(\Delta, h)$.
2. *A set* $\mathscr{F} \subseteq \mathbb{R}^X$ *is* $\mathsf{UEB}(X)$ *if and only if there are a pseudometric* $\Delta \in \mathsf{UP}(X)$ *and* $r \in \mathbb{R}^+$ *such that* $\mathscr{F} \subseteq r\mathsf{BLip_b}(\Delta)$.

Proof. If $\mathscr{F} \subseteq \mathbb{R}^X$ is uniformly equicontinuous and

$$\Delta(x,y) := \sup_{f\in\mathscr{F}} |f(x) - f(y)|, \; x,y\in X,$$

$$h(x) := \sup_{f\in\mathscr{F}} |f(x)|, \; x\in X,$$

then $\Delta \in \mathsf{UP}(X)$, h is Δ-1-Lipschitz and $\mathscr{F} \subseteq \mathsf{BLip}(\Delta,h)$. $\square$

1.4 Uniformizable Topological Spaces

The results in this section show that a topological space has a compatible uniformity if and only if it is completely regular. Moreover, for every completely regular topology there exists the finest compatible uniformity.

Theorem 1.21. *Let X be a uniform space and $x\in X$.*

1. *For every $\Delta \in \mathsf{UP}(X)$, the function $\diagdown_y\Delta(x,y)$ is uniformly continuous on X.*
2. *If $V \subseteq X$ is an open set and $x\in V$, then there exists $f\in \mathsf{U_b}(X)$ such that $0 \leq f \leq 1$, $f(x) = 0$ and f is 1 on $X \setminus V$.*

Proof. 1. Let $\varphi := \diagdown_y\Delta(x,y)$. Then $\overleftarrow{\varphi}\Delta_{\mathbb{R}} \in \mathsf{UP}(X)$ because $\overleftarrow{\varphi}\Delta_{\mathbb{R}} \leq \Delta$.

2. V is a union of basic open sets. Thus $x\in \odot[x_0,\Delta] \subseteq V$ for some $x_0\in X$ and $\Delta \in \mathsf{UP}(X)$, and $\odot[x,\Delta/r] \subseteq V$ where $r := 1 - \Delta(x,x_0) > 0$. The function f defined by

$$f(y) := 1 \wedge \frac{1}{r}\Delta(y,x)$$

is as required. $\square$

Corollary 1.22. *The topology of any uniform space X is the coarsest topology for which every function in $\mathsf{U_b}(X)$ is continuous.* $\square$

Corollary 1.23. *The topology of every uniform space is completely regular.* $\square$

Conversely, every completely regular topology is the topology of a uniform space:

Theorem 1.24. *Let T be a completely regular topological space and let $\mathscr{U}$ be the set of all pseudometrics Δ on T such that for each $x\in T$ the function $\diagdown_y\Delta(x,y)$ is continuous on T. Then $\mathscr{U}$ is a uniformity compatible with T.*

Proof. Clearly $\mathscr{U}$ has properties (U1) and (U2) in Definition 1.1.

By the definition of $\mathscr{U}$, every set $\odot[x,\Delta]$, where $x\in X$ and $\Delta \in \mathscr{U}$, is open in the topology of T. On the other hand, if V is an open subset of T and $x_0\in V$, there is a function $f\in \mathsf{C_b}(T)$ such that $f(x_0) = 0$ and f is 1 on $T \setminus V$. Define the pseudometric $\Delta \in \mathscr{U}$ by $\Delta(x,y) := |f(x) - f(y)|$, $x,y\in T$. Then $\odot[x_0,\Delta] \subseteq V$. Hence $\mathscr{U}$ has property (U3) and is compatible with T. $\square$

Definition 1.25. When T is completely regular topological space, the *fine uniformity of T* is the uniformity $\mathscr{U}$ in Theorem 1.24. The *fine uniform space of T*, denoted $\mathsf{F}T$, is the set T with the fine uniformity of T.

Since every uniform space is a completely regular topological space, this also defines the fine uniform space $\mathsf{F}X$ for every uniform space X. A uniform space X is *fine* iff $X = \mathsf{F}X$.   ∎

Theorem 1.26. *Let X be a uniform space.*

1. *If Y is a uniform space and φ is a continuous mapping from X to Y, then φ is uniformly continuous from $\mathsf{F}X$ to Y.*
2. *Every uniformity on the set X that is compatible with the topology of X is coarser than $\mathsf{F}X$.*

Proof. To prove Part 1, take any continuous mapping $\varphi : X \to Y$ and any $\Delta \in \mathsf{UP}(Y)$. For every $x \in X$ the function $\setminus_y \Delta(\varphi(x), y)$ is continuous on Y; hence the function $\setminus_z \overleftarrow{\varphi} \Delta(x,z) = \setminus_z \Delta(\varphi(x), \varphi(z))$ is continuous on X. Thus $\overleftarrow{\varphi}\Delta \in \mathsf{UP}(\mathsf{F}X)$ by the definition of $\mathsf{F}X$.

Part 2 is a special case of Part 1 for the identity mapping φ on the set X.   □

By Theorem 1.26, the fine uniformity of a completely regular space T is the finest uniformity compatible with the topology of T. Two uniform spaces X and Y on the same set of points are compatible with same topology if and only if $\mathsf{F}X = \mathsf{F}Y$.

Exercise 1.27. Show that $\mathsf{U}(\mathsf{F}X) = \mathsf{C}(X)$ and $\mathsf{U_b}(\mathsf{F}X) = \mathsf{C_b}(X)$ for every uniform space X.   ∎

1.5 Notes for Chap. 1

In his memoir published in 1937, Weil [179] defined uniform spaces and proved their basic properties. Since then, the theory has grown far beyond the brief account here. The foundations of uniform spaces are presented in most books on general topology, such as Bourbaki [12], Čech [24], Császár [31] and Engelking [47]. Isbell's monograph [100] is a classic of the field, covering a number of advanced topics. Bentley et al. [6] trace the history of uniform concepts and include pointers to major developments in uniform spaces up to the early 1990s.

A uniform structure may be defined in several ways, all equivalent to Weil's original definition. The three most popular approaches are those using entourages, uniform covers and pseudometrics. Definition 1.1 is an instance of the latter, adapted from Choquet [25] and Gillman and Jerison [81, 15.3].

Cardinal reflections in Sect. 1.2 are taken from Isbell [100], where $\mathsf{p}_\alpha X$ is written as $X_{\aleph_\alpha}$ and $\mathsf{p}_1 X$ is also written as eX.

Chapter 2
Induced Uniform Structures

In this chapter, I deal with a frequently used method for constructing uniformities: Given a point-separating set of mappings from a set S to uniform spaces, one uniformity on S is the coarsest for which the mappings are uniformly continuous. This is the induced uniformity on S, also known as the projectively induced or projectively generated uniformity. Special cases of this construction are the uniform subspace, the uniform and semiuniform product and various "weak" uniform structures.

2.1 General Properties

Let $\mathfrak{X}$ be a non-empty set of uniformities on the same set S of points and $\mathscr{U} := \bigcap \mathfrak{X}$. Clearly $\mathscr{U}$ has properties (U1) and (U2) in Definition 1.1. Thus there is a uniformity on S coarser than all uniformities in $\mathfrak{X}$ if and only if $\mathscr{U}$ has property (U3), and in that case, $\mathscr{U}$ is the finest among the uniformities that are coarser than every uniformity in $\mathfrak{X}$.

Now let $\mathscr{P}$ be a set of pseudometrics on S, and assume that $\mathscr{P}$ separates the points of S. Let $\mathfrak{X}$ be the set of all uniformities on the set S that contain $\mathscr{P}$. The set $\mathfrak{X}$ is non-empty because it contains the discrete uniformity. The set $\mathscr{U} := \bigcap \mathfrak{X}$ has property (U3) because $\mathscr{P}$ does, and $\mathscr{U}$ is the coarsest uniformity such that $\mathscr{U} \supseteq \mathscr{P}$.

Next, let Φ be a set of mappings where each $\varphi \in \Phi$ maps S to a uniform space X_φ. For $\varphi \in \Phi$, define the set $\overleftarrow{\varphi} \mathsf{UP}(X_\varphi) := \{ \overleftarrow{\varphi} \Delta \mid \Delta \in \mathsf{UP}(X_\varphi) \}$ of pseudometrics on S. If the set Φ of mappings separates the points in S, then so does the set $\bigcup_{\varphi \in \Phi} \overleftarrow{\varphi} \mathsf{UP}(X_\varphi)$ of pseudometrics.

Definition 2.1. Let S be a non-empty set, and $\mathscr{P}$ a set of pseudometrics on S that separates the points of S. The *uniformity induced by* $\mathscr{P}$ is the coarsest uniformity $\mathscr{U}$ on S such that $\mathscr{U} \supseteq \mathscr{P}$.

J. Pachl, *Uniform Spaces and Measures*, Fields Institute Monographs 30,
DOI 10.1007/978-1-4614-5058-0_3,
© Springer Science+Business Media New York 2013

Let Φ be a set of mappings $\varphi \colon S \to X_\varphi$ to uniform spaces X_φ, and assume that Φ separates the points in S. The *uniformity induced by* Φ is the uniformity on S induced by $\bigcup_{\varphi \in \Phi} \overleftarrow{\varphi} \mathsf{UP}(X_\varphi)$.   ∎

Thus the uniformity induced by Φ is the coarsest uniformity on S for which every mapping in Φ is uniformly continuous.

Theorem 2.2. *The uniformity of every uniform space is induced by a set of uniformly continuous mappings to metric spaces.*

Proof. Let X be any uniform space. For every $\Delta \in \mathsf{UP}(X)$, consider the canonical mapping $\chi_\Delta \colon X \to X/\Delta$ from X onto the associated metric space X/Δ with metric $\Delta^\bullet$ (Sect. P.2). Then $\Delta = \overleftarrow{\chi_\Delta}(\Delta^\bullet)$ and χ_Δ is uniformly continuous from X to X/Δ. Thus

$$\mathsf{UP}(X) = \{\overleftarrow{\chi_\Delta}(\Delta^\bullet) \mid \Delta \in \mathsf{UP}(X)\} \subseteq \bigcup_{\Delta \in \mathsf{UP}(X)} \overleftarrow{\chi_\Delta}\mathsf{UP}(X/\Delta) \subseteq \mathsf{UP}(X)$$

and the uniformity of X is induced by the set $\{\chi_\Delta \mid \Delta \in \mathsf{UP}(X)\}$.   □

Example 2.3. Let Σ be a σ-algebra of subsets of a non-empty set S, and assume that Σ separates the points of S. Let $\mathbb{N}$ be the discrete uniform space on the set $\omega := \{0,1,2,\dots\}$, and let $\mathscr{U}(\Sigma)$ be the uniformity on S induced by the set of the mappings $\varphi \colon S \to \mathbb{N}$ such that $\varphi^{-1}(j) \in \Sigma$ for every $j \in \mathbb{N}$.

For another set S' with a point-separating σ-algebra Σ', a mapping $\varphi \colon S \to S'$ is measurable with respect to the σ-algebras Σ and Σ' if and only if it is uniformly continuous with respect to the uniformities $\mathscr{U}(\Sigma)$ and $\mathscr{U}(\Sigma')$. Moreover, a function $f \colon S \to \mathbb{R}$ is measurable with respect to the σ-algebras Σ and $\mathsf{Bo}(\mathbb{R})$ if and only if f is uniformly continuous with respect to $\mathscr{U}(\Sigma)$.

In this sense, every measurable space (i.e. a set with a σ-algebra) is a uniform space. Note that the uniform space X of this form satisfies $X = \mathsf{p}_1 X$.   ∎

Definition 2.4. When E is a locally convex space, *the additive uniformity of* E is induced by the pseudometrics of the continuous seminorms on E, in other words, by the pseudometrics of the form $\Delta(x,y) = \alpha(x - y)$, $x,y \in E$, where α is a continuous seminorm on E.   ∎

Exercise 2.5. Let E be a locally convex space. Show that the additive uniformity of E is compatible with the topology of E. Show that a net in E is Cauchy in the sense of the definition in Sect. P.3 if and only if it is Cauchy in the additive uniformity of E. Show that the space E is complete in the sense of the definition in Sect. P.3 if and only if E is complete with its additive uniformity.   ∎

On several occasions, I shall need a more explicit form of the uniform structure induced by a set of pseudometrics. When $\mathscr{P}$ is a non-empty set of pseudometrics on the same set, let $\mathsf{cl}^\vee \mathscr{P}$ denote the set of pseudometrics of the form $\Delta_0 \vee \Delta_1 \vee \cdots \vee \Delta_k$ where $k \geq 0$ and $\Delta_i \in \mathscr{P}$ for $0 \leq i \leq k$. Thus $\mathsf{cl}^\vee \mathscr{P}$ is the smallest set that contains $\mathscr{P}$ and is closed under the operation $\vee$.

Theorem 2.6. *Let S be a non-empty set, $\mathscr{P}$ a set of pseudometrics on S that separates the points of S, and let $\mathscr{U}$ be the uniformity induced by $\mathscr{P}$.*

1. *A pseudometric Δ on S is in $\mathscr{U}$ if and only if $\Delta \ll \mathrm{cl}^\vee \mathscr{P}$.*
2. *If a bounded pseudometric Δ on S is in the uniformity induced by $\mathscr{P}$, then for every $\varepsilon > 0$ there exist $\Delta' \in \mathrm{cl}^\vee \mathscr{P}$ and $r > 0$ such that $\Delta < \max(\varepsilon, r\Delta')$.*

Proof. Since $\mathscr{P} \subseteq \mathscr{U}$, it follows that $\mathrm{cl}^\vee \mathscr{P} \subseteq \mathrm{cl}^\vee \mathscr{U} = \mathscr{U}$.

1. If $\Delta \ll \mathrm{cl}^\vee \mathscr{P}$, then $\Delta \ll \mathscr{U}$ and thus $\Delta \in \mathscr{U}$. Conversely, let $\mathscr{U}'$ be the set of all pseudometrics Δ on S such that $\Delta \ll \mathrm{cl}^\vee \mathscr{P}$. Then $\mathscr{U}'$ has properties (U1), (U2) and (U3) in Definition 1.1 and $\mathscr{P} \subseteq \mathscr{U}'$; hence $\mathscr{U} \subseteq \mathscr{U}'$.
2. Let $\Delta \in \mathscr{U}$ and $\varepsilon > 0$. Without loss of generality, assume $\Delta \leq 1$. By Part 1, there exist $\Delta' \in \mathrm{cl}^\vee \mathscr{P}$ and $\theta > 0$ such that: If $x, y \in X$, $\Delta'(x,y) \leq \theta$, then $\Delta(x,y) < \varepsilon$. It follows that $\Delta < \max(\varepsilon, \Delta'/\theta)$. $\qquad\qquad\square$

When the set $\mathscr{P}$ of pseudometrics on S is determined by a set of mappings from S to uniform spaces, it is useful to know whether $\mathrm{cl}^\vee \mathscr{P}$ may be replaced by $\mathscr{P}$ in Theorem 2.6. A frequently encountered sufficient condition is the following:

Definition 2.7. Let S be a non-empty set and let Φ be a set of mappings $\varphi \colon S \to X_\varphi$ to uniform spaces X_φ. Say that the set Φ is UCUD (*uniformly continuous upwards directed*) iff for every two mappings $\varphi_i \in \Phi$, $i = 0, 1$, there are $\varphi \in \Phi$ and uniformly continuous mappings $\psi_i \colon X_\varphi \to X_{\varphi_i}$, $i = 0, 1$, such that $\varphi_i = \psi_i \circ \varphi$ for $i = 0, 1$. $\qquad\blacksquare$

Corollary 2.8. *Let X be a uniform space whose uniformity is induced by an UCUD set Φ of mappings $\varphi \colon X \to X_\varphi$ to uniform spaces X_φ. If $\Delta \in \mathrm{UP_b}(X)$, then for every $\varepsilon > 0$ there exist $\varphi \in \Phi$ and $\Delta' \in \mathrm{UP}(X_\varphi)$ such that $\Delta < \overleftarrow{\varphi}\Delta' + \varepsilon$.*

Proof. Let $\mathscr{P} := \bigcup_{\varphi \in \Phi} \overleftarrow{\varphi}\, \mathrm{UP}(X_\varphi)$. If Φ is UCUD, then $\mathrm{cl}^\vee \mathscr{P} = \mathscr{P}$. $\qquad\square$

Theorem 2.9. *Let α be an ordinal and let X be a uniform space whose uniformity is induced by a set of mappings $\varphi \colon X \to X_\varphi$ to uniform spaces X_φ, $\varphi \in \Phi$. If $X_\varphi = \mathrm{p}_\alpha X_\varphi$ for all $\varphi \in \Phi$, then $X = \mathrm{p}_\alpha X$.*

Proof. If a pseudometric Δ on X_φ has property (i) in Theorem 1.8, then so does $\overleftarrow{\varphi}\Delta$. Apply Lemma 1.10 and Part 1 of Theorem 2.6. $\qquad\square$

Theorem 2.10. *For any uniform space X, the uniformity of $\mathrm{p}X$ is induced by $\mathrm{U_b}(X)$.*

Proof. Let $\mathscr{P}$ be the set of pseudometrics $\overleftarrow{f}\Delta_{\mathbb{R}}$ where $f \in \mathrm{U_b}(X)$. Since every $\Delta \in \mathscr{P}$ satisfies the conditions in Theorem 1.8 with $\kappa = \aleph_0$, the uniformity induced by $\mathscr{P}$ is coarser than $\mathrm{p}X$ by Lemma 1.10 and Part 1 of Theorem 2.6.

Conversely, to prove that the uniformity induced by $\mathscr{P}$ is finer than $\mathrm{p}X$, take any $\varepsilon > 0$ and $\Delta \in \mathrm{UP}(\mathrm{p}X)$. Then $\Delta \in \mathrm{UP}(X)$, and there is a finite set $D \subseteq X$ such that $\Delta(x, D) < \varepsilon$ for all $x \in X$. It follows that the Δ-diameter of X is finite. By Theorem 1.21, for each $z \in D$ the function $f_z := \backslash_x \Delta(x, z)$ is in $\mathrm{U_b}(X)$.

The pseudometric $\Delta'(x,y) := \max_{z \in D} |f_z(x) - f_z(y)|$, $x,y \in X$, is in $\mathrm{cl}^\vee \mathscr{P}$. If $x, y \in X$, there is $z \in D$ such that $\Delta(x,z) < \varepsilon$, and then

$$\Delta(x,y) < f_z(y) + \varepsilon < |f_z(x) - f_z(y)| + 2\varepsilon \le \Delta'(x,y) + 2\varepsilon \,,$$

which shows that $\Delta \ll \mathrm{cl}^\vee \mathscr{P}$.   □

Definition 2.11. For any uniform space X, define cX to be the uniform space that has the same points as X and the uniformity induced by $\mathsf{U}(X)$.   ■

Clearly $\mathsf{U}(X) = \mathsf{U}(cX)$, $\mathsf{U}_\mathsf{b}(X) = \mathsf{U}_\mathsf{b}(pX)$, $\mathsf{p}_1 X$ is finer than cX, and cX is finer than pX.

Exercise 2.12. Find a uniform space X such that $\mathsf{p}_1 X \ne cX \ne pX$.   ■

Two familiar classes of completely regular topological spaces may be defined by means of the uniformity of cX. A completely regular space T is *realcompact* iff the uniform space $c_\mathsf{F} T$ is complete; T is *pseudocompact* iff $c_\mathsf{F} T$ is precompact. The space T is compact if and only if it is both realcompact and pseudocompact, by Theorem 1.15.

Exercise 2.13. This exercise is for the readers familiar with categories and functors. Extend the operations U, U_b, F, p_α and c so that they apply not only to uniform spaces but also to uniformly continuous mappings. Show that p_α and c are reflective functors, and F is a coreflective functor, from the category of uniform spaces and uniformly continuous mappings to itself.   ■

Definition 2.14. Let $\mathfrak{S}$ be a set of subsets of a non-empty set S, $\mathfrak{S} \nearrow S$, and let a set $\mathscr{F} \subseteq \mathbb{R}^S$ be such that every function $f \in \mathscr{F}$ is bounded on every $A \in \mathfrak{S}$. The $\mathfrak{S}$-*uniformity* on the set $\mathscr{F}$ is the uniformity induced by the pseudometrics of the seminorms $\|\cdot\|_A$, $A \in \mathfrak{S}$.   ■

Exercise 2.15. When S, $\mathfrak{S}$ and $\mathscr{F}$ are as in Definition 2.14, the $\mathfrak{S}$-uniformity on $\mathscr{F}$ is compatible with the $\mathfrak{S}$-topology (Sect. P.3). If $\mathscr{F}$ is a vector subspace of $\mathbb{R}^S$ equipped with the $\mathfrak{S}$-topology, then the $\mathfrak{S}$-uniformity on $\mathscr{F}$ is the additive uniformity of $\mathscr{F}$.   ■

Several instances of the $\mathfrak{S}$-uniformity have their own names, which agree with the names of the corresponding $\mathfrak{S}$-topologies in Sect. P.3. When $\mathfrak{S}$ is the set of all finite subsets of S, the $\mathfrak{S}$-uniformity is called the S-*pointwise uniformity*. If E and F are two vector spaces in duality, then the F-*weak uniformity* on E is the F-pointwise uniformity obtained by identifying E with a subspace of $\mathbb{R}^F$, and $\mathsf{w}_F E$ denotes the uniform space on the set E with the F-weak uniformity. When E is a locally convex space, $\mathsf{w}E$ denotes the uniform space on the set E with the E^*-weak uniformity.

2.2   Uniform Subspaces

Definition 2.16. Let X be a uniform space and $\emptyset \neq Y \subseteq X$. The *subspace uniformity* on Y, denoted $\mathsf{UP}(X){\upharpoonright}Y$, is the uniformity on Y induced by the inclusion mapping $Y \hookrightarrow X$.

Say that a uniform space Y is a *uniform subspace* of a uniform space X, and write $Y \Subset X$, iff the set Y is a subset of the set X and the uniformity of Y is the subspace uniformity. ∎

When $\emptyset \neq Y \subseteq X$ and Δ is a pseudometric on X, let $\Delta{\upharpoonright}Y$ denote the restriction of Δ to Y. Thus $\Delta{\upharpoonright}Y := \overleftarrow{\iota}\,\Delta$ where $\iota : Y \hookrightarrow X$ is the inclusion mapping. When $Y \Subset X$, Theorem 2.6 yields that a pseudometric Δ' on the set Y is in $\mathsf{UP}(Y)$ if and only if $\Delta' \ll \{\Delta{\upharpoonright}Y \mid \Delta \in \mathsf{UP}(X)\}$.

Theorem 2.17. *Let X be a uniform space and $Y \Subset X$.*

1. *Y is a topological subspace of X.*
2. *If Y is complete, then it is closed in X.*
3. *If X is complete and Y is closed in X, then Y is complete.*

Proof. 1. The inclusion mapping $Y \hookrightarrow X$ is uniformly continuous and therefore continuous. On the other hand, if $\odot[x, \Delta']$ is a basic open set in Y, $x \in Y$ and $\Delta' \in \mathsf{UP}(Y)$, then there is $\Delta \in \mathsf{UP}(X)$ such that $\odot[x, \Delta] \cap Y \subseteq \odot[x, \Delta']$.
2. Let $\{y_\gamma\}_\gamma$ be a net in Y that converges in X. Then $\{y_\gamma\}_\gamma$ is Cauchy in X and therefore also in Y, and its limit is in Y because Y is complete.
3. Let $\{y_\gamma\}_\gamma$ be a Cauchy net in Y. Then $\{y_\gamma\}_\gamma$ is also Cauchy in X because $\Delta{\upharpoonright}Y \in \mathsf{UP}(Y)$ for every pseudometric $\Delta \in \mathsf{UP}(X)$. Thus $\{y_\gamma\}_\gamma$ converges in X, and its limit is in Y because Y is closed. □

Exercise 2.18. If X is a metric space with metric Δ and $Y \Subset X$, then the uniformity of Y is induced by $\Delta{\upharpoonright}Y$. ∎

A remarkable property of uniform subspaces is that if $Y \Subset X$, then every pseudometric in $\mathsf{UP_b}(Y)$ is a restriction of a pseudometric in $\mathsf{UP_b}(X)$, and every function in $\mathsf{U_b}(Y)$ is a restriction of a function in $\mathsf{U_b}(X)$. The proof below relies on the following combinatorial lemma:

Lemma 2.19. *Let X and Y be two sets, $\emptyset \neq Y \subseteq X$, let Δ be a pseudometric on Y and Δ' a pseudometric on X. If $\Delta \leq \Delta'{\upharpoonright}Y$, then there is a pseudometric $\widetilde{\Delta}$ on X such that $\Delta = \widetilde{\Delta}{\upharpoonright}Y$ and $\widetilde{\Delta} \leq \Delta'$.*

Proof. Let $p : X \times X \to \mathbb{R}$ be the function defined by $p(x,y) := \Delta(x,y)$ when $x, y \in Y$ and $p(x,y) := \Delta'(x,y)$ otherwise. When $s = (s_0, s_1, \ldots, s_k)$ is a finite sequence of elements $s_i \in X$, $k \geq 1$, write $s : s_0 \rightsquigarrow s_k$ and

$$p^*(s) := \sum_{i=1}^{k} p(s_{i-1}, s_i).$$

For $x, y \in X$, define

$$\tilde{\Delta}(x,y) := \inf\{p^*(s) \mid s : x \rightsquigarrow y\}.$$

Clearly $\tilde{\Delta}$ is a pseudometric on X and $\tilde{\Delta} \leq \Delta'$. It remains to be proved that $\Delta = \tilde{\Delta} \upharpoonright Y$.

Let $s = (s_0, s_1, \ldots, s_k)$ be any sequence in X such that $s_0, s_k \in Y$ and $k \geq 2$. Consider two cases:

1. $s_i \in Y$ for all i. Let s' be the sequence obtained from s by deleting the element s_1. Then $s' : s_0 \rightsquigarrow s_k$ and

$$p^*(s) - p^*(s') = p(s_0, s_1) + p(s_1, s_2) - p(s_0, s_2)$$
$$= \Delta(s_0, s_1) + \Delta(s_1, s_2) - \Delta(s_0, s_2) \geq 0.$$

2. There is i such that $s_i \notin Y$. Let s' be the sequence obtained from s by deleting s_i. Then $s' : s_0 \rightsquigarrow s_k$ and

$$p^*(s) - p^*(s') = p(s_{i-1}, s_i) + p(s_i, s_{i+1}) - p(s_{i-1}, s_{i+1})$$
$$= \Delta'(s_{i-1}, s_i) + \Delta'(s_i, s_{i+1}) - p(s_{i-1}, s_{i+1})$$
$$\geq \Delta'(s_{i-1}, s_{i+1}) - p(s_{i-1}, s_{i+1}) \geq 0.$$

Thus in either case, there is a shorter s' such that $s' : s_0 \rightsquigarrow s_k$ and $p^*(s) \geq p^*(s')$. It follows that if $x, y \in Y$ and $s : x \rightsquigarrow y$, then $p^*(s) \geq p^*((x,y)) = \Delta(x,y)$, and therefore $\tilde{\Delta}(x,y) = \Delta(x,y)$.   □

Theorem 2.20. *Let X be a uniform space, $Y \Subset X$ and $\Delta \in \mathsf{UP}_b(Y)$. Then there exists $\tilde{\Delta} \in \mathsf{UP}_b(X)$ such that $\Delta = \tilde{\Delta} \upharpoonright Y$.*

Proof. Without loss of generality, assume that $\Delta \leq 1$. Let $\mathscr{P}$ be the set of all restrictions to Y of pseudometrics in $\mathsf{UP}(X)$. The uniformity of Y is induced by $\mathscr{P}$, and $\Delta \ll \mathscr{P}$ by Theorem 2.6. Hence there are $\Delta_j \in \mathsf{UP}(X)$ for $j \in \omega$ such that

$$\forall x, y \in Y \quad \left[\Delta_j(x,y) < 1 \Rightarrow \Delta(x,y) < \frac{1}{2^{j+1}} \right].$$

Define the pseudometric Δ' on X by

$$\Delta'(x,y) := \sum_{j \in \omega} \frac{1}{2^j} (\Delta_j(x,y) \wedge 1).$$

If $x, y \in Y$ and $j \in \omega$ are such that $\Delta'(x,y) < 1/2^j$, then $\Delta_j(x,y) < 1$, and therefore $\Delta(x,y) < 1/2^{j+1}$. It follows that $\Delta \leq \Delta' \upharpoonright Y$. By Lemma 2.19, there exists a pseudometric $\tilde{\Delta}$ on X such that $\tilde{\Delta} \upharpoonright Y = \Delta$ and $\tilde{\Delta} \leq \Delta'$. Clearly $\Delta' \in \mathsf{UP}_b(X)$, and $\tilde{\Delta} \in \mathsf{UP}_b(X)$ because $\tilde{\Delta} \leq \Delta'$.   □

Lemma 2.21. *Let Δ be a pseudometric on a set X and $\emptyset \neq Y \subseteq X$. For every function f on Y that is 1-Lipschitz for $\Delta\lfloor Y$, there exists a function $\widetilde{f}$ on X that is 1-Lipschitz for Δ and such that $f(y) = \widetilde{f}(y)$ for all $y \in Y$.*

Proof. Define $\widetilde{f}(x) := \sup\{f(y) - \Delta(x,y) \mid y \in Y\}$ for $x \in X$.   □

Theorem 2.22. *Let X be a uniform space and $Y \Subset X$. For every $f \in \mathsf{U_b}(Y)$ there exists $\widetilde{f} \in \mathsf{U_b}(X)$ such that $f(y) = \widetilde{f}(y)$ for all $y \in Y$.*

Proof. When $f \in \mathsf{U_b}(Y)$, the pseudometric defined by $\Delta(y,y') := |f(y) - f(y')|$, $y,y' \in Y$, is in $\mathsf{UP_b}(Y)$. By Theorem 2.20, there exists $\widetilde{\Delta} \in \mathsf{UP_b}(X)$ such that $\Delta = \widetilde{\Delta}\lfloor Y$, and by Lemma 2.21, there exists $\widetilde{f} \in \mathsf{BLip_b}(\widetilde{\Delta})$ that extends f.   □

Exercise 2.23. Consider the uniform subspace $\mathbb{N} := \{0,1,\dots\}$ of $\mathbb{R}$. Show that there is a function in $\mathsf{U}(\mathbb{N})$ that cannot be extended to a function in $\mathsf{U}(\mathbb{R})$. Show that there is a pseudometric in $\mathsf{UP}(\mathbb{N})$ that cannot be extended to a pseudometric in $\mathsf{UP}(\mathbb{R})$.   ∎

Compare Theorem 2.20 and Exercise 2.23 with Corollary 2.25 below.

Theorem 2.24. *Let X, Y and Z be uniform spaces, $Y \Subset X$ and let Y be dense in X. If Z is complete and $\varphi: Y \to Z$ is a uniformly continuous mapping, then there is a uniformly continuous mapping $\widetilde{\varphi}: X \to Z$ that extends φ.*

Proof. Let E be the closure of the graph of φ in $X \times Z$:

$$E := \{(x,z) \in X \times Z \mid \exists \text{ net } \{y_\gamma\}_\gamma \text{ in } Y : \lim_\gamma y_\gamma = x \text{ and } \lim_\gamma \varphi(y_\gamma) = z\}.$$

To prove that E is the graph of a uniformly continuous mapping $\widetilde{\varphi}: X \to Z$ that extends φ, it suffices to establish the following.

(a) For every $x \in X$ there is $z \in Z$ such that $(x,z) \in E$.
(b) For every $\Delta' \in \mathsf{UP}(Z)$ and $\varepsilon > 0$, there exists $\Delta \in \mathsf{UP}(X)$ such that

$$\forall\, x,x' \in X \,\forall\, z,z' \in Z\, [\,\Delta(x,x') < 1 \text{ and } (x,z),(x',z') \in E \;\Rightarrow\; \Delta'(z,z') < 3\varepsilon\,].$$

To show (a), take any $x \in X$. There is a net $\{y_\gamma\}_\gamma$ in Y such that $\lim_\gamma y_\gamma = x$. The net $\{y_\gamma\}_\gamma$ is Cauchy in Y; hence the net $\{\varphi(y_\gamma)\}_\gamma$ is Cauchy in Z and there is $z \in Z$ such that $\lim_\gamma \varphi(y_\gamma) = z$.

To show (b), take any $\Delta' \in \mathsf{UP}(Z)$ and $\varepsilon > 0$. Since $\overleftarrow{\varphi}\Delta' \in \mathsf{UP}(Y)$ and $Y \Subset X$, there is $\Delta \in \mathsf{UP}(X)$ such that

$$\forall\, y,y' \in Y\, [\,\Delta(y,y') < 3 \;\Rightarrow\; \overleftarrow{\varphi}\Delta'(y,y') < \varepsilon\,].$$

Now if $(x,z),(x',z') \in E$, $\Delta(x,x') < 1$, then there are nets $\{y_\beta\}_\beta$ and $\{y'_\gamma\}_\gamma$ in Y such that $\lim_\beta y_\beta = x$, $\lim_\gamma y'_\gamma = x'$, $\lim_\beta \varphi(y_\beta) = z$ and $\lim_\gamma \varphi(y'_\gamma) = z'$. Choose β, γ so that $\Delta(y_\beta,x) < 1$, $\Delta(y'_\gamma,x') < 1$, $\Delta'(\varphi(y_\beta),z) < \varepsilon$ and $\Delta'(\varphi(y'_\gamma),z') < \varepsilon$. Straightforward estimates show that $\Delta(y_\beta,y'_\gamma) < 3$ and $\Delta'(z,z') < 3\varepsilon$, concluding the proof of (b).   □

Corollary 2.25. *Let X be a uniform space, Y $\in$ X. If Y is dense in X, then for every pseudometric $\Delta \in$ UP(Y) there exists $\widetilde{\Delta} \in$ UP(X) such that $\Delta = \widetilde{\Delta} \lceil Y$.*

Proof. Fix any $y_0 \in Y$, and

$$\mathscr{F} := \left\{ f: Y \to \mathbb{R} \ \middle| \ f(y_0) = 0 \text{ and } |f(y) - f(y')| \leq \Delta(y, y') \text{ for all } y, y' \in Y \right\}.$$

Let $\varphi: Y \to \ell_\infty(\mathscr{F})$ be the mapping defined by $\varphi(y)(f) := f(y)$ for $y \in Y$, $f \in \mathscr{F}$. Let Δ' be the metric of $\|\cdot\|_{\mathscr{F}}$; that means $\Delta'(z, z') := \|z - z'\|_{\mathscr{F}}$ for $z, z' \in \ell_\infty(\mathscr{F})$.

For every $y' \in Y$, the function $\backslash_y (\Delta(y, y') - \Delta(y_0, y'))$ belongs to $\mathscr{F}$; hence

$$\Delta(y, y') = \sup\{|f(y) - f(y')| \ | \ f \in \mathscr{F}\} \quad \text{for all } y, y' \in Y.$$

That means that $\overleftarrow{\varphi} \Delta' = \Delta$, and the mapping φ is uniformly continuous from Y to $\ell_\infty(\mathscr{F})$ with Δ'. By virtue of Theorem 2.24, there is a uniformly continuous mapping $\varphi': X \to \ell_\infty(\mathscr{F})$ such that $\varphi'(y) = \varphi(y)$ for all $y \in Y$. The pseudometric $\widetilde{\Delta} = \overleftarrow{\varphi'} \Delta'$ is in UP(X) because φ' is uniformly continuous, and obviously $\Delta = \widetilde{\Delta} \lceil Y$. $\qquad \square$

Note the correspondence between pseudometrics and certain uniformly equicontinuous sets of functions in the last proof. Variants of this important construction appear more than once further on.

2.3   Uniform Structures on Products

Definition 2.26. The *(uniform) product* of uniform spaces X_γ, $\gamma \in \Gamma$, is the uniform space whose set of points is the product set $\prod_{\gamma \in \Gamma} X_\gamma$ and whose uniformity (the *product uniformity*) is induced by the set of canonical projections onto the spaces X_γ, $\gamma \in \Gamma$.

Let $\prod_{\gamma \in \Gamma} X_\gamma$ denote the product of uniform spaces X_γ. Let $X \times Y$ denote the product of two spaces X and Y. $\qquad \square$

Lemma 2.27. *Let X be a uniform space and Δ a pseudometric on the set X. Then $\Delta \in$ UP(X) if and only if Δ is a uniformly continuous function on the product uniform space $X \times X$.*

Proof. If $\Delta \in$ UP(X), then $\Delta \in$ U(X × X) because

$$|\Delta(x, y) - \Delta(x', y')| \leq |\Delta(x, y) - \Delta(x', y)| + |\Delta(x', y) - \Delta(x', y')|$$
$$\leq \Delta(x, x') + \Delta(y, y')$$

for $(x, y), (x', y') \in X \times X$. If $\Delta \in$ U(X × X), then $\Delta \in$ UP(X) because by Theorem 2.6, for every $\varepsilon > 0$ there are $\Delta_\varepsilon \in$ UP(X) and $\theta > 0$ such that

$$[\Delta_\varepsilon(x, x') < \theta \text{ and } \Delta_\varepsilon(y, y') < \theta] \Rightarrow |\Delta(x, y) - \Delta(x', y')| < \varepsilon$$

for $(x, y), (x', y') \in X \times X$, and in particular $\Delta_\varepsilon(x, y) < \theta \Rightarrow \Delta(x, y) < \varepsilon$. $\qquad \square$

Exercise 2.28. Let X, Y and Z be three uniform spaces. Prove that a mapping $\varphi \colon X \times Y \to Z$ is uniformly continuous if and only if two conditions hold:

(i) The set $\{\backslash_x \varphi(x,y) \mid y \in Y\}$ of mappings from X to Z is uniformly equicontinuous.
(ii) The set $\{\backslash_y \varphi(x,y) \mid x \in X\}$ of mappings from Y to Z is uniformly equicontinuous.

■

Exercise 2.29. If $Y_\gamma \Subset X_\gamma$ for every $\gamma \in \Gamma$, then $\prod_{\gamma \in \Gamma} Y_\gamma \Subset \prod_{\gamma \in \Gamma} X_\gamma$. ■

Exercise 2.30. The topology of the uniform product $\prod_{\gamma \in \Gamma} X_\gamma$ is the product of the topologies of X_γ. ■

Exercise 2.31. The uniform product of complete spaces is complete. ■

For certain applications, the product uniformity on $X \times Y$ is not fine enough (see Exercise 3.11). A finer uniformity, such as the one defined next, is needed.

Theorem 2.32. *Let X and Y be two uniform spaces. Let $\mathcal{U}$ be the set of all pseudometrics Δ on the set $X \times Y$ for which:*

• *The pseudometric on X defined by*

$$(x,x') \mapsto \sup_{y \in Y} \left(1 \wedge \Delta((x,y),(x',y))\right), \ x,x' \in X,$$

belongs to $\mathsf{UP}(X)$.
• *For every $x \in X$, the pseudometric on Y defined by*

$$(y,y') \mapsto \Delta((x,y),(x,y')), \ y,y' \in Y,$$

belongs to $\mathsf{UP}(Y)$.

Then $\mathcal{U}$ has properties (U1), (U2) and (U3) in Definition 1.1, and the uniformity $\mathcal{U}$ is finer than the product uniformity.

Proof. $\mathcal{U}$ has properties (U1) and (U2) because $\mathsf{UP}(X)$ and $\mathsf{UP}(Y)$ do.

Let π_X and π_Y be the canonical projections from $X \times Y$ onto X and Y, respectively. Then $\overleftarrow{\pi_X}\Delta \in \mathcal{U}$ for every $\Delta \in \mathsf{UP}(X)$ and $\overleftarrow{\pi_Y}\Delta \in \mathcal{U}$ for every $\Delta \in \mathsf{UP}(Y)$. It follows that $\mathcal{U}$ has property (U3) and is finer than the product uniformity. □

Definition 2.33. Let X and Y be two uniform spaces. The *semiuniform product $X*Y$* of X and Y (in this order) is the uniform space whose set of points is the product set $X \times Y$ and whose uniformity is the set $\mathcal{U}$ in Theorem 2.32. ■

Theorem 2.34. *For any uniform spaces X, Y and Z, a set Φ of mappings from $X*Y$ to Z is uniformly equicontinuous if and only if these two conditions hold:*

(i) *The set $\{\backslash_x \varphi(x,y) \mid \varphi \in \Phi, y \in Y\}$ of mappings from X to Z is uniformly equicontinuous.*
(ii) *For every $x \in X$, the set $\{\backslash_y \varphi(x,y) \mid \varphi \in \Phi\}$ of mappings from Y to Z is uniformly equicontinuous.*

Proof. With $\Psi := \{\backslash_x \varphi(x,y) \mid \varphi \in \Phi, y \in Y\}$ and $\Psi_x := \{\backslash_y \varphi(x,y) \mid \varphi \in \Phi\}$ for $x \in X$, conditions (i) and (ii) mean that $\overleftarrow{\Psi}\Delta' \in \mathrm{UP}(X)$ and $\overleftarrow{\Psi_x}\Delta' \in \mathrm{UP}(Y)$ for all $\Delta' \in \mathrm{UP_b}(Z)$ and $x \in X$. We have

$$\overleftarrow{\Psi}\Delta'(x,x') = \sup\left\{\overleftarrow{\varphi}\Delta'((x,y),(x',y)) \mid \varphi \in \Phi, y \in Y\right\}$$

$$= \sup_{y \in Y} \overleftarrow{\Phi}\Delta'((x,y),(x',y)) \quad \text{for } x, x' \in X;$$

$$\overleftarrow{\Psi_x}\Delta'(y,y') = \sup\left\{\overleftarrow{\varphi}\Delta'((x,y),(x,y')) \mid \varphi \in \Phi\right\}$$

$$= \overleftarrow{\Phi}\Delta'((x,y),(x,y')) \quad \text{for } y, y' \in Y,$$

hence (i) and (ii) hold if and only if $\overleftarrow{\Phi}\Delta' \in \mathrm{UP}(X*Y)$ for every $\Delta' \in \mathrm{UP_b}(Z)$. $\square$

Corollary 2.35. *For any uniform spaces X, Y and Z, a mapping $\varphi\colon X*Y \to Z$ is uniformly continuous if and only if these two conditions hold:*

(i) The set $\{\backslash_x \varphi(x,y) \mid y \in Y\}$ of mappings from X to Z is uniformly equicontinuous.
(ii) The mapping $\backslash_y \varphi(x,y)$ from Y to Z is uniformly continuous for every $x \in X$. $\square$

Lemma 2.36. *For any three uniform spaces X_0, X_1 and X_2, the natural mapping $\psi\colon (X_0*X_1)*X_2 \to X_0*(X_1*X_2)$ defined by $\psi(((x_0,x_1),x_2)) := (x_0,(x_1,x_2))$ is a uniform isomorphism from $(X_0*X_1)*X_2$ onto $X_0*(X_1*X_2)$.*

Proof. I shall prove that a mapping $\varphi\colon X_0*(X_1*X_2) \to Y$ to a uniform space Y is uniformly continuous if and only if the mapping $\varphi \circ \psi$ is uniformly continuous on $(X_0*X_1)*X_2$.

By Theorem 2.34, $\varphi \circ \psi$ is uniformly continuous on $(X_0*X_1)*X_2$ if and only if:

(a) The set $\{\backslash_{x_0} \varphi \circ \psi((x_0,x_1),x_2) \mid x_1 \in X_1, x_2 \in X_2\}$ of mappings from X_0 to Y is uniformly equicontinuous on X_0.
(b) For every $x_0 \in X_0$, the set $\{\backslash_{x_1} \varphi \circ \psi((x_0,x_1),x_2) \mid x_2 \in X_2\}$ of mappings from X_1 to Y is uniformly equicontinuous on X_1.
(c) For every $x_0 \in X_0$ and $x_1 \in X_1$, the mapping $\backslash_{x_2} \varphi \circ \psi((x_0,x_1),x_2)$ from X_2 to Y is uniformly continuous on X_2.

Also by Theorem 2.34, φ is uniformly continuous on $X_0*(X_1*X_2)$ if and only if:

(a$'$) The set $\{\backslash_{x_0} \varphi(x_0,(x_1,x_2)) \mid x_1 \in X_1, x_2 \in X_2\}$ of mappings from X_0 to Y is uniformly equicontinuous on X_0.
(b$'$) For every $x_0 \in X_0$, the set $\{\backslash_{x_1} \varphi(x_0,(x_1,x_2)) \mid x_2 \in X_2\}$ of mappings from X_1 to Y is uniformly equicontinuous on X_1.
(c$'$) For every $x_0 \in X_0$ and $x_1 \in X_1$, the mapping $\backslash_{x_2} \varphi(x_0,(x_1,x_2))$ from X_2 to Y is uniformly continuous on X_2.

Clearly (a), (b) and (c) are equivalent to (a$'$), (b$'$) and (c$'$). $\square$

Exercise 2.37. For any uniform spaces X and Y, the topology of $X*Y$ is the product of the topologies of X and Y. ∎

2.4 Notes for Chap. 2

The references in Sect. 1.5 deal with basic properties of induced uniformities, using entourages or uniform covers. Realcompact and pseudocompact spaces are discussed by Császár [31] and Gillman and Jerison [81].

The uniformity in Example 2.3 and similar more general constructions were studied by Frolík [65] and Hager [89]. The survey by Rice [157] includes a discussion of these constructions and their connections to descriptive set theory.

Extension results for pseudometrics and functions in Theorems 2.20 and 2.22 were proved by Isbell [99] and Katětov [106], respectively. The construction in the proof of Theorem 2.24 is adapted from Kelley [109, 6.26], and the semiuniform product in Definition 2.33 from Isbell [100, Ch.III].

2.6 Notes for Chap. 2

Chapter 3
Uniform Structures on Semigroups

This chapter begins with basic properties of the right uniformity in semitopological semigroups and in topological groups. Then I abstract the key property of the right uniformity in the definition of a semiuniform semigroup, which will provide a framework for the study of convolution in Chap. 9. I also define ambitable semigroups and prove that many topological groups are ambitable.

3.1 Semitopological Semigroups

Let Δ be a pseudometric on a semigroup S. Say that Δ is *right non-expansive* iff $\Delta(x,y) \geq \Delta(xz, yz)$ for all $x, y, z \in S$. Note that if S is a group, then a pseudometric on S is right non-expansive if and only if it is right-invariant.

Definition 3.1. A *semitopological semigroup* is a semigroup S with a topology such that for every $x \in S$ the mappings $y \mapsto xy$ and $y \mapsto yx$ are continuous from S to itself.

Let S be a semitopological semigroup. Let $\mathsf{RP}(S)$ denote the set of the right non-expansive pseudometrics Δ on S such that for every $x \in S$ the function $\setminus_y \Delta(x,y)$ from S to $\mathbb{R}$ is continuous. Let $\mathsf{LUC}(S)$ denote the set of functions $f \in \mathsf{C_b}(S)$ such that the mapping $x \mapsto \setminus_y f(xy)$ is continuous from S to $\mathsf{C_b}(S)$ with the $\|\cdot\|_S$ norm.

Let S be a semitopological semigroup such that $\mathsf{RP}(S)$ separates the points of S. The *right uniformity* of S is the uniformity induced by $\mathsf{RP}(S)$. The uniform space rS is the set S with the right uniformity. ∎

Exercise 3.2. Let E be a locally convex space. Show that the additive uniformity of E (Definition 2.4) is the right uniformity of E when E is considered as a semitopological semigroup with its addition operation $+$. ∎

Lemma 3.3. *Let S be a semitopological semigroup whose points are separated by $\mathsf{RP}(S)$. For every $\Delta \in \mathsf{UP_b}(rS)$ there exists $\Delta' \in \mathsf{RP}(S)$ such that $\Delta \leq \Delta'$.*

J. Pachl, *Uniform Spaces and Measures*, Fields Institute Monographs 30,
DOI 10.1007/978-1-4614-5058-0_4,
© Springer Science+Business Media New York 2013

Proof. The construction is similar to that in the proof of Theorem 2.20. Take any $\Delta \in UP_b(rS)$. It is enough to consider the case $\Delta \leq 1$. As $RP(S)$ is closed under the operation $\vee$, we have $\Delta \ll RP(S)$ by Theorem 2.6, and there are $\Delta_j \in RP(S)$ for $j \in \omega$ such that

$$\forall x,y \in S \left[\Delta_j(x,y) < 1 \Rightarrow \Delta(x,y) < \frac{1}{2^{j+1}} \right].$$

Define the pseudometric $\Delta' \in RP(S)$ by

$$\Delta'(x,y) := \sum_{j \in \omega} \frac{1}{2^j} (\Delta_j(x,y) \wedge 1).$$

If $x,y \in X$ and $j \in \omega$ are such that $\Delta'(x,y) < 1/2^j$, then $\Delta_j(x,y) < 1$ and therefore $\Delta(x,y) < 1/2^{j+1}$. It follows that $\Delta \leq \Delta'$.   □

Exercise 3.4. For the additive group of integers with the discrete topology, show that Lemma 3.3 does not hold when Δ is not required to be bounded.   ∎

Lemma 3.5. *Let S be a semitopological semigroup.*

1. $RP(S)$ *separates the points of S if and only if* $LUC(S)$ *does.*
2. *If* $RP(S)$ *separates the points of S, then* $LUC(S) = U_b(rS)$.

Proof. If $f \in LUC(S)$, then the pseudometric Δ defined by

$$\Delta(x,y) := |f(x) - f(y)| \vee \sup_{z \in S} |f(xz) - f(yz)| \quad \text{for } x,y \in S$$

is in $RP(S)$, and $f \in Lip_b(\Delta)$. Hence if $LUC(S)$ separates the points of S, then so does $RP(S)$ and $LUC(S) \subseteq U_b(rS)$.

Now assume that $RP(S)$ separates the points of S, and take any $f \in U_b(rS)$, $\|f\|_S \leq 1$. By Lemma 3.3, there exists $\Delta \in RP(S)$ such that $f \in BLip_b(\Delta)$. Then $|f(x) - f(y)| \leq \Delta(x,y)$ for $x,y \in S$ and $|f(xz) - f(yz)| \leq \Delta(xz,yz) \leq \Delta(x,y)$ for $x,y,z \in S$, which shows that $f \in LUC(S)$. Thus $U_b(rS) \subseteq LUC(S)$ and $LUC(S)$ separates the points of S.   □

In a general semitopological semigroup S, even one whose topology is Hausdorff, the set $RP(S)$ need not separate the points of S. But then little is lost by passing to the quotient space S' whose points are separated by $RP(S')$:

Exercise 3.6. Let S be a semitopological semigroup. For $x,x' \in S$, let $x \sim x'$ iff $\Delta(x,x') = 0$ for every $\Delta \in RP(S)$. Let S' be the set of equivalence classes of $\sim$. If $x,x',y,y' \in S$, $x \sim x'$ and $y \sim y'$, then $xy \sim x'y'$. Thus there is a unique semigroup operation on S' for which the quotient mapping $\chi : S \to S'$ is a homomorphism. With the quotient topology, S' is a semitopological semigroup whose points are separated by $RP(S')$, and $LUC(S) = \{f \circ \chi \mid f \in LUC(S')\}$.   ∎

Theorem 3.7. *Let S and S' be two semitopological semigroups whose points are separated by* RP(S) *and* RP(S'). *If* $\varphi: S \to S'$ *is a continuous homomorphism, then* φ *is uniformly continuous from rS to rS'.*

Proof. In view of Lemma 3.3, it is enough to show that $\overleftarrow{\varphi}\Delta \in \mathrm{UP}(rS)$ for every $\Delta \in \mathrm{RP}(S')$. If $\Delta \in \mathrm{RP}(S')$, then the pseudometric $\overleftarrow{\varphi}\Delta$ is right non-expansive, and the function $\backslash_y \overleftarrow{\varphi}\Delta(x,y)$ is continuous for every $x \in S$. Therefore $\overleftarrow{\varphi}\Delta \in \mathrm{RP}(S)$. □

Theorem 3.8. *For any semitopological semigroup S whose points are separated by* RP(S), *the semigroup operation* $(x,y) \mapsto xy$ *is uniformly continuous from the semiuniform product rS*rS to rS.*

Proof. Set $\varphi(x,y) := xy$ for $x,y \in S$. If $\Delta \in \mathrm{RP}(S)$, then

$$\Delta(\varphi(x,z), \varphi(y,z)) = \Delta(xz, yz) \le \Delta(x,y)$$

for $x,y,z \in S$, which proves condition (i) in Corollary 2.35.

To prove condition (ii) in 2.35, take any $\Delta \in \mathrm{RP}(S)$. For every $z \in S$ the pseudometric Δ_z defined by

$$\Delta_z(x,y) := \Delta(\varphi(z,x), \varphi(z,y)) = \Delta(zx, zy)$$

for $x,y \in S$ is right non-expansive and the function $\backslash_y \Delta_z(x,y)$ is continuous for every $x \in S$, hence $\Delta_z \in \mathrm{RP}(G)$. Condition (ii) follows. □

3.2 Topological Groups

Evidently a pseudometric Δ on a topological group G is in RP(G) if and only if Δ is right-invariant and the function $y \mapsto \Delta(e_G, y)$ is continuous on G.

Theorem 3.9. *Let G be a topological group.*

1. *The set* RP(G) *separates the points of G.*
2. *The right uniformity of G is compatible with the topology of G.*

Proof. Part 1 follows from Theorem P.3.

In view of Lemma 3.3, the sets $\odot[x, \Delta]$ where $x \in G$ and $\Delta \in \mathrm{RP}(G)$ form a base of the topology of rG. Every such set $\odot[x, \Delta]$ is open in the topology of G by the continuity of the function $\backslash_y \Delta(x,y)$.

Conversely, if V is an open set in the topology of G and $x_0 \in V$, then Vx_0^{-1} is a neighbourhood of e_G, and by Theorem P.3, there is a pseudometric $\Delta \in \mathrm{RP}(G)$ such that $\odot[x_0, \Delta] \subseteq V$. □

Theorem 3.10. *Let G' be a topological group and G its topological subgroup. Then* $rG \Subset rG'$.

Proof. Let $\mathcal{U}$ be the restriction of the uniformity of rG' to G. By Theorem 3.7, the uniformity of rG is finer than $\mathcal{U}$.

To prove that $\mathscr{U}$ is finer than the uniformity of rG, take any $\Delta \in \mathrm{RP}(G)$, $\varepsilon > 0$. Then $\odot[e_G, \Delta/\varepsilon]$ is a neighbourhood of e_G in G. By Lemma 3.3 and Theorem 3.9, there is $\Delta' \in \mathrm{RP}(G')$ such that $\odot[e_{G'}, \Delta'] \cap G \subseteq \odot[e_G, \Delta/\varepsilon]$. Since both Δ and Δ' are right-invariant, if $x, y \in G$, $\Delta'(x, y) < 1$, then $\Delta(x, y) < \varepsilon$. That proves $\Delta \in \mathscr{U}$. $\square$

Exercise 3.11. Prove that, for a topological group G, the mapping $(x, y) \mapsto xy$ is uniformly continuous from $rG \times rG$ to rG if and only if the mapping $x \mapsto x^{-1}$ is uniformly continuous from rG to rG. Find a topological group G for which these two mappings are not uniformly continuous. ∎

Definition 3.12. Let G be a topological group and κ an infinite cardinal. A subset A of G is κ-*bounded* iff for every neighbourhood V of e_G there exists a set $H \subseteq G$ such that $|H| \leq \kappa$ and $A \subseteq VH$.

The group G is κ-*bounded* iff it is κ-bounded as a subset of itself; it is *locally* κ-*bounded* iff there is a κ-bounded neighbourhood of e_G. ∎

Every locally compact group is locally $\aleph_0$-bounded and therefore also locally κ-bounded for any $\kappa \geq \aleph_0$.

Exercise 3.13. Let G be any topological group and α any ordinal. Show that G is $\aleph_\alpha$-bounded if and only if $rG = \mathrm{p}_{\alpha+1} rG$. ∎

3.3 Semiuniform Semigroups

Definition 3.14. A *semiuniform semigroup* is a semigroup X with a uniform structure such that the semigroup operation $(x, y) \mapsto xy$, $x, y \in X$, is uniformly continuous from the semiuniform product X^*X to X. ∎

A more precise term would be a *right semiuniform semigroup*, since we may define a *left semiuniform semigroup* by requiring that the mapping $(x, y) \mapsto yx$ be uniformly continuous from X^*X to X. However, here I only consider right semiuniform semigroups and omit the qualifier *right*.

If S is a semitopological semigroup whose points are separated by $\mathrm{LUC}(S)$, then rS is a semiuniform semigroup and $\mathrm{LUC}(S) = \mathrm{U_b}(rS)$ by Lemma 3.5 and Theorem 3.8.

Exercise 3.15. Every semiuniform semigroup X is a semitopological semigroup when considered with the topology of its uniformity, so that $\mathrm{LUC}(X)$ is defined. Find a semiuniform semigroup X such that $\mathrm{LUC}(X) \neq \mathrm{U_b}(X)$. ∎

Exercise 3.16. Let A be a normed algebra and $0 < r \leq 1$. Show that both the open and the closed ball of radius r in A with the algebra multiplication and the uniformity defined by the norm of A are semiuniform semigroups. ∎

Exercise 3.17. Let X be a semiuniform semigroup and $\Delta \in \mathsf{UP}_\mathsf{b}(X)$. Prove that there exists a right non-expansive pseudometric $\Delta' \in \mathsf{UP}_\mathsf{b}(X)$ such that $\Delta \leq \Delta'$ and $\Delta(xz, yz) \leq \Delta'(x, y)$ for all $x, y, z \in X$. ∎

Definition 3.18. Let f be a function on a semigroup S and $y \in S$. The *right translation of f by y* is the function $\backslash_x f(xy)$. The *right S-orbit of f*, or simply the *orbit of f*, is the set of all right translations of f:

$$\mathsf{orb}(f) := \{\backslash_x f(xy) \mid y \in S\}.$$

The *right S-orbit closure of f*, or simply the *orbit closure of f*, denoted by $\overline{\mathsf{orb}}(f)$, is the S-pointwise closure of $\mathsf{orb}(f)$ in the space $\mathbb{R}^S$. ∎

Lemma 3.19. *Let X be a semiuniform semigroup. For every function $f \in \mathsf{U}(X)$, the set $\overline{\mathsf{orb}}(f)$ is uniformly equicontinuous on X.*

Proof. For $f \in \mathsf{U}(X)$, the function $(x, y) \mapsto f(xy)$ is uniformly continuous on $X * X$. By Corollary 2.35, there is $\Delta \in \mathsf{UP}(X)$ such that $1 \wedge |g(x) - g(x')| \leq \Delta(x, x')$ for all $x, x' \in X$ and $g \in \mathsf{orb}(f)$. Then the same holds for all $g \in \overline{\mathsf{orb}}(f)$. □

Thus $\overline{\mathsf{orb}}(f)$ is a $\mathsf{UEB}(X)$ set for every $f \in \mathsf{U}_\mathsf{b}(X)$. The next definition singles out the semiuniform semigroups in which, conversely, every UEB set is included in $\overline{\mathsf{orb}}(f)$ for some $f \in \mathsf{U}_\mathsf{b}(X)$. This property has an important role in the study of topological centres in Chap. 9.

Definition 3.20. A semiuniform semigroup X is *ambitable* iff for every $\mathsf{UEB}(X)$ set $\mathscr{F} \subseteq \mathsf{U}_\mathsf{b}(X)$ there exists $f \in \mathsf{U}_\mathsf{b}(X)$ such that $\mathscr{F} \subseteq \overline{\mathsf{orb}}(f)$.

A semitopological semigroup, and in particular a topological group, is *ambitable* iff it is ambitable as a semiuniform semigroup with its right uniformity. ∎

For every pseudometric Δ we have $\mathsf{BLip}_\mathsf{b}(\Delta) \subseteq 2\mathsf{BLip}_\mathsf{b}(\Delta)^+ - 1$. By Lemma 1.20, a semiuniform semigroup X is ambitable if and only if for every $\Delta \in \mathsf{UP}(X)$, there is $f \in \mathsf{U}_\mathsf{b}(X)$ such that $\mathsf{BLip}_\mathsf{b}(\Delta)^+ \subseteq \overline{\mathsf{orb}}(f)$. And for semitopological semigroups it is enough to consider right non-expansive pseudometrics:

Lemma 3.21. *A semitopological semigroup S is ambitable if and only if for every pseudometric $\Delta \in \mathsf{RP}(S)$ there exists $f \in \mathsf{U}_\mathsf{b}(S)$ such that $\mathsf{BLip}_\mathsf{b}(\Delta)^+ \subseteq \overline{\mathsf{orb}}(f)$.*

Proof. By Lemma 3.3, for every pseudometric $\Delta \in \mathsf{UP}(rS)$ there is $\Delta' \in \mathsf{RP}(S)$ such that $\Delta \wedge 1 \leq \Delta'$, and then $\mathsf{BLip}_\mathsf{b}(\Delta)^+ \subseteq \mathsf{BLip}_\mathsf{b}(\Delta')^+$. □

Definition 3.22. Let X be a semiuniform semigroup and Y a uniform space. A *semiuniform action of X on Y* is a uniformly continuous mapping $(x, y) \mapsto x \cdot y$ from $X * Y$ to Y such that $(xx') \cdot y = x \cdot (x' \cdot y)$ for $x, x' \in X$, $y \in Y$. We say that X *acts semiuniformly on Y* by such a mapping. ∎

Every semiuniform semigroup acts on itself by the action $x \cdot y = xy$.

3.4   Ambitable Groups

The main result of this section (Theorem 3.35) states that many topological groups
are ambitable. But first I show that precompact groups are not.

A topological group G is said to be *precompact* iff the uniform space rG is
precompact.

Exercise 3.23. Prove that these three properties of a topological group G are
equivalent:

(i) G is precompact.
(ii) For every neighbourhood V of e_G there exists a finite set $D \subseteq G$ such that
     $G \subseteq VD$.
(iii) For every neighbourhood V of e_G there exists a finite set $D \subseteq G$ such that
     $G \subseteq DV$.   ∎

Theorem 3.24. *No precompact topological group is ambitable.*

Proof. Let G be a precompact group, and fix any $f \in U_b(rG)$. Since the orbit $\mathrm{orb}(f)$
is uniformly equicontinuous (Lemma 3.19), there are $\Delta \in UP(rG)$ and $\theta > 0$ such
that if $x, x', y \in G$ and $\Delta(x, x') < \theta$, then $|f(xy) - f(x'y)| < 1/3$. As G is precompact,
there is a finite set $D \subseteq G$ such that for every $x \in G$ there is $z \in D$ with $\Delta(x, z) < \theta$,
and thus $|f(xy) - f(zy)| < 1/3$ for every $y \in G$.

Consider the constant functions 0 and 1. If $0 \in \overline{\mathrm{orb}}(f)$, then there is $y \in G$ such that
$f(zy) < 1/3$ for every $z \in D$; hence $f(xy) < 2/3$ for every $x \in G$. Thus $f(x) < 2/3$
for every $x \in G$, and $1 \notin \overline{\mathrm{orb}}(f)$. This proves that there is no $f \in U_b(rG)$ for which
$0, 1 \in \overline{\mathrm{orb}}(f)$.   □

Definition 3.25. For every right-invariant pseudometric Δ on a group G, define
three cardinal numbers:

- $d(\Delta)$, the Δ-*density* of G, is the least cardinality of a Δ-dense subset of G.
- $\eta^{\sharp}(\Delta)$ is the least cardinality of a set $A \subseteq G$ such that $G = \odot[e_G, \Delta] A$.
- $\eta(\Delta)$ is the least cardinality of a set $A \subseteq G$ for which there exists a finite set
  $D \subseteq G$ such that $G = D \odot [e_G, \Delta] A$.   ∎

Lemma 3.26. *Let Δ be a right-invariant pseudometric on a group G such that
$\eta(\Delta) \geq \aleph_0$. Let Γ be a set of cardinality $\eta(\Delta)$, and for each $\gamma \in \Gamma$ let D_γ be a
non-empty finite subset of G. Then there are elements $x_\gamma \in G$ for $\gamma \in \Gamma$ such that
$\Delta(D_\beta x_\beta, D_\gamma x_\gamma) \geq 1$ whenever $\beta, \gamma \in \Gamma$, $\beta \neq \gamma$.*

Proof. Without loss of generality, assume that Γ is the set of ordinals smaller than
$\eta(\Delta)$. The construction of x_γ proceeds by transfinite recursion.

Start with any $x_0 \in G$. When $\alpha \in \Gamma$, $\alpha > 0$, assume that $x_\gamma \in G$ for $\gamma < \alpha$ have
already been constructed for which $\Delta(D_\beta x_\beta, D_\gamma x_\gamma) \geq 1$ whenever $\beta, \gamma < \alpha$, $\beta \neq \gamma$.
Since $\eta(\Delta) \geq \aleph_0$ and $\alpha < \eta(\Delta)$, the definition of $\eta(\Delta)$ gives

$$G \neq D_\alpha^{-1} \odot [e_G, \Delta] \bigcup_{\gamma < \alpha} D_\gamma x_\gamma.$$

Thus there exists $x_\alpha \in G$ such that $\Delta(y, zx_\alpha) \geq 1$ for all $z \in D_\alpha$ and all $y \in D_\gamma x_\gamma$, $\gamma < \alpha$. That means $\Delta(D_\gamma x_\gamma, D_\alpha x_\alpha) \geq 1$ for all $\gamma < \alpha$. $\qquad\square$

Lemma 3.27. *Let G be a topological group, $\Delta \in \mathrm{RP}(G)$ and $\eta(\Delta) \geq \aleph_0$. If $\mathcal{W}$ is a collection of non-empty G-pointwise open subsets of $\mathrm{BLip_b}(\Delta)^+$ and $|\mathcal{W}| \leq \eta(\Delta)$, then there exists $f \in \mathrm{BLip_b}(\Delta)^+$ such that $\mathrm{orb}(f)$ intersects every set in $\mathcal{W}$.*

Proof. Without loss of generality, assume that every set in $\mathcal{W}$ is a basic neighbourhood of the form

$$U = \{ g \in \mathrm{BLip_b}(\Delta)^+ \mid |g(x) - h_U(x)| < \varepsilon_U \text{ for all } x \in D_U \}$$

where $D_U \subseteq G$ is a non-empty finite set, $h_U \in \mathrm{BLip_b}(\Delta)^+$ and $\varepsilon_U > 0$.

By Lemma 3.26 with $\mathcal{W}$ in place of Γ, there are elements $x_U \in X$ for $U \in \mathcal{W}$ such that $\Delta(D_U x_U, D_V x_V) \geq 1$ whenever $U, V \in \mathcal{W}$, $U \neq V$. Define $f : G \to \mathbb{R}$ by

$$f(x) := \sup_{V \in \mathcal{W}} \max_{z \in D_V} (h_V(z) - \Delta(x, zx_V))^+ \quad \text{for } x \in G.$$

For every $V \in \mathcal{W}$ and every $z \in D_V$, the function $x \mapsto (h_V(z) - \Delta(x, zx_V))^+$ is in $\mathrm{BLip_b}(\Delta)^+$, hence so is f.

Take any $U \in \mathcal{W}$ and $x \in D_U$. From the definition of f we get $f(xx_U) \geq h_U(x)$. To prove the opposite inequality, take any $V \in \mathcal{W}$ and $z \in D_V$.

If $V = U$ then from

$$h_U(z) - h_U(x) \leq |h_U(z) - h_U(x)| \leq \Delta(x, z) = \Delta(xx_U, zx_U)$$

and $h_U = h_V$, $x_U = x_V$, we get $(h_V(z) - \Delta(xx_U, zx_V))^+ \leq h_U(x)$. If $V \neq U$, then we get the same inequality from $\Delta(xx_U, zx_V) \geq 1$.

Thus $(h_V(y) - \Delta(xx_U, zx_V))^+ \leq h_U(x)$ in both cases, and $f(xx_U) \leq h_U(x)$ follows from the definition of f.

In conclusion, $f(xx_U) = h_U(x)$ for every $U \in \mathcal{W}$ and $x \in D_U$, which means that the right translation of f by x_U is in U. $\qquad\square$

Lemma 3.28. *Let G be a topological group, $\Delta \in \mathrm{RP}(G)$ and $\mathrm{d}(\Delta) = \eta(\Delta) \geq \aleph_0$. Then there exists $f \in \mathrm{BLip_b}(\Delta)^+$ such that $\mathrm{BLip_b}(\Delta)^+ = \overline{\mathrm{orb}}(f)$.*

Proof. Let A be a Δ-dense subset of G of cardinality $\mathrm{d}(\Delta)$. The restriction to $\mathrm{BLip_b}(\Delta)^+$ of the canonical projection $\mathbb{R}^G \to \mathbb{R}^A$ is a homeomorphism with respect to the pointwise topologies. Hence the G-pointwise topology on $\mathrm{BLip_b}(\Delta)^+$ has a base of cardinality at most $\eta(\Delta)$, and by Lemma 3.27 there is $f \in \mathrm{BLip_b}(\Delta)^+$ whose orbit intersects every non-empty open set in $\mathrm{BLip_b}(\Delta)^+$. $\qquad\square$

Exercise 3.29. Let E be a normed space, $E \neq \{0\}$. Show that E, considered as a topological group with its addition operation $+$ and the norm topology, is ambitable.

$\qquad\qquad\qquad\qquad\qquad\qquad\qquad\qquad\qquad\qquad\qquad\qquad\qquad\qquad\qquad\qquad\blacksquare$

Lemma 3.28 becomes more useful when the cardinal $\eta(\Delta)$ is replaced by $\eta^\#(\Delta)$. The following estimate makes that possible.

Theorem 3.30. *Let G be a topological group, let κ be an infinite cardinal, and $\Delta \in \mathrm{RP}(G)$. If $\eta(2\Delta) < \kappa$, then $\eta^{\#}(\Delta) < \kappa$.*

The proof comes after a combinatorial lemma.

Lemma 3.31. *Let G be a group, $j \in \omega$, $A \subseteq G$, $B_i \subseteq G$ for $0 \leq i \leq j$, and $|A| < \kappa$, where κ is an infinite cardinal. If $G = \bigcup_{i=0}^{j} B_i A$, then there are a set $A' \subseteq G$ and i, $0 \leq i \leq j$, such that $G = B_i^{-1} B_i A'$ and $|A'| < \kappa$.*

Proof. Proceed by induction. When $j = 0$, the statement holds with $i = 0$ and $A' = A$.

For the induction step, assume that the statement in the lemma is true for j, and let $A \subseteq G$, $\kappa \geq \aleph_0$ and $B_0, B_1, \ldots, B_{j+1} \subseteq G$ be such that $G = \bigcup_{i=0}^{j+1} B_i A$ and $|A| < \kappa$.

If $G = B_{j+1}^{-1} B_{j+1} A$, then set $i = j + 1$ and $A' = A$.

On the other hand, if $G \neq B_{j+1}^{-1} B_{j+1} A$, then there is $x \in G \setminus B_{j+1}^{-1} B_{j+1} A$. Thus $B_{j+1} x \cap B_{j+1} A = \emptyset$, and $B_{j+1} \subseteq \bigcup_{i=0}^{j} B_i A x^{-1}$. It follows that

$$G = \bigcup_{i=0}^{j} B_i (A \cup A x^{-1} A)$$

and clearly $|A \cup A x^{-1} A| < \kappa$. By the induction hypothesis, there are $A' \subseteq G$ and i, $0 \leq i \leq j$, such that $G = B_i^{-1} B_i A'$ and $|A'| < \kappa$.

Thus in either case the statement holds with $j + 1$ in place of j.   □

Proof of Theorem 3.30. Let $B := \odot[e_G, 2\Delta]$. For $x, y \in B$ we have

$$\Delta(e_G, x^{-1} y) = \Delta(y^{-1}, x^{-1}) \leq \Delta(y^{-1}, e_G) + \Delta(x^{-1}, e_G) = \Delta(e_G, y) + \Delta(e_G, x) < 1$$

and thus $B^{-1} B \subseteq \odot[e_G, \Delta]$.

By the definition of $\eta(2\Delta)$, there are sets $A, D \subseteq G$ such that $|A| = \eta(2\Delta) < \kappa$, $|D| < \aleph_0$ and $G = DBA$.

By Lemma 3.31, there are $x \in D$ and $A' \subseteq G$ such that $|A'| < \kappa$ and

$$G = (xB)^{-1} x B A' = B^{-1} B A' \subseteq \odot[e_G, \Delta] A' ,$$

which shows that $\eta^{\#}(\Delta) \leq |A'| < \kappa$.   □

Corollary 3.32. *A topological group G is precompact if and only if $\eta(\Delta)$ is finite for every $\Delta \in \mathrm{RP}(G)$.*

Proof. By definition, G is precompact if and only if $\eta^{\#}(\Delta)$ is finite for every $\Delta \in \mathrm{RP}(G)$. Apply Theorem 3.30.   □

Corollary 3.33. *Let G be a topological group and $\Delta \in \mathrm{RP}(G)$. If $\mathrm{d}(\Delta) > \aleph_0$ then*

$$\mathrm{d}(\Delta) = \lim_{j \in \omega} \eta(j\Delta).$$

Proof. Clearly $\mathrm{d}(\Delta) = \lim_{j \in \omega} \eta^{\#}(j\Delta)$. Apply Theorem 3.30.   □

Corollary 3.34. *Let κ be an infinite cardinal. The following properties of a topological group G are equivalent:*

(i) G is κ-bounded.
(ii) $d(\Delta) \leq \kappa$ for every $\Delta \in RP(G)$.
(iii) $\eta^{\sharp}(\Delta) \leq \kappa$ for every $\Delta \in RP(G)$.
(iv) $\eta(\Delta) \leq \kappa$ for every $\Delta \in RP(G)$.

Proof. The sets $\odot[e_G, \Delta]$, $\Delta \in RP(G)$, form a base of the topology of G. Therefore (i)$\Leftrightarrow$(iii) by the definition of $\eta^{\sharp}(\Delta)$. Implications (ii)$\Rightarrow$(iii)$\Rightarrow$(iv) follow from $\eta(\Delta) \leq \eta^{\sharp}(\Delta) \leq d(\Delta)$, and (iv) $\Rightarrow$ (ii) follows from Corollary 3.33. □

Theorem 3.35. *Let G be a topological group and κ an infinite cardinal.*

1. *If G is locally κ-bounded and if there exists $\Delta_0 \in RP(G)$ such that $\eta^{\sharp}(\Delta_0) \geq \kappa$ then G is ambitable.*
2. *If G is locally κ^+-bounded and not κ-bounded, then it is ambitable.*
3. *If κ is the least infinite cardinal for which G is κ-bounded and κ is a successor cardinal, then G is ambitable.*
4. *If G is locally $\aleph_k$-bounded for some $k \in \omega$ and not precompact, then it is ambitable.*
5. *If for every $\Delta \in RP(G)$ there is $\Delta' \in RP(G)$ with $\Delta \leq \Delta'$ and $cf(d(\Delta')) > \aleph_0$, then G is ambitable.*

By Part 4 of the theorem, every $\aleph_0$-bounded group is either precompact or ambitable, and every locally compact group is either compact or ambitable.

Proof. To prove Part 1, let $\Delta_0, \Delta_1 \in RP(G)$ be such that $\eta^{\sharp}(\Delta_0) \geq \kappa$ and the set $\odot[e_G, \Delta_1]$ is κ-bounded. Take any $\Delta \in RP(G)$, and $\Delta' := \Delta \vee \Delta_0 \vee \Delta_1$.

There is $B \subseteq G$ such that $G = \odot[e_G, \Delta']B$ and $|B| = \eta^{\sharp}(\Delta') \geq \eta^{\sharp}(\Delta_0) \geq \kappa$. As the set $\odot[e_G, \Delta']$ is κ-bounded, there are sets $A_j \subseteq G$, $j \in \omega$, such that $|A_j| \leq \kappa$ and $\odot[e_G, \Delta'] \subseteq \odot[e_G, j\Delta']A_j$ for all j. The set $\bigcup_{j \in \omega} A_j B$ is Δ'-dense in G, hence

$$d(2\Delta') = d(\Delta') \leq \kappa \cdot \aleph_0 \cdot \eta^{\sharp}(\Delta') = \eta^{\sharp}(\Delta') \leq \eta(2\Delta'),$$

where the last inequality holds by Theorem 3.30. By Lemma 3.28, there exists $f \in BLip_b(2\Delta')^+$ such that $BLip_b(\Delta)^+ \subseteq BLip_b(2\Delta')^+ = \overline{orb}(f)$.

Part 2 follows from Part 1 and Corollary 3.34. Part 3 follows from Part 2.

To prove Part 4, let k be the smallest integer for which G is locally $\aleph_k$-bounded. If $k \geq 1$, then the statement follows from Part 3. If $k = 0$ and G is not precompact, then there is $\Delta \in RP(G)$ such that $\eta^{\sharp}(\Delta) \geq \aleph_0$, and G is ambitable by Part 1.

To prove Part 5, take any $\Delta \in RP(G)$ and $\Delta' \in RP(G)$ such that $\Delta \leq \Delta'$ and $cf(d(\Delta')) > \aleph_0$. Since $d(\Delta') \geq cf(d(\Delta')) > \aleph_0$, we get $d(\Delta') = \lim_{j \in \omega} \eta(j\Delta')$ by Corollary 3.33. Since $cf(d(\Delta')) > \aleph_0$, it follows that $d(j\Delta') = d(\Delta') = \eta(j\Delta')$ for some $j \in \omega$, and $BLip_b(\Delta)^+ \subseteq BLip_b(j\Delta')^+ = \overline{orb}(f)$ for some $f \in BLip_b(j\Delta')^+$ by Lemma 3.28. □

Exercise 3.36. Let X be an infinite semigroup, and assume that X is cancellative; that is, the mappings $x \mapsto xy$ and $x \mapsto yx$ from X to itself are injective for every $y \in X$. Show that X with the discrete uniformity is ambitable.   ■

3.5 Notes for Chap. 3

Function spaces on semitopological semigroups, including the space $\mathsf{LUC}(S)$, are covered in detail by Berglund et al. [8] (where $\mathsf{LUC}(S)$ is written as $\mathcal{LC}(S)$). The right uniformity in Definition 3.1 generalizes the right uniformity as commonly defined for topological groups; it coincides with a particular case of the *left strong uniformity* of Pym [150, 1.5]. Megrelishvili [129, 4.14] established several characterizations of the topological semigroups in which the right uniformity is compatible with the topology.

Section 3.1 is the one place in this treatise where it would be advantageous to allow non-Hausdorff uniform structures in Definition 1.1. We would then have $\mathsf{LUC}(S) = \mathsf{U}_b(rS)$ for every semitopological semigroup S, without having to require $\mathsf{RP}(S)$ to separate the points of S.

For topological groups, the definition and key properties of the right and the left uniformity are due to Weil [179]. These and other uniform structures are among the basic tools in analysis on topological groups, as developed, for example, by Hewitt and Ross [97]. Right and left uniformities and related constructions have an important role in dynamics of infinite-dimensional groups, as presented by Pestov [143]. The monograph by Roelcke and Dierolf [159] is a comprehensive study of uniform structures on topological groups.

The two conditions in Exercise 3.11 characterize the groups for which the right and the left uniformities coincide [159, 2.23].

The notion of a κ-bounded topological group (Definition 3.12) was introduced by Guran [87]. See also Arkhangel'skiĭ [4].

The conditions that define semiuniform semigroups (Definition 3.14) were used by Hindman and Strauss [98, 21.43] in extending the operation in a semigroup to its uniform compactification. Every uniform semigroup as defined by Marxen [126] is a semiuniform semigroup. The theory of semiuniform semigroups remains to be developed.

Ambitable semigroups are those having a certain factorization property that was employed by Neufang [131] and Ferri and Neufang [55] in their study of topological centres. Orbits and orbit closures are key notions in topological dynamics. Definition 3.18 is adapted from de Vries [176, E.5]. The adjective *ambitable* comes from the term *ambit* [176, IV.4.1]; the sets $\overline{\mathrm{orb}}(f)$ are ambits of a special kind. Roelcke and Dierolf [159, Ch.4] described group actions on uniform spaces.

The content of Sect. 3.4 appears in my paper [140]. Corollary 3.32 is essentially a result of Uspenskij [171, p. 338][172, p. 1581]. Bouziad and Troallic [14] derived it from the case $\kappa = \aleph_0$ of Lemma 3.31 which they proved by modifying an idea of Neumann [133]. Ferri and Neufang [55] proved Corollary 3.32 using a result of Protasov [146].

Research Problem 1. This is related to Problem 3 in Sect. 9.5. Characterize, or at least find large interesting classes of, ambitable semiuniform semigroups. In particular, is there a succinct characterization of the discrete semigroups that are ambitable? Is every topological group either precompact or ambitable?

Theorem 3.35 is a partial answer to the last question. The result in Exercise 3.36 holds even with somewhat weaker assumptions [139]. Investigations by Dales et al. [33] regarding topological centres indicate what further generalizations may be possible for discrete semigroups.   ∎

Chapter 4
Some Notable Classes of Uniform Spaces

In this chapter I describe several classes of uniform spaces that will find their use in the study of uniform measures in Parts II and III:

- Inversion-closed spaces, and related properties.
- Supercomplete spaces.
- Spaces with the (ℓ_1) property.

4.1 Inversion-Closed and Alexandroff Spaces

Definition 4.1. A uniform space X is *inversion-closed* iff $1/f \in U(X)$ for every function $f \in U(X)$ such that $f(x) > 0$ for all $x \in X$.

A uniform space X is *sub-inversion-closed* iff there is an inversion-closed space Y such that $X \Subset Y$. ∎

If $U(X) = C(X)$, then X is inversion-closed. In particular, for every completely regular space T, the fine uniform space $\mathsf{F}T$ is inversion-closed.

Exercise 4.2. Prove that if X is an inversion-closed space and $f, g \in U(X)$, then $fg \in U(X)$. ∎

Exercise 4.3. Find an inversion-closed space X such that $U(X) \neq C(X)$. ∎

When f is a real-valued function on a set S, write $\operatorname{coz}(f) := \{x \in S \mid f(x) \neq 0\}$.

Definition 4.4. Let X be uniform space. A *cozero set in X* is a subset of X of the form $\operatorname{coz}(g)$ where $g \in U_b(X)$. A *zero set in X* is a complement of a cozero set. A function $f : X \to \mathbb{R}$ is *cozero-continuous* iff $f^{-1}(V)$ is a cozero set in X for every open set $V \subseteq \mathbb{R}$ (iff $f^{-1}(Z)$ is a zero set for every closed set $Z \subseteq \mathbb{R}$).

A uniform space X is an *Alexandroff space* iff every bounded cozero-continuous function on X is in $U_b(X)$. ∎

J. Pachl, *Uniform Spaces and Measures*, Fields Institute Monographs 30,
DOI 10.1007/978-1-4614-5058-0_5,
© Springer Science+Business Media New York 2013

Lemma 4.5. *Let X be a uniform space.*

1. *If Z_0, Z_1 are zero sets in X, then so is $Z_0 \cup Z_1$.*
2. *If Z_j, $j \in \omega$, are zero sets in X, then so is $\bigcap_{j \in \omega} Z_j$.*

Proof. Let $g_j \in \mathsf{U}_b(X)$ be such that $0 \leq g_j \leq 1$ and $Z_j = g_j^{-1}(0)$ for $j \in \omega$. The functions $f := g_0 \wedge g_1$ and $g := \sum_j g_j / 2^j$ are in $\mathsf{U}_b(X)$ by Theorem 1.18, $Z_0 \cup Z_1 = f^{-1}(0)$ and $\bigcap_j Z_j = g^{-1}(0)$. $\qquad\square$

Theorem 4.6. *These three properties of a uniform space X are equivalent:*

(i) *X is an Alexandroff space.*
(ii) *If $\{f_j\}_j$ and $\{g_j\}_j$ are two sequences of functions in $\mathsf{U}_b(X)$ and $h: X \to \mathbb{R}$ is a function such that $f_j \nearrow h$ and $g_j \searrow h$, then $h \in \mathsf{U}_b(X)$.*
(iii) *If Z, Z' are zero sets in X, $Z \cap Z' = \emptyset$, then there exists $h \in \mathsf{U}_b(X)$ that is 0 on Z and 1 on Z'.*

Proof. Assume that X is an Alexandroff space and take any two sequences $\{f_j\}_j$ and $\{g_j\}_j$ of functions in $\mathsf{U}_b(X)$ such that $f_j \nearrow h$ and $g_j \searrow h$. For every $r \in \mathbb{R}$, we have

$$\{x \in X \mid h(x) > r\} = \bigcup_{j \in \omega} \{x \in X \mid f_j(x) > r\}$$

$$\{x \in X \mid h(x) < r\} = \bigcup_{j \in \omega} \{x \in X \mid g_j(x) < r\}$$

and by Lemma 4.5, the function h is cozero-continuous. That proves (i)$\Rightarrow$(ii).

To prove (ii)$\Rightarrow$(iii), take any two disjoint zero sets $Z = f^{-1}(0)$ and $Z' = g^{-1}(0)$ where $f, g \in \mathsf{U}_b(X)$, and assume $f \geq 0$, $g \geq 0$. Define recursively

$$f_0 := 0$$
$$g_0 := 1$$
$$f_j := jf \wedge g_{j-1}$$
$$g_j := (1 - jg) \vee f_j$$

for $j \in \omega$, $j \geq 1$. Then

$$0 = f_0 \leq f_1 \leq \cdots \leq g_1 \leq g_0 = 1$$

and $\lim_j f_j(x) = \lim_j g_j(x)$ for all $x \in X$. The pointwise limit $h := \lim_j f_j$ is 0 on Z and 1 on Z', and if X has property (ii), then $h \in \mathsf{U}_b(X)$.

To prove (iii)$\Rightarrow$(i), take any cozero-continuous function $f: X \to [0,1]$. If (iii) holds, then there are $g_{jk} \in \mathsf{U}_b(X)$ for $j, k \in \omega$, $0 \leq j < k$, such that $0 \leq g_{jk} \leq 1$ and such that each g_{jk} is 0 on the zero set $\{x \in X \mid f(x) \leq j/k\}$ and 1 on the zero set $\{x \in X \mid f(x) \geq (j+1)/k\}$. The functions $f_k := \sum_{j=0}^{k-1} g_{jk}/k$ are in $\mathsf{U}_b(X)$ and $\lim_k \|f_k - f\|_X = 0$. Hence $f \in \mathsf{U}_b(X)$ by Lemma 1.17. $\qquad\square$

Theorem 4.7. *These four properties of a uniform space X are equivalent:*

(i) X is inversion-closed.

(ii) Every cozero-continuous function $f: X \to \mathbb{R}$ is in $\mathsf{U}(X)$.

(iii) If $\{f_j\}_j$ is a sequence of functions $f_j \in \mathsf{U_b}(X)$ such that $f_j \searrow 0$, then the set $\{f_j \mid j \in \omega\}$ is uniformly equicontinuous.

(iv) If $f: X \to \mathbb{R}^+$ is a function such that $f \wedge j \in \mathsf{U}(X)$ for all $j \in \omega$, then $f \in \mathsf{U}(X)$.

Proof. Assume that X is inversion-closed. First note that X has property (iii) in Theorem 4.6. Indeed, if $Z = f^{-1}(0)$ and $Z' = g^{-1}(0)$ are disjoint zero sets, $f, g \in \mathsf{U_b}(X)^+$, then the function $f/(f+g)$ is 0 on Z and 1 on Z'. To see that $f/(f+g) \in \mathsf{U_b}(X)$, observe that $(f+g)/(f+1) \in \mathsf{U}(X)$ by Theorem 1.18, and apply the inversion-closed property twice to obtain $(f+1)/(f+g) \in \mathsf{U}(X)$ and $1/(f+g) \in \mathsf{U}(X)$.

Now take any cozero-continuous function $f: X \to \mathbb{R}$. The functions $1/(f^+ + 1)$ and $1/(f^- + 1)$ are cozero-continuous and bounded, therefore uniformly continuous by Theorem 4.6. Hence the functions $f^+ + 1$ and $f^- + 1$ are uniformly continuous, and so is $f = f^+ - f^-$. That proves (i)$\Rightarrow$(ii).

Next assume (ii), and take any sequence of functions $f_j \in \mathsf{U_b}(X)$ such that $f_j \searrow 0$. Without loss of generality, assume that $f_0 = 1$. Let $g_j := f_j + 1/(j+1)$ for $j \in \omega$, so that $g_j(x) > g_{j+1}(x)$ for all $j \in \omega$ and $x \in X$, $g_0 = 2$ and $g_j \searrow 0$. Extend the sequence $\{g_j\}_j$ to a net $\{g_r\}_r$ of functions indexed by $r \in \mathbb{R}^+$ using linear interpolation on each interval $[j, j+1]$:

$$g_{j+t}(x) := t g_{j+1}(x) + (1-t) g_j(x) \text{ for } j \in \omega, \, 0 \le t < 1, \, x \in X.$$

The functions g_r are in $\mathsf{U_b}(X)$, $g_r \searrow 0$, and we have $g_r(x) > g_s(x)$ when $x \in X$, $r, s \in \mathbb{R}^+$, $r < s$. Moreover, for every $x \in X$, the function $r \mapsto g_r(x)$ satisfies the Lipschitz condition $|g_r(x) - g_s(x)| \le 2|r - s|$ for all $r, s \in \mathbb{R}^+$.

For $0 < t \le 2$ and $x \in X$ define $h_t(x) := \inf\{r \mid g_r(x) \le t\}$. Since the functions $r \mapsto g_r(x)$ are continuous and strictly decreasing, we have

$$h_t(x) > r \text{ if and only if } g_r(x) > t$$

$$h_t(x) < r \text{ if and only if } g_r(x) < t$$

for $0 < t \le 2$, $r \in \mathbb{R}$ and $x \in X$, which along with Lemma 4.5 shows that the functions h_t are cozero-continuous, hence uniformly continuous by (ii).

Now take any $\varepsilon > 0$ of the form $\varepsilon = 1/k$, $k \in \omega$. There is $\Delta \in \mathsf{UP}(X)$ such that if $x, y \in X$, $\Delta(x, y) < 1$, then $|h_{i/k}(x) - h_{i/k}(y)| \le \varepsilon$ for $i = 1, 2, \dots, 2k$.

Next, take any $r \in \mathbb{R}^+$ and $x, y \in X$ such that $\Delta(x, y) < 1$ and $g_r(y) > g_r(x)$. Since $0 \le g_r(x) \le 2$, there is $i \in \{1, 2, \dots, 2k\}$ such that $(i-1)/k \le g_r(x) \le i/k$. Then $h_{i/k}(x) \le r$, and therefore $h_{i/k}(y) \le s := r + \varepsilon$. But that means $g_s(y) \le i/k$, and from the previously mentioned Lipschitz condition we get

$$g_r(y) - g_r(x) \le g_s(y) + |g_r(y) - g_s(y)| - g_r(x) \le \frac{i}{k} + 2\varepsilon - \frac{i-1}{k} = 3\varepsilon.$$

In particular, for every $j \in \omega$, we get $|f_j(y) - f_j(x)| = |g_j(y) - g_j(x)| \leq 3\varepsilon$ whenever $x, y \in X$, $\Delta(x,y) < 1$. That proves the set $\{f_j \mid j \in \omega\}$ to be uniformly equicontinuous, thus concluding the proof of the implication (ii)$\Rightarrow$(iii).

To prove (iii)$\Rightarrow$(iv), take any function $f \colon X \to \mathbb{R}^+$ such that $f \wedge j \in U(X)$ for all $j \in \omega$. The functions $f_j := f \wedge (j+1) - f \wedge j$, where $j \in \omega$, are in $U_b(X)$ and $1 \geq f_j \searrow 0$. If (iii) holds, then there is $\Delta \in UP(X)$ such that $f_j \in BLip_b(\Delta)$ for all j. Then $1 \wedge |f(x) - f(y)| \leq 2\Delta(x,y)$ for $x, y \in X$, and it follows that $f \in U(X)$.

To prove (iv)$\Rightarrow$(i), take any $f \in U(X)$ such that $f(x) > 0$ for all $x \in X$. Then $(1/f) \wedge j = 1/(f \vee (1/j)) \in U(X)$ for every $j \in \omega$ by Theorem 1.18, and if X has property (iv), then $1/f \in U(X)$. $\qquad\square$

The property in the next definition is used in Sect. 10.3.

Definition 4.8. The *CDE property* of a uniform space X is defined as follows:

- For every $\Delta \in UP(X)$ and for every sequence of functions $g_j \in U_b(X)^+$, $j \in \omega$, if $coz(g_j) \neq \emptyset$ for all j and $\Delta(coz(g_i), coz(g_j)) \geq 1$ for all i, j, $i \neq j$, then the set $\{g_j \mid j \in \omega\}$ is $UE(X)$. $\qquad\blacksquare$

Theorem 4.9. *Every sub-inversion-closed uniform space has the CDE property.*

Proof. Take any inversion-closed space X, $Y \Subset X$, $\Delta \in UP(Y)$ and $g_j \in U_b(Y)^+$, $j \in \omega$, such that $coz(g_j) \neq \emptyset$ and $\Delta(coz(g_i), coz(g_j)) \geq 1$ for $i \neq j$. Assume without loss of generality that Δ is bounded and $g_j \in Lip_b(\Delta)$ for all j (if not, then replace Δ by $1 \wedge \Delta + \sum_j \Delta_j/2^j$, where $\Delta_j \in UP(Y)$, $\Delta_j \leq 1$ and $g_j \in Lip_b(\Delta_j)$).

By Theorem 2.20, there is $\widetilde{\Delta} \in UP_b(X)$ such that $\Delta = \widetilde{\Delta} \restriction Y$. For every $j \in \omega$ there is $r_j \geq 3\|g_j\|_X$ such that g_j is 1-Lipschitz for $r_j\Delta$. Define

$$f_j(x) := \sup \{(g_j(y) - r_j\widetilde{\Delta}(x,y))^+ \mid y \in Y\} \quad \text{for } j \in \omega,\ x \in X.$$

Each f_j is in $U_b(X)^+$ and extends g_j, and $\widetilde{\Delta}(coz(f_i), coz(f_j)) < 1/3$ for $i \neq j$. Let

$$A := \{x \in X \mid \widetilde{\Delta}(x, \bigcup_{j\in\omega} coz(f_j)) > 1/4\},$$

$$A_j := \{x \in X \mid \widetilde{\Delta}(x, coz(f_j)) < 1/3\} \quad \text{for } j \in \omega.$$

Then A and A_j are cozero sets and $X = A \cup \bigcup_j A_j$. Define the real-valued function f on X by $f(x) := \max_j f_j(x)$, $x \in X$. Thus f agrees with f_j on A_j for each $j \in \omega$, and f is 0 on A. If $V \subseteq \mathbb{R}$ is an open set, then

$$f^{-1}(V) = (A \cap f^{-1}(V)) \cup \bigcup_{j\in\omega} (A_j \cap f_j^{-1}(V));$$

hence $f^{-1}(V)$ is a cozero set by Lemma 4.5. Thus f is cozero-continuous, and $f \in U(X)$ by Theorem 4.7.

Now let $g(y) := \max_j g_j(y)$ for $y \in Y$. As g is a restriction of f to Y, it is in $U(Y)$. It follows that the set $\{g_j \mid j \in \omega\}$ is $UE(X)$ because

$$\sup_j |g_j(y) - g_j(y')| = |g(y) - g(y')|$$

whenever $y, y' \in Y$, $\Delta(y, y') \leq 1$. □

I conclude this section with an easy but useful lemma.

Lemma 4.10. *Let X be a uniform space and Y a dense uniform subspace of X. Then Y has the CDE property if and only if X does.*

Proof. By Theorem 2.24, every $g \in U(Y)$ is the restriction of a function $\widetilde{g} \in U(X)$ to Y. By density, the extension $\widetilde{g}$ is unique. By Corollary 2.25, every $\Delta \in UP(Y)$ is the restriction of a pseudometric $\widetilde{\Delta} \in UP(X)$ to Y, and again $\widetilde{\Delta}$ is unique. It follows that a set $\mathscr{F} \subseteq U(Y)$ is $UE(Y)$ if and only if the set $\{\widetilde{g} \mid g \in \mathscr{F}\}$ is $UE(X)$. For every $g \in U(Y)$ we have $\mathrm{coz}(\widetilde{g}) \subseteq \overline{\mathrm{coz}(g)}$ (the closure taken in X); therefore $\widetilde{\Delta}(\mathrm{coz}(\widetilde{f}), \mathrm{coz}(\widetilde{g})) = \Delta(\mathrm{coz}(f), \mathrm{coz}(g))$. □

4.2 Supercomplete Spaces

Complete metric spaces have some remarkable properties that do not hold for general complete uniform spaces. For certain results, such as those in Sect. 7.3, the right generalization of a complete metric space is a supercomplete uniform space.

Definition 4.11. When Δ is a bounded pseudometric on a set X, the pseudometric Δ^H on the set of non-empty subsets of X is defined by

$$\Delta^H(A, B) := \max\left(\sup_{a \in A} \Delta(a, B), \ \sup_{b \in B} \Delta(A, b) \right) \quad \text{for } \emptyset \neq A, B \subseteq X.$$

For any uniform space X, the *hyperspace* HX of X is a uniform space whose points are the closed non-empty subsets of X. The uniformity of HX is induced by the pseudometrics Δ^H, where $\Delta \in UP_b(X)$.

The space X is *supercomplete* iff the space HX is complete. ■

The mapping $x \mapsto \{x\}$ is a uniform isomorphism from X onto a uniform subspace of HX. By the next exercise, every supercomplete space is complete.

Exercise 4.12. Let X be a uniform space. Prove that the image of X in HX under the mapping $x \mapsto \{x\}$ is closed in HX. ■

Theorem 4.13. *Let X be a uniform space. If a net $\{A_\gamma\}_\gamma$ of non-empty closed subsets of X converges in the topology of HX, then*

$$\lim_\gamma A_\gamma = \bigcap_\beta \overline{\bigcup_{\gamma \geq \beta} A_\gamma}.$$

Proof. Set $L := \lim_\gamma A_\gamma$, and $B_\beta := \bigcup_{\gamma \geq \beta} A_\gamma$.

To prove $L \supseteq \bigcap_\beta \overline{B_\beta}$, take any $x \in \bigcap_\beta \overline{B_\beta}$ and any $\Delta \in \mathsf{UP}(X)$. By the definition of $\mathsf{H}X$, there exists β such that $\Delta(y,L) \leq 1$ for every $y \in B_\beta$. Since $\Delta(x,B_\beta) = 0$, it follows that $\Delta(x,L) \leq 1$. Since the set L is closed, $x \in L$ by Theorem 1.3.

To prove $L \subseteq \bigcap_\beta \overline{B_\beta}$, take any $x \notin \bigcap_\beta \overline{B_\beta}$. There is β for which $x \notin \overline{B_\beta}$, and there is $\Delta \in \mathsf{UP}(X)$ for which $\Delta(x,B_\beta) > 1$, so that $\Delta(x,A_\gamma) > 1$ for all $\gamma \geq \beta$. Thus $\Delta(x,L) > 1/2$, and hence $x \notin L$.   □

For decreasing nets in $\mathsf{H}X$, convergence has a particularly simple form:

Corollary 4.14. *Let X be a uniform space, and let $\{A_\gamma\}_\gamma$ be a net of non-empty closed subsets of X such that $A_\gamma \searrow A$. If the net $\{A_\gamma\}_\gamma$ converges in the topology of $\mathsf{H}X$, then $\lim_\gamma A_\gamma = A$.*   □

Definition 4.15. A uniform space X is *uniformly locally compact* iff there exists $\Delta \in \mathsf{UP}(X)$ such that for every $x \in X$ the set $\overline{\odot[x,\Delta]}$ is compact.   ∎

A prototypical example of a uniformly locally compact space is a locally compact group with its right (or left) uniformity (see Sect. 3.2).

Theorem 4.16. *Every uniformly locally compact uniform space is supercomplete.*

Proof. Let X be uniformly locally compact and let $\Delta_0 \in \mathsf{UP}(X)$ be such that for every $x \in X$ the set $\overline{\odot[x,\Delta_0]}$ is compact. Take any Cauchy net $\{A_\gamma\}_\gamma$ in $\mathsf{H}X$, and let $B_\beta := \bigcup_{\gamma \geq \beta} A_\gamma$ for every β, and $B := \bigcap_\beta \overline{B_\beta}$.

To prove that $\lim_\gamma A_\gamma = B$, take any $\Delta \in \mathsf{UP}(X)$ such that $\Delta \geq \Delta_0$. There is γ_0 such that $\Delta^{\mathsf{H}}(A_\beta, A_\gamma) < 1$ for all $\beta, \gamma \geq \gamma_0$, and therefore $\Delta^{\mathsf{H}}(\overline{B_{\gamma_0}}, A_\gamma) < 1$ for all $\gamma \geq \gamma_0$. Now fix $\gamma \geq \gamma_0$.

If $x \in B \subseteq \overline{B_{\gamma_0}}$, then $\Delta(x,A_\gamma) \leq \Delta^{\mathsf{H}}(\overline{B_{\gamma_0}}, A_\gamma) < 1$.

If $y \in A_\gamma$, then $\Delta(A_\beta, y) \leq \Delta^{\mathsf{H}}(A_\beta, A_\gamma) < 1$ and $A_\beta \cap \odot[y,\Delta] \neq \emptyset$ for all $\beta \geq \gamma_0$. The set $\overline{\odot[y,\Delta]}$ is compact; hence $B \cap \overline{\odot[y,\Delta]} \neq \emptyset$, and $\Delta(B,y) \leq 1$.

Thus $\Delta^{\mathsf{H}}(B, A_\gamma) \leq 1$ for all $\gamma \geq \gamma_0$.   □

Corollary 4.17. *For every locally compact group G, the uniform space $\mathsf{r}G$ is supercomplete.*   □

Exercise 4.18. The hyperspace of any metrizable uniform space is metrizable. A metrizable uniform space is complete if and only if it is supercomplete.   ∎

Exercise 4.19. The hyperspace of a compact uniform space is compact.   ∎

The following theorem is used in only one place further on (in Sect. 8.4), and no other results in this treatise depend on it. I omit the proof, which may be found in [100, VII.41] along with a characterization of supercomplete spaces.

Theorem 4.20. *Let T be a completely regular space. The uniform space $\mathsf{F}T$ is supercomplete if and only if T is paracompact.*   □

4.3 Partitions of Unity

The key concept in this section is the (ℓ_1) property, to be used in Chap. 12.

Definition 4.21. A *partition of unity* on a uniform space X is a mapping φ from X to $\ell_\infty(S)$ where S is a non-empty set, $0 \le \varphi(x)(s) \le 1$ for all $x \in X$ and $s \in S$, and $\sum_{s \in S} \varphi(x)(s) = 1$ for every $x \in X$. Write $\varphi_s(x) := \varphi(x)(s)$; thus φ_s is a function on X with values in the interval $[0,1]$. ∎

Note that a partition of unity $\varphi : X \to \ell_\infty(S)$ is uniformly continuous from X to $\ell_\infty(S)$ with the $\|\cdot\|_S$ norm if and only if $\{\varphi_s \mid s \in S\}$ is a UEB subset of $\mathsf{U}_b(X)$.

Exercise 4.22. Show that for every uniform space X and every $\Delta \in \mathsf{UP}(X)$, there exists a partition of unity φ on X that is uniformly continuous from X to $\ell_\infty(S)$ with the $\|\cdot\|_S$ norm and such that $\Delta\text{-diam}(\text{coz}(\varphi_s)) \le 1$ for every $s \in S$. ∎

When $\varphi : X \to \ell_\infty(S)$ is a partition of unity on X, its range is included in the space $\ell_1(S) \subseteq \ell_\infty(S)$. For certain applications, it is important to know whether φ may be chosen to be uniformly continuous from X to $\ell_1(S)$ with the $\|\cdot\|_1$ norm.

Definition 4.23. A uniform space X has the (ℓ_1) *property* iff for every pseudometric $\Delta \in \mathsf{UP}(X)$ there exists a partition of unity φ on X that is uniformly continuous from X to $\ell_1(S)$ with the $\|\cdot\|_1$ norm and such that $\Delta\text{-diam}(\text{coz}(\varphi_s)) \le 1$ for every $s \in S$. ∎

In view of Theorem 2.20, if a uniform space X has the (ℓ_1) property, then so does every uniform subspace of X. That along with the next theorem shows that uniform subspaces of finite-dimensional normed spaces have the (ℓ_1) property.

Theorem 4.24. *Let G be a locally compact group. Then the uniform space rG has the (ℓ_1) property.*

Proof. Take any $\Delta \in \mathsf{UP}_b(rG)$. By Lemma 3.3, no generality is lost by assuming that $\Delta \in \mathsf{RP}(G)$ and the set $\odot[e_G, \Delta]$ is compact.

Let S be a maximal subset of G with the property that $\Delta(s, s') \ge 1/4$ for all $s, s' \in S$, $s \ne s'$. Since Δ is right-invariant, by compactness there is $k \in \omega$ such that $|Sx \cap \odot[e_G, 2\Delta]| \le k$ for every $x \in G$. For every $s \in S$, define

$$f_s(x) := \left(\frac{1}{2} - \Delta(s, x) \right)^+ \quad \text{for } x \in G.$$

Let $S_x := \{ s \in S \mid f_s(x) > 0 \}$ for $x \in G$; then $S_x x^{-1} \subseteq Sx^{-1} \cap \odot[e_G, 2\Delta]$, and thus $|S_x| \le k$. On the other hand, for every $x \in G$ there is $s \in S$ such that $\Delta(x, s) < 1/4$; hence $f_s(x) > 1/4$. Therefore $f(x) := \sum_{s \in S} f_s(x)$ defines a function f on G for which $1/4 < f \le k/2$, and $\varphi_s := f_s/f$ defines a partition of unity $\varphi : rG \to \ell_\infty(S)$. Clearly $\Delta\text{-diam}(\text{coz}(\varphi_s)) = \Delta\text{-diam}(\text{coz}(f_s)) \le 1$ for $s \in S$.

For $x,y \in G$, we have $|f(x) - f(y)| \leq \sum_{s \in S} |f_s(x) - f_s(y)| \leq 2k\Delta(x,y)$; hence

$$|\varphi_s(x) - \varphi_s(y)| \leq \left| \frac{f_s(x)}{f(x)} - \frac{f_s(y)}{f(x)} \right| + \left| \frac{f_s(y)}{f(x)} - \frac{f_s(y)}{f(y)} \right|$$

$$\leq 4|f_s(x) - f_s(y)| + f_s(y) \cdot \frac{|f(x) - f(y)|}{f(x)f(y)}$$

$$\leq 4\Delta(x,y) + 16k\Delta(x,y)$$

$$\sum_{s \in S} |\varphi_s(x) - \varphi_s(y)| \leq 8k(4k+1)\Delta(x,y)$$

which demonstrates that φ is uniformly continuous from rG to $\ell_1(S)$. $\square$

4.4   Notes for Chap. 4

Function space properties in Sect. 4.1 have a long history. Theorems 4.6 and 4.7 include only few of the many relationships between spaces of real-valued functions and uniform structures. For more and references to original sources, see Gillman and Jerison [81], Hager [92] and Rice [156], [157].

Deaibes [36], Frolík [66], [71] and Hager [88], [90], [91] include many equivalent properties of Alexandroff and inversion-closed spaces. The implication (ii) $\Rightarrow$ (iii) in Theorem 4.7 is due to Preiss (unpublished) and Zahradník [182]. Rice [156, 2.8] characterizes sub-inversion-closed spaces.

The spaces in Definition 4.8 have some noteworthy characterizations. They are known as spaces of type (σ_1^∞) in the work of Deaibes and Pupier [40] and as $H(\omega)- t_f$ (or hedgehog–topologically-fine spaces) in the work of Frolík et al. [77], [78].

The definition and basic properties of supercomplete spaces are in Isbell [100], with a correction in [101]. Other characterizations of supercompleteness of X, in terms of ideals and sublattices in the lattice $U_b(X)$, are due to Fedorova [51], [52]. Theorem 4.20 is a special case of Isbell's theorem [100, VII.41]: A uniform space is supercomplete if and only if it is paracompact and its locally fine coreflection is fine.

The result about partitions of unity in Exercise 4.22 is due to Isbell [100, IV.11]. The (ℓ_1) property was introduced by Frolík [63], [68], [75]. Since every paracompact, and therefore every metrizable, topological space has enough continuous locally finite partitions of unity [31, 8.3] [47, 5.1], it follows that for every completely regular space T, the fine space $\mathsf{F}T$ has the (ℓ_1) property. On the other hand, Zahradník [183] proved that no infinite-dimensional normed space with its metric uniformity has the (ℓ_1) property.

Theorem 4.24 is implicit in Frolík's seminar notes [68]. The proof here is from our joint work with Christensen [28, L.4.1]. Van Handel [94, B.1] described a concrete Lipschitz partition of unity on $\mathbb{R}^k$.

Part II
Uniform Measures

In this part I develop the theory of uniform measures and their core properties. Chapter 5 deals with tight (also known as Radon) measures on complete metric spaces in duality with bounded uniformly continuous functions. Uniform measures are then introduced in Chap. 6 as a natural generalization to arbitrary uniform spaces.

Chapter 7 includes conditions for uniform measures to be represented as integrals with respect to various types of measures. In Chap. 8 several familiar spaces of measures are obtained as spaces of uniform measures, by an appropriate choice of a uniform structure in the underlying space. Chapter 9 connects uniform measures to continuity properties of convolution.

Chapter 5
Measures on Complete Metric Spaces

After several definitions and preliminary results that apply to arbitrary uniform spaces, this chapter deals with tight measures on complete metric spaces. Four results provide a basis for much that follows in subsequent chapters:

(1) (Theorem 5.28) The functionals on $U_b(X)$ represented by tight Borel measures on X are exactly those that are sup-norm continuous and X-pointwise continuous on $\mathsf{BLip}_b(\Delta)$.
(2) (Corollaries 5.17 and 5.37) Three topologies on the space of functionals in (1), namely two "weak" topologies and the topology of uniform convergence on $\mathsf{BLip}_b(\Delta)$, coincide on the positive cone and on spheres of the $\|\cdot\|$ norm.
(3) (Theorem 5.41) Two of the three topologies in (2) have the same compact sets.
(4) (Theorem 5.45) In duality with $U_b(X)$, the space of functionals in (1) is weakly sequentially complete.

5.1   Tight Measures on Uniform Spaces

Let X be a uniform space, and consider $U_b(X)$, the space of bounded uniformly continuous functions on X. With the vector operations and partial order defined pointwise at every $x \in X$, $U_b(X)$ is a Riesz subspace of $\ell_\infty(X)$. Moreover, $U_b(X)$ is a Banach sublattice of $\ell_\infty(X)$ with the sup norm $\|\cdot\|_X$ (Theorem 1.18).

By Theorem P.21, the order-bounded dual $U_b(X)^\sim$ of $U_b(X)$ is also its normed-space dual $U_b(X)^*$, and the space $U_b(X)^\sim = U_b(X)^*$ is a Banach lattice.

Definition 5.1. Let X be a uniform space.

1. $\mathfrak{M}_b(X)$ is the dual $U_b(X)^\sim = U_b(X)^*$ of $U_b(X)$, with the dual partial order $\leq$ and the dual norm $\|\cdot\|$.
2. $\mathfrak{M}_t(X)$ is the space of the linear functionals on $U_b(X)$ whose restriction to the $\|\cdot\|_X$-unit ball in $U_b(X)$ is continuous in the compact–open topology on $U_b(X)$.

∎

J. Pachl, *Uniform Spaces and Measures*, Fields Institute Monographs 30,
DOI 10.1007/978-1-4614-5058-0_6,
© Springer Science+Business Media New York 2013

Bounded Lipschitz functions (Definition P.16) have an important role in the study of $\mathfrak{M}_b(X)$ and its subspaces. For any pseudometric Δ, $\mathrm{Lip}_b(\Delta)$ is a Riesz subspace of $\ell_\infty(S)$. With the X-pointwise topology and partial order, $\mathrm{BLip}_b(\Delta)$ is a compact lattice. If X is a uniform space and $\Delta \in \mathrm{UP}(X)$, then $\mathrm{Lip}_b(\Delta) \subseteq \mathrm{U}_b(X)$.

Lemma 5.2. *Let X be any uniform space and let $\Delta \in \mathrm{UP}(X)$. If $\mathfrak{m} \in \mathfrak{M}_t(X)$, then $\mathfrak{m} \in \mathfrak{M}_b(X)$ and $\mathfrak{m}$ is continuous on $\mathrm{BLip}_b(\Delta)$ with the X-pointwise topology.*

Proof. The $\|\cdot\|_X$ norm topology on $\mathrm{U}_b(X)$ is finer than the compact–open topology. Thus every $\mathfrak{m} \in \mathfrak{M}_t(X)$ is in $\mathfrak{M}_b(X)$.

The compact–open topology and the X-pointwise topology coincide on $\mathrm{BLip}_b(\Delta)$. Thus every $\mathfrak{m} \in \mathfrak{M}_t(X)$ is X-pointwise continuous on $\mathrm{BLip}_b(\Delta)$.   □

Clearly $\mathfrak{M}_t(X)$ is a vector subspace of $\mathfrak{M}_b(X)$, closed in the $\|\cdot\|$ norm. By Corollary 5.4 below, $\mathfrak{M}_t(X)$ is even a band in the Banach lattice $\mathfrak{M}_b(X)$.

When X is a uniform space, every pseudometric $\Delta \in \mathrm{UP}(X)$ defines the seminorm $\|\cdot\|_{\mathrm{BLip}_b(\Delta)}$ on $\mathfrak{M}_b(X)$, which will be written simply as $\|\cdot\|_\Delta$. Since $\mathrm{BLip}_b(\Delta)$ is contained in the $\|\cdot\|_X$ unit ball in $\mathrm{U}_b(X)$, it follows that $\|\cdot\|_\Delta \leq \|\cdot\|$ on $\mathfrak{M}_b(X)$.

The next theorem and its corollary show that, up to a Banach-lattice isomorphism, the space $\mathfrak{M}_t(X)$ with the norm $\|\cdot\|$ depends only on the topology of X. The theorem refers to tight Borel measures on X, which are defined in Sect. P.5 for every topological space, and hence also for every uniform space.

Theorem 5.3. *Let X be a uniform space. The following properties of a linear functional $\mathfrak{m} \colon \mathrm{U}_b(X) \to \mathbb{R}$ are equivalent:*

(i) $\mathfrak{m} \in \mathfrak{M}_t(X)$.
(ii) There exists a tight Borel measure μ on X such that $\mathfrak{m}(f) = \int f \, d\mu$ for $f \in \mathrm{U}_b(X)$.

If (ii) holds, then there is a unique such tight Borel measure μ, and moreover, $\mathfrak{m}^+(f) = \int f \, d\mu^+$, $\mathfrak{m}^-(f) = \int f \, d\mu^-$ and $|\mathfrak{m}|(f) = \int f \, d|\mu|$ for every $f \in F$, and $\|\mathfrak{m}\| = \|\mu\|_{\mathrm{TV}}$.

Proof. By Corollary 1.22, the topology of X is the coarsest topology for which every function in $\mathrm{U}_b(X)$ is continuous. Apply Theorem P.30.   □

Corollary 5.4. *For any uniform space X, the space $\mathfrak{M}_t(X)$ is a band in the Banach lattice $\mathfrak{M}_b(X)$.*

Proof. Apply Theorem 5.3 with Theorem P.21.   □

Corollary 5.5. *Let X be a uniform space. For every $\mathfrak{m} \in \mathfrak{M}_t(X)$ there exists a unique $\widetilde{\mathfrak{m}} \in \mathfrak{M}_t(\mathsf{F}X)$ such that $\mathfrak{m}(f) = \widetilde{\mathfrak{m}}(f)$ for all $f \in \mathrm{U}_b(X)$. Moreover, $\mathfrak{m}^+(f) = \widetilde{\mathfrak{m}}^+(f)$, $\mathfrak{m}^-(f) = \widetilde{\mathfrak{m}}^-(f)$, and $|\mathfrak{m}|(f) = |\widetilde{\mathfrak{m}}|(f)$ for $f \in \mathrm{U}_b(X)$, and $\|\mathfrak{m}\| = \|\widetilde{\mathfrak{m}}\|$.*

Proof. Since the topology of X is the same as that of $\mathsf{F}X$, Borel sets and tight Borel measures are the same for the two uniformities.   □

If X and Y are two uniform spaces with the same set of points and the same topology, then $\mathsf{F}X = \mathsf{F}Y$, and by Corollary 5.5, there is a norm-preserving

Banach-lattice isomorphism between $\mathfrak{M}_t(X)$ and $\mathfrak{M}_t(Y)$. The isomorphism is produced by extending each $\mathsf{m} \in \mathfrak{M}_t(X)$ to a unique $\widetilde{\mathsf{m}} \in \mathfrak{M}_t(\mathsf{F}X)$ and then restricting $\widetilde{\mathsf{m}}$ to $\mathsf{U}_b(Y)$.

Exercise 5.6. Prove Corollary 5.4 without using representation by Borel measures (Theorem P.30). ∎

Example 5.7. Let X be the metric space on the set $\omega := \{0, 1, 2, \dots\}$ with the metric Δ defined by $\Delta(x, y) := 2$ for $x \neq y$. Then $\mathsf{U}_b(X) = \ell_\infty$ and $\mathrm{BLip}_b(\Delta)$ is the norm unit ball in ℓ_∞. Every $\mathsf{m} \in \mathfrak{M}_t(X)$ is represented by a tight Borel measure μ on X, and the function $j \mapsto \mu(\{j\})$, $j \in \omega$, is in ℓ_1. It follows that if ℓ_1 is identified with its image under the natural embedding into $\ell_1^{**} = \ell_\infty^* = \mathfrak{M}_b(X)$ then $\ell_1 = \mathfrak{M}_t(X)$. ∎

Theorem 5.3 justifies the following definition.

Definition 5.8. A linear functional on $\mathsf{U}_b(X)$ is *tight* iff it belongs to $\mathfrak{M}_t(X)$. A set $\mathfrak{A} \subseteq \mathfrak{M}_t(X)$ is *uniformly tight* iff the set of measures that represent the elements of $\mathfrak{A}$ as in Theorem 5.3 is uniformly tight. ∎

Lemma 5.9. *For any uniform space, a set $\mathfrak{A} \subseteq \mathfrak{M}_t(X)$ is uniformly tight if and only if the set of restrictions of the functionals in $\mathfrak{A}$ to the $\|\cdot\|_X$-unit ball in $\mathsf{U}_b(X)$ is equicontinuous in the compact–open topology.*

Proof. Apply Theorem P.29. □

5.2 Point Masses and Molecular Measures

Point masses and molecular measures are defined as functionals on the space $\mathbb{R}^S$ of *all* functions on a set S, although in this treatise they are usually considered as functionals on spaces of uniformly continuous functions.

Definition 5.10. Let S be a non-empty set.

1. For every $x \in S$, the *point mass at x* is the linear functional $\partial_S(x)$ on $\mathbb{R}^S$ defined by $\partial_S(x)(f) := f(x)$ for $f \in \mathbb{R}^S$. Thus ∂_S is a mapping from S to the space of linear functionals on $\mathbb{R}^S$, written simply as ∂ when S is understood from context.
2. A linear functional on $\mathbb{R}^S$ is a *molecular measure on S* iff it is a finite linear combination of point masses. The space of molecular measures on S is denoted by $\mathrm{Mol}(S)$. ∎

Although the elements of $\mathrm{Mol}(S)$ are linear functionals, not measures, the term molecular measure is commonly used, and I use it here as well. Evidently, every functional in $\mathrm{Mol}(S)$ is represented by a measure. Scalar multiples of point masses are sometimes called atomic measures. That explains the term molecular measure for a finite combination of point masses.

If X is a uniform space, then $\mathsf{U}_b(X)$ separates the points of X. It follows that every molecular measure is uniquely determined by its values on $\mathsf{U}_b(X)$, so that the natural mapping from $\mathrm{Mol}(X)$ to $\mathfrak{M}_b(X)$ is injective, and we may identify $\mathrm{Mol}(X)$ with its image in $\mathfrak{M}_b(X)$. Similarly, we may identify $\mathrm{Mol}(X)$ with its image in the space of linear functionals on $U(X)$. Accordingly, say that a linear functional on the space $U_b(X)$ or $U(X)$ is a *molecular measure on X* iff it is the restriction of a molecular measure on the point set of X (a functional on $\mathbb{R}^X$) to the space $U_b(X)$ or $U(X)$.

Lemma 5.11. *Let X be a uniform space. A linear functional on $U_b(X)$ is a molecular measure if and only if it is continuous on $U_b(X)$ in the X-pointwise topology.*

Proof. The condition is obviously necessary. To prove it is sufficient, take any linear functional $\mathfrak{m}$ on $U_b(X)$ that is continuous in the X-pointwise topology. There is a finite set $D \subseteq X$ such that if $f \in U_b(X)$ and $\|f\|_D = 0$, then $|\mathfrak{m}(f)| < 1$. Therefore also $\mathfrak{m}(f) = 0$ whenever $\|f\|_D = 0$; in other words, $\mathfrak{m}(f) = 0$ whenever $\partial_X(x)(f) = 0$ for each $x \in D$. By Lemma P.4, $\mathfrak{m}$ is a linear combination of the functionals $\partial_X(x)$, $x \in D$. $\qquad\square$

The mapping $\partial_X \colon X \to \mathrm{Mol}(X) \subseteq \mathfrak{M}_b(X)$ is a valuable tool in several constructions in this and later chapters. It is often convenient to identify X with its image $\partial_X(X) \subseteq \mathfrak{M}_b(X)$ by means of ∂_X.

Lemma 5.12. *Let X be a uniform space and $\Delta \in \mathrm{UP}(X)$. If $x, y \in X$, then*

$$\|\partial_X(x) - \partial_X(y)\|_\Delta = 2 \wedge \Delta(x, y).$$

Proof. For every $x \in X$, the function $f_x \colon y \mapsto (1 \wedge (\Delta(x, y) - 1))$ is in $\mathrm{BLip}_b(\Delta)$, and $|f_x(x) - f_x(y)| = 2 \wedge \Delta(x, y)$. Thus

$$\|\partial_X(x) - \partial_X(y)\|_\Delta = \sup\{|f(x) - f(y)| \mid f \in \mathrm{BLip}_b(\Delta)\} \geq 2 \wedge \Delta(x, y).$$

The opposite inequality follows from the definition of $\mathrm{BLip}_b(\Delta)$. $\qquad\square$

The constant 2 in Lemma 5.12 comes from the condition $\|f\|_X \leq 1$ imposed on functions $f \in \mathrm{BLip}_b(\Delta)$. The lemma is a simple special case of a general formula in Theorem 11.3.

5.3 Two Weak Topologies

Let X be a uniform space. Every $\mathfrak{m} \in \mathfrak{M}_t(\mathsf{F}X)$ is a functional on the space $C_b(X)$, and $U_b(X) \subseteq C_b(X) = U_b(\mathsf{F}X)$. If $\mathfrak{m}(f) = 0$ for all $f \in U_b(X)$, then $\mathfrak{m}(f) = 0$ also for all $f \in C_b(X)$, by Corollary 5.5. Thus there are two "weak" Hausdorff topologies on $\mathfrak{M}_t(\mathsf{F}X)$, namely, the $U_b(X)$-weak and the $C_b(X)$-weak topology.

Alternatively and equivalently, the same two topologies may be considered on the space $\mathfrak{M}_t(X)$. By Corollary 5.5, every $\mathfrak{m} \in \mathfrak{M}_t(X)$ uniquely extends to $\widetilde{\mathfrak{m}} \in \mathfrak{M}_t(\mathsf{F}X)$, and the extension defines a Banach-lattice isomorphism. Thus $\mathfrak{M}_t(X)$ is in a natural duality with $C_b(X)$, namely, $\langle m, f \rangle := \tilde{m}(f)$ for $m \in \mathfrak{M}_t(X)$, $f \in C_b(X)$, and that defines the $C_b(X)$-weak topology on $\mathfrak{M}_t(X)$.

Clearly the $U_b(X)$-weak and the $C_b(X)$-weak topologies coincide on the whole space $\mathfrak{M}_t(X)$ if and only if $U_b(X) = C_b(X)$. Next I now show that the two topologies always coincide on the positive cone $\mathfrak{M}_t(X)^+$ and on $\|\cdot\|$ spheres in $\mathfrak{M}_t(X)$. The proof relies on two lemmas.

Lemma 5.13. *Let X be a uniform space, $f \in C_b(X)$ and $x \in X$. Then*

$$f(x) = \sup\{\, g(x) \mid g \in U_b(X) \text{ and } g \le f \,\}$$
$$= \inf\{\, g(x) \mid g \in U_b(X) \text{ and } g \ge f \,\}.$$

Proof. Let $\varepsilon > 0$. There is a neighbourhood V of x such that $f(y) > f(x) - \varepsilon$ for all $y \in V$, and there is $\Delta \in UP(X)$ such that if $\Delta(x, y) < 1$, then $y \in V$. Define $g \in \mathsf{Lip}_b(\Delta)$ by

$$g(y) := (f(x) - \varepsilon - 2\|f\|_X \Delta(x, y)) \vee (-\|f\|_X).$$

Then $g \le f$ and $g(x) \ge f(x) - \varepsilon$. This proves the first equality in the lemma. The second equality follows by symmetry. □

Lemma 5.14. *Let K be a compact subset of a uniform space X; let $f \in C_b(X)$, and $\theta > 0$. Then there are functions $g, g' \in U_b(X)$ such that $g \le f \le g'$ and $\|g' - g\|_K < \theta$.*

Proof. By Lemma 5.13, there are nets $\{g_\gamma\}_\gamma$ and $\{g'_\gamma\}_\gamma$ of functions in $U_b(X)$ such that $g_\gamma \le f \le g'_\gamma$ for every γ, $g_\gamma \nearrow f$ and $g'_\gamma \searrow f$. The nets $\{g_\gamma\}_\gamma$ and $\{g'_\gamma\}_\gamma$ converge uniformly on every compact set; hence $\|g'_\gamma - g_\gamma\|_K < \theta$ for some γ. □

Exercise 5.15. Prove Corollary 5.5 from Lemma 5.14, without using representation by Borel measures (Theorem P.30). ∎

Theorem 5.16. *Let X be a uniform space, $m \in \mathfrak{M}_t(FX)^+$, and let $\{m_\gamma\}_\gamma$ be a net in $\mathfrak{M}_b(FX)^+$. If $\lim_\gamma m_\gamma(g) = m(g)$ for every $g \in U_b(X)$, then $\lim_\gamma m_\gamma(f) = m(f)$ for every $f \in C_b(X)$.*

Proof. Take any $\varepsilon > 0$. There are a compact set $K \subseteq X$ and $\theta > 0$ such that if $h \in C_b(X)$, $\|h\|_X \le 2$ and $\|h\|_K < \theta$, then $|m(h)| < \varepsilon$.

Let $f \in C_b(X)$ and assume, without loss of generality, that $\|f\|_X \le 1$. In view of Lemma 5.14, there are $g, g' \in U_b(X)$ such that $g \le f \le g'$ and $\|g' - g\|_K < \theta$. Replace g by $g \vee -1$ and g' by $g' \wedge 1$, so that $-1 \le g \le f \le g' \le 1$ and $\|f - g\|_X \le 2$, $\|g' - f\|_X \le 2$. Thus $m(f - g) < \varepsilon$, $m(g' - f) < \varepsilon$ and

$$m(f) - \varepsilon < m(g) = \lim_\gamma m_\gamma(g) \le \liminf_\gamma m_\gamma(f)$$

$$\le \limsup_\gamma m_\gamma(f) \le \lim_\gamma m_\gamma(g') = m(g') < m(f) + \varepsilon,$$

which proves $\lim_\gamma m_\gamma(f) = m(f)$. □

Corollary 5.17. *For any uniform space X, the $\mathsf{U}_b(X)$-weak topology coincides with the $\mathsf{C}_b(X)$-weak topology on $\mathfrak{M}_t(X)^+$.*   □

Corollary 5.18. *Let X be a uniform space, $\mathfrak{m} \in \mathfrak{M}_t(\mathsf{F}X)$, and let $\{\mathfrak{m}_\gamma\}_\gamma$ be a net in $\mathfrak{M}_b(\mathsf{F}X)$. If $\lim_\gamma \mathfrak{m}_\gamma(g) = \mathfrak{m}(g)$ for every $g \in \mathsf{U}_b(X)$ and $\lim_\gamma \|\mathfrak{m}_\gamma\| \leq \|\mathfrak{m}\|$, then $\lim_\gamma \mathfrak{m}_\gamma(f) = \mathfrak{m}(f)$ for every $f \in \mathsf{C}_b(X)$.*

Proof. Apply Lemma P.22 with $S = X$ and $F = \mathsf{U}_b(X)$. Then apply Theorem 5.16 twice: Once with $\mathfrak{m}^+$ and $\mathfrak{m}_\gamma^+$ in place of $\mathfrak{m}$ and $\mathfrak{m}_\gamma$, and once with $\mathfrak{m}^-$ and $\mathfrak{m}_\gamma^-$ in place of $\mathfrak{m}$ and $\mathfrak{m}_\gamma$.   □

Corollary 5.19. *For any uniform space X and any $r \in \mathbb{R}^+$, the $\mathsf{U}_b(X)$-weak topology coincides with the $\mathsf{C}_b(X)$-weak topology on the set*

$$\{\mathfrak{m} \in \mathfrak{M}_t(X) \mid \|\mathfrak{m}\| = r\}.$$ □

5.4   Tight Measures on Complete Metric Spaces

The proof of Lemma 5.13 shows that on a metric space continuous functions are pointwise approximated by Lipschitz functions. The next lemma describes a stronger approximation of uniformly continuous functions.

Lemma 5.20. *Let X be a metric space with metric Δ.*

1. *If $\mathscr{F} \subseteq \mathsf{U}_b(X)$ is a $\mathsf{UEB}(X)$ set, then*

$$\forall \varepsilon > 0 \, \exists r \in \mathbb{R}^+ \, \forall f \in \mathscr{F} \, \exists g \in r\mathsf{BLip}_b(\Delta) \, [\, \|f - g\|_X \leq \varepsilon \,].$$

2. *The space $\mathsf{Lip}_b(\Delta)$ is $\|\cdot\|_X$-dense in the space $\mathsf{U}_b(X)$.*

Proof. The construction of g for Part 1 is similar to the proof of Lemma 5.13. Fix $\varepsilon > 0$. By Theorem 2.6, there is $r \in \mathbb{R}^+$ such that

$$\sup_{f \in \mathscr{F}} |f(x) - f(y)| < r\Delta(x,y) + \varepsilon$$

for all $x, y \in X$. For any $f \in \mathscr{F}$, define $g \in r\mathsf{BLip}_b(\Delta)$ by

$$g(y) := \sup_{x \in X} (f(x) - \varepsilon - r\Delta(x,y)), \quad y \in X.$$

Then $f - \varepsilon \leq g \leq f$.

Part 2 follows from Part 1.   □

Corollary 5.21. *Let X be a metric space with metric Δ. Then the seminorm $\|\cdot\|_\Delta$ on the space $\mathfrak{M}_b(X)$ is a norm.*   □

Corollary 5.22. *Let X be a metric space with metric Δ. Then the $\mathsf{Lip}_b(\Delta)$-weak topology coincides with the $\mathsf{U}_b(X)$-weak topology on every $\|\cdot\|$-bounded subset of $\mathfrak{M}_b(X)$.* □

Next I show that the $\|\cdot\|_\Delta$ dual of $\mathfrak{M}_t(X)$ naturally identifies with $\mathsf{Lip}_b(\Delta)$.

Lemma 5.23. *Let X be a metric space with metric Δ. Every linear functional on $\mathfrak{M}_t(X)$ continuous in the $\|\cdot\|_\Delta$ norm is of the form $\mathfrak{m} \mapsto \mathfrak{m}(f)$, $\mathfrak{m} \in \mathfrak{M}_t(X)$, for some $f \in \mathsf{Lip}_b(\Delta)$.*

Proof. By Lemma 5.2, the X-pointwise topology and the $\mathfrak{M}_t(X)$-weak topology coincide on $r\mathsf{BLip}_b(\Delta)$ for every $r \in \mathbb{R}^+$. Apply the Mackey–Arens Theorem P.8 with $E = \mathfrak{M}_t(X)$ and $F = \mathsf{Lip}_b(\Delta)$. □

Corollary 5.24. *Let X be a metric space with metric Δ and $\mathfrak{m} \in \mathfrak{M}_t(X)$. Then for every $\varepsilon > 0$ there is $\mathfrak{n} \in \mathsf{Mol}(X)$ such that $\|\mathfrak{n}\| \leq \|\mathfrak{m}\|$ and $\|\mathfrak{m} - \mathfrak{n}\|_\Delta < \varepsilon$.*

Proof. Take any $r \in \mathbb{R}^+$ such that $\mathfrak{m}$ is not in the $\|\cdot\|_\Delta$ norm closure of the set $\mathfrak{B}(r) := \{\mathfrak{n} \in \mathsf{Mol}(X) \mid \|\mathfrak{n}\| \leq r\}$. By virtue of Theorem P.5 and Lemma 5.23, there is $f \in \mathsf{Lip}_b(\Delta)$ such that $\mathfrak{m}(f) > r' := \sup\{|\mathfrak{n}(f)| \mid \mathfrak{n} \in \mathfrak{B}(r)\}$. Since $r\partial_X(x) \in \mathfrak{B}(r)$ for every $x \in X$, we get $r\|f\|_X \leq r' < \mathfrak{m}(f) \leq \|\mathfrak{m}\| \cdot \|f\|_X$ and $\|\mathfrak{m}\| > r$.

It follows that $\mathfrak{m}$ is in the $\|\cdot\|_\Delta$ closure of $\mathfrak{B}(\|\mathfrak{m}\|)$. □

Now we come to one of the central notions in this treatise, X-pointwise continuity of linear functionals on sets of the form $\mathsf{BLip}_b(\Delta)$. I start with a simple lemma that simplifies several proofs further on. In agreement with the general notation in Sect. P.4, $\mathsf{BLip}_b(\Delta)^+$ is the set $\{f \in \mathsf{BLip}_b(\Delta) \mid f \geq 0\}$.

Lemma 5.25. *Let X be a uniform space and $\Delta \in \mathsf{UP}(X)$. The following properties of a set $\mathfrak{A}$ of linear functionals on the space $\mathsf{U}_b(X)$ are equivalent:*

(i) *The set of restrictions of the functionals in $\mathfrak{A}$ to $\mathsf{BLip}_b(\Delta)$ is X-pointwise equicontinuous on $\mathsf{BLip}_b(\Delta)$.*

(ii) *The set of restrictions of the functionals in $\mathfrak{A}$ to $\mathsf{BLip}_b(\Delta)^+$ is X-pointwise equicontinuous at 0.*

Proof. Assume that (ii) holds. By linearity, the restrictions of the functionals in $\mathfrak{A}$ to $2\mathsf{BLip}_b(\Delta)^+$ are also X-pointwise equicontinuous at 0. If $\{f_\gamma\}_\gamma$ is a net in $\mathsf{BLip}_b(\Delta)$ that converges X-pointwise to $f \in \mathsf{BLip}_b(\Delta)$, then the nets $\{(f_\gamma - f)^+\}_\gamma$ and $\{(f_\gamma - f)^-\}_\gamma$ in $2\mathsf{BLip}_b(\Delta)^+$ converge X-pointwise to 0, and

$$\limsup_{\gamma} \sup_{\mathfrak{m} \in \mathfrak{A}} |\mathfrak{m}(f_\gamma) - \mathfrak{m}(f)| \leq \limsup_{\gamma} \sup_{\mathfrak{m} \in \mathfrak{A}} |\mathfrak{m}(f_\gamma - f)^+| + \limsup_{\gamma} \sup_{\mathfrak{m} \in \mathfrak{A}} |\mathfrak{m}(f_\gamma - f)^-| = 0$$

by (ii). Thus (ii) implies (i). □

The next theorem characterizes uniformly tight sets of positive measures, and in particular positive tight Borel measures, on complete metric spaces.

Theorem 5.26. *Let X be a complete metric space with metric Δ. The following properties of a set $\mathfrak{A} \subseteq \mathfrak{M}_b(X)^+$ are equivalent:*

(i) $\mathfrak{A}$ is uniformly tight.
(ii) The set of restrictions of the functionals in $\mathfrak{A}$ to $\mathsf{BLip}_b(\Delta)$ is X-pointwise equicontinuous.

The proof of the implication (ii)$\Rightarrow$(i) uses the following technical lemma:

Lemma 5.27. *Let X be a complete metric space with metric Δ, $\emptyset \neq B_j \subseteq X$ for $j = 1,2,\ldots$ and $K \subseteq X$. Let each B_j be a finite union of closed sets of Δ-diameter at most $1/j$, and $B_j \searrow K$. Then $K \neq \emptyset$, K is compact and*

$$\Delta(x,K) = \lim_j \Delta(x,B_j)$$

for every $x \in X$.

Proof. Take any $x \in X$ and $\varepsilon > 0$. There are $y_j \in B_j$ for $j = 1,2,\ldots$ such that $\Delta(x,y_j) < \Delta(x,B_j) + \varepsilon$. By the assumption about B_j, the sequence $\{y_j\}_j$ has a Cauchy subsequence $\{y_{j(i)}\}_i$, and $\lim_i y_{j(i)} = y$ for some $y \in X$ because X is complete. Each set B_j is closed, hence $y \in B_j$ for $j = 1,2,\ldots$. Thus $y \in K$, $K \neq \emptyset$ and

$$\Delta(x,K) \leq \Delta(x,y) = \lim_i \Delta(x,y_{j(i)}) \leq \lim_i \Delta(x,B_{j(i)}) + \varepsilon = \lim_j \Delta(x,B_j) + \varepsilon$$

so that $\Delta(x,K) \leq \lim_j \Delta(x,B_j)$. The opposite inequality follows from $K \subseteq B_j$.

The set K with the metric Δ is precompact because for every $\varepsilon > 0$ it is covered by finitely many sets of diameter at most ε, and K is complete because it is a closed subset of the complete space X. Hence K is compact by Theorem 1.15. $\square$

Proof of Theorem 5.26. Since $\mathsf{BLip}_b(\Delta)$ is a subset of the $\|\cdot\|_X$ unit ball in $\mathsf{U}_b(X)$, the implication (i)$\Rightarrow$(ii) follows from Lemma 5.9.

To prove (ii)$\Rightarrow$(i), take any set $\mathfrak{A} \subseteq \mathfrak{M}_b(X)^+$ that satisfies (ii). The equicontinuity of $\mathfrak{A}$ implies that the values $\|m\| = m(1)$, $m \in \mathfrak{A}$, are bounded. Take any $\varepsilon > 0$ such that there exists $m_0 \in \mathfrak{A}$ for which $\varepsilon < \|m_0\|/2$.

There are non-empty finite sets $D_j \subseteq X$, $j = 1,2,\ldots$, such that $|m(h)| < \varepsilon/j2^j$ when $m \in \mathfrak{A}$, $h \in \mathsf{BLip}_b(\Delta)$, $\|h\|_{D_j} = 0$. In particular, this means that $m(h_j) < \varepsilon/2^j$ for all $m \in \mathfrak{A}$ and the functions h_j defined by

$$h_j(x) := 1 \wedge j\Delta(x,D_j) \ \text{ for } x \in X.$$

Write

$$A_j := \left\{ x \mid h_j(x) \leq \tfrac{1}{2} \right\}, \quad B_j := \bigcap_{i=1}^{j} A_i, \quad K := \bigcap_{i=1}^{\infty} A_i.$$

If $x \in X$ and j are such that $x \notin B_j$, then there is $i \leq j$ with $x \notin A_i$, and thus $2h_i(x) > 1$. It follows that $B_j \neq \emptyset$ for all j; indeed, if $B_j = \emptyset$, then $1 \leq \max_{1 \leq i \leq j} 2h_i$ and therefore $m_0(1) \leq \sum_{i=1}^{j} m_0(2h_i) < 2\varepsilon$, contradicting the choice of ε. Hence K is non-empty and compact by Lemma 5.27.

Now let f be any function in $U_b(X)$ such that $\|f\|_X \leq 1$ and $\|f\|_K < \varepsilon$. I shall derive an estimate for $|m(f)|$, $m \in \mathfrak{A}$, that will conclude the proof.

By Lemma 5.20, $\|f - f'\|_X < \varepsilon$ for some $f' \in \mathsf{Lip}_b(\Delta)$. Take $r \geq 1$ for which $f' \in r\mathsf{BLip}_b(\Delta)$, and define functions $g, g_j \in r\mathsf{BLip}_b(\Delta)$, $j = 1, 2, \ldots$, by

$$g(x) := 1 \wedge r\Delta(x, K) \, , \quad g_j(x) := 1 \wedge r\Delta(x, B_j).$$

By Lemma 5.27, we have $g_j \nearrow g$. As $\mathfrak{A}$ is X-pointwise equicontinuous on $r\mathsf{BLip}_b(\Delta)$, there is j such that $m(g) - m(g_j) < \varepsilon$ for all $m \in \mathfrak{A}$. If $g_j(x) > 0$, then $x \notin B_j$; hence $g_j \leq \max_{1 \leq i \leq j} 2h_i \leq \sum_{i=1}^{j} 2h_i$ and for all $m \in \mathfrak{A}$ we have

$$m(g) = m(g) - m(g_j) + m(g_j) < \varepsilon + 2 \sum_{i=1}^{j} m(h_i) < 3\varepsilon.$$

The function $(|f'| - 2\varepsilon)^+$ belongs to $r\mathsf{BLip}_b(\Delta)$, is 0 on K and $(|f'| - 2\varepsilon)^+ \leq 1$. Therefore $(|f'| - 2\varepsilon)^+ \leq g$ and

$$|m(f)| \leq |m(f) - m(f')| + |m(f')| \leq \|m\|\varepsilon + m(|f'|)$$
$$\leq \|m\|\varepsilon + m((|f'| - 2\varepsilon)^+) + 2\|m\|\varepsilon$$
$$\leq 3\|m\|\varepsilon + m(g) < 3\|m\|\varepsilon + 3\varepsilon$$

for all $m \in \mathfrak{A}$. Since the set $\mathfrak{A}$ is $\|\cdot\|$-bounded, this estimate shows that the set of restrictions of the functionals in $\mathfrak{A}$ to the $\|\cdot\|_X$ unit ball in $U_b(X)$ is equicontinuous in the compact–open topology. $\qquad\square$

Theorem 5.28. *Let X be a complete metric space with metric Δ. The following properties of a functional $m \in \mathfrak{M}_b(X)$ are equivalent:*

(i) $m \in \mathfrak{M}_t(X)$.
(ii) For every $\varepsilon > 0$, there is $n \in \mathsf{Mol}(X)$ such that $\|n\| \leq \|m\|$ and $\|m - n\|_\Delta < \varepsilon$.
(iii) The restriction of m to $\mathsf{BLip}_b(\Delta)$ is X-pointwise continuous.
(iv) The restrictions of m^+ and m^- to $\mathsf{BLip}_b(\Delta)$ are X-pointwise continuous.

Proof. (i)$\Rightarrow$(ii) by Corollary 5.24, clearly (ii)$\Rightarrow$(iii), and the implication (iv)$\Rightarrow$(i) is a special case of Theorem 5.26.

To prove (iii)$\Rightarrow$(iv), assume (iii), take any $\varepsilon > 0$, and let $\{f_\gamma\}_\gamma$ be a net of functions $f_\gamma \in \mathsf{BLip}_b(\Delta)$ such that $\lim_\gamma f_\gamma(x) = 0$ for every $x \in X$. By Lemma P.20 with $F = U_b(X)$ and $h = 1$, there is $g \in U_b(X)^+$ such that $m^+(f_\gamma^+) < m(g \wedge f_\gamma^+) + \varepsilon$ and $m^+(f_\gamma^-) < m(g \wedge f_\gamma^-) + \varepsilon$ for all γ.

By Lemma 5.20, there is $g' \in \mathsf{Lip}_b(\Delta)^+$ for which $\|g - g'\|_X \leq \varepsilon$. It follows that $\|g \wedge f_\gamma^+ - g' \wedge f_\gamma^+\|_X \leq \varepsilon$ and

$$m^+(f_\gamma^+) < m(g \wedge f_\gamma^+) + \varepsilon$$
$$\leq m(g' \wedge f_\gamma^+) + \|g \wedge f_\gamma^+ - g' \wedge f_\gamma^+\|_X \|m\| + \varepsilon$$
$$\leq m(g' \wedge f_\gamma^+) + \varepsilon\|m\| + \varepsilon$$

for every γ. Now $g' \in r\mathsf{BLip}_b(\Delta)^+$ for some $r \geq 1$, and $\mathfrak{m}$ is continuous on $r\mathsf{BLip}_b(\Delta)$. Since $g' \wedge f_\gamma^+ \in r\mathsf{BLip}_b(\Delta)$ and $\lim_\gamma g' \wedge f_\gamma^+(x) = 0$ for every $x \in X$, it follows that

$$\limsup_\gamma \mathfrak{m}^+(f_\gamma^+) \leq \varepsilon \|\mathfrak{m}\| + \varepsilon$$

and similarly

$$\limsup_\gamma \mathfrak{m}^+(f_\gamma^-) \leq \varepsilon \|\mathfrak{m}\| + \varepsilon.$$

Thus

$$\limsup_\gamma \mathfrak{m}^+(f_\gamma) \leq 2\varepsilon \|\mathfrak{m}\| + 2\varepsilon$$

for every $\varepsilon > 0$. This proves that the restriction of $\mathfrak{m}^+$ to $\mathsf{BLip}_b(\Delta)$ is continuous at 0, and by Lemma 5.25 also at every point in $\mathsf{BLip}_b(\Delta)$. Hence $\mathfrak{m}^- = \mathfrak{m}^+ - \mathfrak{m}$ is also continuous on $\mathsf{BLip}_b(\Delta)$. □

Exercise 5.29. Show that the implication (iv)$\Rightarrow$(i) in Theorem 5.28 does not hold if the metric space X is not complete. ∎

Theorem 5.30. *Let X be a metric space with metric Δ, and let $\mathfrak{B}$ be the closed $\|\cdot\|$ unit ball in $\mathfrak{M}_b(X)$.*

1. *$\mathfrak{B}$ is complete in the $\|\cdot\|_\Delta$ norm.*
2. *If X is complete, then $\mathfrak{M}_t(X)$ is $\|\cdot\|_\Delta$ closed in $\mathfrak{M}_b(X)$.*
3. *If X is complete, then $\mathfrak{B} \cap \mathfrak{M}_t(X)$ is complete in the $\|\cdot\|_\Delta$ norm.*

Proof. Let $\{\mathfrak{m}_j\}_j$ be a $\|\cdot\|_\Delta$ Cauchy sequence in $\mathfrak{B}$. For every $f \in \mathsf{Lip}_b(\Delta)$, the sequence $\{\mathfrak{m}_j(f)\}_j$ in $\mathbb{R}$ is Cauchy and therefore has a limit $\mathfrak{m}'(f)$. That defines a linear functional $\mathfrak{m}'$ on $\mathsf{Lip}_b(\Delta)$ and $|\mathfrak{m}'(f)| \leq 1$ when $f \in \mathsf{Lip}_b(\Delta)$, $\|f\|_X \leq 1$. Use Lemma 5.20 to extend $\mathfrak{m}'$ to $\mathfrak{m} \in \mathfrak{M}_b(X)$ such that $\|\mathfrak{m}\| \leq 1$. Since the sequence $\{\mathfrak{m}_j\}_j$ is $\|\cdot\|_\Delta$ Cauchy, it converges to $\mathfrak{m}$ in the $\|\cdot\|_\Delta$ norm.

Part 2 follows from Theorem 5.28. Part 3 follows from Parts 1 and 2. □

Exercise 5.31. Show that the statement in Part 3 of Theorem 5.30 does not hold when the metric space X is not complete. ∎

Exercise 5.32. Let X be the unit interval $[0,1]$ and Δ the usual metric on X. Show that the space $\mathfrak{M}_b(X) = \mathfrak{M}_t(X)$ is not complete in the $\|\cdot\|_\Delta$ norm. ∎

Exercise 5.33. Let X be a metric space with metric Δ, and let $\mathfrak{A}$ be the space of those linear functionals on $\mathsf{Lip}_b(\Delta)$ that are X-pointwise continuous on $\mathsf{BLip}_b(\Delta)$. Abbreviate again the norm $\|\cdot\|_{\mathsf{BLip}_b(\Delta)}$ on $\mathfrak{A}$ as $\|\cdot\|_\Delta$. Identify each $\mathfrak{m} \in \mathfrak{M}_b(X)$ with its restriction to $\mathsf{Lip}_b(\Delta)$ so that $\mathsf{Mol}(X) \subseteq \mathfrak{M}_t(X) \subseteq \mathfrak{A}$. Prove the following:

1. The $\|\cdot\|_\Delta$ norm dual of $\mathfrak{A}$ is $\mathsf{Lip}_b(\Delta)$.
2. In the $\|\cdot\|_\Delta$ norm, the space $\mathfrak{A}$ is complete and $\mathsf{Mol}(X)$ is dense in $\mathfrak{A}$.
3. If X is complete, then $\mathfrak{M}_t(X) = \mathfrak{A} \cap \mathfrak{M}_b(X)$. ∎

The reader likely knows that every metric space is isometrically embedded into a complete metric space. It is instructive to see how this result follows directly from Part 1 of Theorem 5.30, using the mapping $\partial \colon X \to \mathfrak{M}_b(X)$ from Sect. 5.2. The following construction appears again in a more general form in Sect. 6.5, where I use it to construct a completion of an arbitrary uniform space.

When X is a metric space with metric Δ, let $\widehat{X}$ be the uniform space whose set of points is the $\|\cdot\|_\Delta$ closure of $\partial(X)$ in $\mathfrak{M}_b(X)$, and whose uniformity is induced by the metric of $\|\cdot\|_\Delta$.

Corollary 5.34. *Let X be a metric space with metric Δ, and identify the set X with $\partial(X) \subseteq \mathfrak{M}_b(X)$ by means of the mapping ∂. Then $X \Subset \widehat{X}$, X is dense in $\widehat{X}$, and the uniform space $\widehat{X}$ is metrizable by a complete metric $\widehat{\Delta}$ such that $\widehat{\Delta}\!\restriction\!X = \Delta$.*

Proof. The metric Δ' of the norm $\|\cdot\|_\Delta$ satisfies $\Delta'\!\restriction\!X = \Delta \wedge 2$ by Lemma 5.12, hence $X \Subset \widehat{X}$. By Part 1 of Theorem 5.30, the unit ball $\mathfrak{B}$ is Δ' complete; hence its closed subspace $\widehat{X}$ is also Δ' complete. By Corollary 2.25, there is $\widehat{\Delta} \in \mathrm{UP}(\widehat{X})$ such that $\Delta = \widehat{\Delta}\!\restriction\!X$, and by continuity (Part 1 of Theorem 1.21), we get $\widehat{\Delta} \wedge 2 = \Delta' \wedge 2$. Thus $\widehat{\Delta}$ is a complete metric inducing the uniformity of $\widehat{X}$. $\square$

Exercise 5.35. Let X be a metric space and let $\mathfrak{m} \in \mathfrak{M}_b(X)$ satisfy condition (iii) in Theorem 5.28. Show that the restriction of $\mathfrak{m}$ to the $\|\cdot\|_X$ unit ball in $\mathrm{U}_b(X)$ is continuous in the topology of uniform convergence on precompact sets in X. ∎

5.5 Topologies on the Positive Cone and on Spheres

By Corollaries 5.17 and 5.19, two weak topologies coincide on the positive cone and on spheres in $\mathfrak{M}_t(X)$. Next I add a third topology to the list when X is a metric space. As in Sect. 5.3, once the result is established for the positive cone, the corresponding result for spheres follows by Lemma P.22.

Theorem 5.36. *Let X be a metric space with metric Δ and $\mathfrak{m} \in \mathfrak{M}_t(X)^+$. The following three conditions are equivalent for a net $\{\mathfrak{m}_\gamma\}_\gamma$ in $\mathfrak{M}_b(X)^+$:*

(i) $\lim_\gamma \mathfrak{m}_\gamma(g) = \mathfrak{m}(g)$ for every $g \in \mathrm{U}_b(X)$
(ii) $\lim_\gamma \mathfrak{m}_\gamma(g) = \mathfrak{m}(g)$ for every $g \in \mathrm{Lip}_b(\Delta)$
(iii) $\lim_\gamma \|\mathfrak{m}_\gamma - \mathfrak{m}\|_\Delta = 0$

Proof. Obviously (i)$\Rightarrow$(ii) and (iii)$\Rightarrow$(ii). The implication (ii)$\Rightarrow$(i) follows from Corollary 5.22, because every $\mathrm{Lip}_b(X)$-weakly convergent net in $\mathfrak{M}_t(X)^+$ is $\|\cdot\|$ bounded.

To prove (ii)$\Rightarrow$(iii), assume that $\lim_\gamma \mathfrak{m}_\gamma(g) = \mathfrak{m}(g)$ for every $g \in \mathrm{Lip}_b(\Delta)$, and take any $\varepsilon > 0$. Since $\mathfrak{m}$ is continuous on $\mathrm{BLip}_b(\Delta)$ with the X-pointwise topology, there are a non-empty finite set $D \subseteq X$ and $\theta > 0$ such that if $g \in \mathrm{BLip}_b(\Delta)$ and $\|g\|_D < \theta$, then $|\mathfrak{m}(g)| < \varepsilon$. As the set $\mathrm{BLip}_b(\Delta)$ is compact in the X-pointwise topology, there is a finite set $\mathscr{F} \subseteq \mathrm{BLip}_b(\Delta)$ such that for every $g \in \mathrm{BLip}_b(\Delta)$ there exists $f \in \mathscr{F}$ with $\|f - g\|_D < \theta$.

Define $h \in \mathsf{BLip}_b(\Delta)$ by $h(x) := (\Delta(x,D) + \theta) \wedge 1$. Since $\lim_\gamma \mathsf{m}_\gamma(g) = \mathsf{m}(g)$ for every $g \in \mathsf{Lip}_b(\Delta)$, there is γ_0 such that if $\gamma \geq \gamma_0$, then $|\mathsf{m}_\gamma(f) - \mathsf{m}(f)| < \varepsilon$ for every $f \in \mathscr{F}$ and $|\mathsf{m}_\gamma(h) - \mathsf{m}(h)| < \varepsilon$.

Let g be any function in $\mathsf{BLip}_b(\Delta)$, and let $f \in \mathscr{F}$ be such that $\|f - g\|_D < \theta$. Since $(f - g) \in 2\mathsf{BLip}_b(\Delta)$, it follows that $|\mathsf{m}(f) - \mathsf{m}(g)| < 2\varepsilon$ and $|f - g| \leq 2h$. From the last inequality, we get

$$\mathsf{m}_\gamma(|f - g|) \leq 2\mathsf{m}_\gamma(h) \leq 2|\mathsf{m}_\gamma(h) - \mathsf{m}(h)| + 2\mathsf{m}(h) < 4\varepsilon$$

for $\gamma \geq \gamma_0$. Putting all the estimates together, we have

$$|\mathsf{m}_\gamma(g) - \mathsf{m}(g)| \leq |\mathsf{m}_\gamma(g) - \mathsf{m}_\gamma(f)| + |\mathsf{m}_\gamma(f) - \mathsf{m}(f)| + |\mathsf{m}(f) - \mathsf{m}(g)|$$
$$< \mathsf{m}_\gamma(|f - g|) + 3\varepsilon < 7\varepsilon$$

for $\gamma \geq \gamma_0$, which proves that $\{\mathsf{m}_\gamma\}_\gamma$ converges to m uniformly on $\mathsf{BLip}_b(\Delta)$. □

Corollary 5.37. *For any metric space X with metric Δ, the $\|\cdot\|_\Delta$ norm topology and the $\mathsf{U}_b(X)$-weak topology coincide on $\mathfrak{M}_t(X)^+$.* □

Theorem 5.38. *Let X be a metric space with metric Δ and $\mathsf{m} \in \mathfrak{M}_t(X)$. For any net $\{\mathsf{m}_\gamma\}_\gamma$ in $\mathfrak{M}_b(X)$ such that $\lim_\gamma \|\mathsf{m}_\gamma\| \leq \|\mathsf{m}\|$, the following three conditions are equivalent:*

(i) $\lim_\gamma \mathsf{m}_\gamma(f) = \mathsf{m}(f)$ *for every* $f \in \mathsf{U}_b(X)$.
(ii) $\lim_\gamma \mathsf{m}_\gamma(f) = \mathsf{m}(f)$ *for every* $f \in \mathsf{Lip}_b(\Delta)$.
(iii) $\lim_\gamma \|\mathsf{m}_\gamma - \mathsf{m}\|_\Delta = 0$.

Proof. Obviously (i)⇒(ii) and (iii)⇒(ii).

To prove (ii)⇒(i), note that $\|\mathsf{m}_\gamma\| \leq 2\|\mathsf{m}\|$ for almost all γ and apply Corollary 5.22.

To prove (ii)⇒(iii), apply Lemma P.22 with $S = X$ and $F = \mathsf{Lip}_b(X)$. Then apply the implication (ii)⇒(iii) in Theorem 5.36 twice: Once with m^+ and m_γ^+ in place of m and m_γ, and once with m^- and m_γ^- in place of m and m_γ. □

Corollary 5.39. *For any metric space X with metric Δ and any $r \in \mathbb{R}^+$, the $\|\cdot\|_\Delta$ norm topology and the $\mathsf{U}_b(X)$-weak topology coincide on the set*

$$\{\mathsf{m} \in \mathfrak{M}_t(X) \mid \|\mathsf{m}\| = r\}.$$ □

In the next section I prove that the two topologies in Corollaries 5.37 and 5.39 coincide also on $\mathsf{U}_b(X)$-weakly countably compact sets.

5.6   Compact Sets and Sequences of Measures

Example 5.7 shows that the $U_b(X)$-weak topology and the $\|\cdot\|_\Delta$ norm topology need not coincide on the whole space $\mathfrak{M}_t(X)$, or even on the unit ball in $\mathfrak{M}_t(X)$. In the example, $\|\cdot\| = \|\cdot\|_\Delta = \|\cdot\|_1$ on $\mathfrak{M}_t(X) = \ell_1$, and by Part 4 of Theorem P.15, the $U_b(X)$-weak topology and the $\|\cdot\|_\Delta$ norm topology do not coincide on the unit ball in $\mathfrak{M}_t(X)$.

Nevertheless, for the particular choice of X in that example, the two topologies do coincide on $U_b(X)$-weakly countably compact sets, by Part 3 of Theorem P.15. That result is generalized to an arbitrary metric space X in the forthcoming Corollary 5.43. Similarly, Part 1 in Theorem P.15, concerning weak sequential completeness, is generalized in Theorem 5.45.

Lemma 5.40. *Let X be a metric space with metric Δ and let a set $\mathfrak{A} \subseteq \mathfrak{M}_t(X)$ be $\|\cdot\|$ bounded and not X-pointwise equicontinuous on $\mathsf{BLip}_b(\Delta)$. Then there exist $\varepsilon > 0$, $r \geq 1$, and $\mathsf{m}_j \in \mathfrak{A}$, $g_j \in r\mathsf{BLip}_b(\Delta)^+$ for $j \in \omega$, such that*

(i) $|\mathsf{m}_j(g_j)| > 2\varepsilon$ for all j;
(ii) $|\mathsf{m}_i(g_j)| < \varepsilon$ for all i, j, $i < j$;
(iii) $g_i \wedge g_j = 0$ for all i, j, $i < j$.

Proof. It is clearly enough to consider the case where $\|\mathsf{m}\| \leq 1$ for all $\mathsf{m} \in \mathfrak{A}$. By Lemma 5.25, $\mathfrak{A}$ is not X-pointwise equicontinuous at 0 on $\mathsf{BLip}_b(\Delta)^+$. Thus

$$\exists \varepsilon > 0 \,\forall \text{ finite } D \subseteq X \,\exists f \in \mathsf{BLip}_b(\Delta)^+ \,\exists \mathsf{m} \in \mathfrak{A} \;[\, \|f\|_D < \varepsilon \text{ and } |\mathsf{m}(f)| > 5\varepsilon\,].$$

Let $r := (1/\varepsilon) \vee 1$, and construct $f_j \in \mathsf{BLip}_b(\Delta)^+$, $\mathsf{m}_j \in \mathfrak{A}$ and non-empty finite sets $D_j \subseteq X$ for $j \in \omega$ such that

(a) $\|f_j\|_{D_i} < \varepsilon$ for all $i < j$;
(b) $|\mathsf{m}_j(f_j)| > 5\varepsilon$;
(c)   if $g \in (r+1)\mathsf{BLip}_b(\Delta)$, $\|g\|_{D_j} = 0$ then $|\mathsf{m}_j(g)| < \varepsilon$.

Define functions $g_j \in r\mathsf{BLip}_b(\Delta)^+$, $j \in \omega$, by

$$g_j(x) := \max_{y \in D_j} \left(f_j(y) - 2\varepsilon - r\Delta(x,y) \right)^+, \quad x \in X.$$

Then $0 \leq g_j \leq (f_j - 2\varepsilon)^+$ because $f_j(y) - r\Delta(x,y) \leq f_j(x)$ for $x, y \in X$.

The function $(f_j - 2\varepsilon)^+ - g_j$ is in $(r+1)\mathsf{BLip}_b(\Delta)$ and vanishes on D_j; therefore $|\mathsf{m}_j(g_j) - \mathsf{m}_j((f_j - 2\varepsilon)^+)| < \varepsilon$ by (c). Since $\|\mathsf{m}_j\| \leq 1$, we get

$$|\mathsf{m}_j(g_j) - \mathsf{m}_j(f_j)| < |\mathsf{m}_j(g_j) - \mathsf{m}_j((f_j - 2\varepsilon)^+)| + |\mathsf{m}_j((f_j - 2\varepsilon)^+) - \mathsf{m}_j(f_j)|$$

$$\leq \varepsilon + 2\varepsilon = 3\varepsilon,$$

and (i) follows from (b).

The function $(f_j - 2\varepsilon)^+$ vanishes on D_i for $i < j$, therefore so does g_j, and (ii) follows from (c).

Finally, to prove (iii), let $i < j$, $x \in X$ and $g_j(x) > 0$. Then $(f_j(x) - 2\varepsilon)^+ > 0$, so that $f_j(x) > 2\varepsilon$. If $y \in D_i$ then $\Delta(x, y) \geq f_j(x) - f_j(y) > \varepsilon$ by (a), $r\Delta(x, y) > r\varepsilon \geq 1$, and $f_i(y) - r\Delta(x, y) \leq 0$. Thus $g_i(x) = 0$ by the definition of g_i.   □

Theorem 5.41. *Let X be a complete metric space with metric Δ. The following properties of a $\|\cdot\|$-bounded set $\mathfrak{A} \subseteq \mathfrak{M}_t(X)$ are equivalent:*

(i) $\mathfrak{A}$ *is X-pointwise equicontinuous on* $\mathsf{BLip}_b(\Delta)$.

(ii) $\mathfrak{A}$ *is* $\|\cdot\|_\Delta$ *precompact.*

(iii) $\mathfrak{A}$ *is relatively* $\|\cdot\|_\Delta$ *compact in* $\mathfrak{M}_t(X)$.

(iv) $\mathfrak{A}$ *is relatively* $\mathsf{U}_b(X)$-*weakly compact in* $\mathfrak{M}_t(X)$.

(v) $\mathfrak{A}$ *is relatively* $\mathsf{U}_b(X)$-*weakly sequentially compact in* $\mathfrak{M}_t(X)$.

(vi) $\mathfrak{A}$ *is relatively* $\mathsf{U}_b(X)$-*weakly countably compact in* $\mathfrak{M}_t(X)$.

(vii) $\mathfrak{A}$ *is relatively* $\mathsf{Lip}_b(\Delta)$-*weakly compact in* $\mathfrak{M}_t(X)$.

(viii) $\mathfrak{A}$ *is relatively* $\mathsf{Lip}_b(\Delta)$-*weakly sequentially compact in* $\mathfrak{M}_t(X)$.

(ix) $\mathfrak{A}$ *is relatively* $\mathsf{Lip}_b(\Delta)$-*weakly countably compact in* $\mathfrak{M}_t(X)$.

(x) *If $\{\mathfrak{m}_j\}_j$ is a sequence in $\mathfrak{A}$ and $\{f_i\}_i$ is a sequence in $\mathsf{BLip}_b(\Delta)$ and the two double limits $\lim_i \lim_j \mathfrak{m}_j(f_i)$ and $\lim_j \lim_i \mathfrak{m}_j(f_i)$ exist, then they are equal.*

The completeness of X is used only in the proof of (ii)⇒(iii). By Corollary 5.43, properties (iii) to (vi) are equivalent in any (not necessarily complete) metric space.

Proof. (i)⇒(ii) follows from the Ascoli Theorem P.14, and (ii)⇒(iii) from Part 3 in Theorem 5.30. Clearly (iii)⇒(iv)⇒(vi)⇒(ix) and (iv)⇒(vii)⇒(ix), as well as (iii)⇒(v)⇒(viii)⇒(ix).

To prove the equivalence of (vii) and (x), let E be the space $\mathfrak{M}_t(X)$ with the $\|\cdot\|_\Delta$ norm and apply Theorem P.13. The closed convex hull of $\mathfrak{A}$ is complete by Part 3 of Theorem 5.30. The dual E^* identifies with $\mathsf{Lip}_b(\Delta)$ by Lemma 5.23, and a set $\mathscr{F} \subseteq \mathsf{Lip}_b(\Delta)$ is equicontinuous in $\|\cdot\|_\Delta$ if and only if $\mathscr{F} \subseteq r\mathsf{BLip}_b(\Delta)$ for some $r \in \mathbb{R}^+$.

To prove (ix)⇒(i), take any $\|\cdot\|$-bounded set $\mathfrak{A} \subseteq \mathfrak{M}_t(X)$ that is not X-pointwise equicontinuous on $\mathsf{BLip}_b(\Delta)$. Let $\varepsilon > 0$, $r \geq 1$ and $\mathfrak{m}_j \in \mathfrak{A}$, $g_j \in r\mathsf{BLip}_b(\Delta)^+$ for $j \in \omega$ be as in Lemma 5.40. The expression $\varphi(\mathfrak{m})(j) := \mathfrak{m}(g_j)/2r$, $\mathfrak{m} \in \mathfrak{M}_t(X)$, $j \in \omega$, defines a linear mapping φ from $\mathfrak{M}_t(X)$ to $\mathbb{R}^\omega$. If $d \in \ell_\infty$, $\|d\|_\omega \leq 1$, then the sequence of functions $\sum_{j=0}^k d(j)g_j/2r$ indexed by $k \in \omega$ is in $\mathsf{BLip}_b(\Delta)$ and converges X-pointwise to $\sum_{j \in \omega} d(j)g_j/2r$, hence

$$\mathfrak{m}\left(\sum_{j \in \omega} d(j)g_j/2r\right) = \lim_k \mathfrak{m}\left(\sum_{j=0}^k d(j)g_j/2r\right) = \lim_k \sum_{j=0}^k d(j)\mathfrak{m}(g_j)/2r = \langle d, \varphi(\mathfrak{m})\rangle.$$

Thus φ maps $\mathfrak{M}_t(X)$ into ℓ_1, $\|\varphi(\mathfrak{m})\|_1 \leq \|\mathfrak{m}\|_\Delta$ for all $\mathfrak{m} \in \mathfrak{M}_t(X)$, and φ is continuous from $\mathfrak{M}_t(X)$ with the $\mathsf{Lip}_b(X)$-weak topology to ℓ_1 with the ℓ_∞-weak topology. The set $\varphi(\mathfrak{A})$ is bounded and not precompact in the norm $\|\cdot\|_1$ on ℓ_1 because for $i < j$ we have

$$\|\varphi(\mathfrak{m}_j) - \varphi(\mathfrak{m}_i)\|_1 \geq |\mathfrak{m}_j(g_j)/2r - \mathfrak{m}_i(g_j)/2r| > \varepsilon/2r$$

by (i) and (ii) in Lemma 5.40. Apply Theorem P.15 to conclude that $\varphi(\mathfrak{A})$ is not relatively ℓ_∞-weakly countably compact in ℓ_1, and therefore $\mathfrak{A}$ is not relatively $\mathrm{Lip}_b(X)$-weakly countably compact in $\mathfrak{M}_t(X)$. $\qquad\square$

Corollary 5.42 (Prokhorov). *Let X be a complete metric space. A subset of $\mathfrak{M}_t(X)^+$ is relatively $C_b(X)$-weakly compact if and only if it is uniformly tight.*

Proof. Combine Corollary 5.17 with Theorems 5.26 and 5.41. $\qquad\square$

Corollary 5.43. *For any metric space X with metric Δ, the $U_b(X)$-weak topology and the $\|\cdot\|_\Delta$ topology coincide on every relatively $U_b(X)$-weakly countably compact subset of $\mathfrak{M}_t(X)$. In particular, the $U_b(X)$-weak topology and the $\|\cdot\|_\Delta$ topology on $\mathfrak{M}_t(X)$ have the same convergent sequences.*

Proof. First note that if $\mathfrak{A}$ is relatively $U_b(X)$-weakly countably compact in $\mathfrak{M}_t(X)$ then it is also relatively $U_b(X)$-weakly countably compact in $\mathfrak{M}_b(X)$, hence $\|\cdot\|$ bounded by the Banach–Steinhaus Theorem P.11.

By Corollary 5.34 there is a complete metric space $\widehat{X}$ with metric $\widehat{\Delta}$ such that $X \subseteq \widehat{X}$, X is dense in $\widehat{X}$ and $\widehat{\Delta}\!\restriction\! X = \Delta$. Define $\iota : \mathfrak{M}_b(X) \to \mathfrak{M}_b(\widehat{X})$ by

$$\iota(\mathfrak{m})(f) := \mathfrak{m}(f\!\restriction\! X) \text{ for } f \in U_b(\widehat{X}),$$

where $f\!\restriction\! X$ is the restriction of f to X (this is a special case of a general construction in Definition 6.7).

By Theorem 2.22 (or 2.24), ι is an isomorphism of the Banach lattices $\mathfrak{M}_b(X)$ and $\mathfrak{M}_b(\widehat{X})$, and also a homeomorphism from $\mathfrak{M}_b(X)$ with the $U_b(X)$-weak topology onto $\mathfrak{M}_b(\widehat{X})$ with the $U_b(\widehat{X})$-weak topology. Evidently, ι is an isometry for the norms $\|\cdot\|_\Delta$ and $\|\cdot\|_{\widehat{\Delta}}$, and $\iota(\mathfrak{M}_t(X)) \subseteq \iota(\mathfrak{M}_t(\widehat{X}))$.

Take any relatively $U_b(\widehat{X})$-weakly countably compact subset $\mathfrak{A}$ of $\mathfrak{M}_t(X)$. Then $\iota(\mathfrak{A})$ is relatively $U_b(\widehat{X})$-weakly countably compact in $\mathfrak{M}_t(\widehat{X})$, therefore relatively $\|\cdot\|_{\widehat{\Delta}}$ compact by Theorem 5.41. Hence the $U_b(\widehat{X})$-weak and the $\|\cdot\|_{\widehat{\Delta}}$ topologies coincide on $\iota(\mathfrak{A})$, and the $U_b(X)$-weak and the $\|\cdot\|_\Delta$ topologies coincide on $\mathfrak{A}$. $\qquad\square$

Exercise 5.44. Show that if a metric space X is not complete, then (i) does not imply (iii) in Theorem 5.41. $\qquad\blacksquare$

Theorem 5.45. *The space $\mathfrak{M}_t(X)$ is $U_b(X)$-weakly sequentially complete for every complete metric space X.*

Proof. Take any $U_b(X)$-weakly Cauchy sequence $\{\mathfrak{m}_j\}_j$ in $\mathfrak{M}_t(X)$. By the Banach–Steinhaus Theorem P.11, the set $\mathfrak{A} := \{\mathfrak{m}_j \mid j \in \omega\}$ is bounded in the $\|\cdot\|$ norm. To prove that $\{\mathfrak{m}_j\}_j$ converges $U_b(X)$-weakly, by Corollary 5.22 and Part 3 of Theorem 5.30, it is enough to prove that $\{\mathfrak{m}_j\}_j$ is Cauchy in the norm $\|\cdot\|_\Delta$.

Suppose that $\{\mathfrak{m}_j\}_j$ is not $\|\cdot\|_\Delta$ Cauchy. Thus there is $\varepsilon > 0$ such that for every $j \in \omega$, there exist $k(j), k'(j) \in \omega$, $k(j), k'(j) \geq j$, for which $\|\mathfrak{m}_{k(j)} - \mathfrak{m}_{k'(j)}\|_\Delta \geq \varepsilon$. Let $\mathfrak{n}_j := \mathfrak{m}_{k(j)} - \mathfrak{m}_{k'(j)}$. As $\{\mathfrak{m}_j\}_j$ is $U_b(X)$-weakly Cauchy, the sequence $\{\mathfrak{n}_j\}_j$ $U_b(X)$-weakly converges to 0. On the other hand, $\{\mathfrak{n}_j\}_j$ does not converge to 0 in $\|\cdot\|_\Delta$, which contradicts Corollary 5.43. $\qquad\square$

5.7   Notes for Chap. 5

Following the notation commonly used in topological measure theory, in these notes I write simply $\mathfrak{M}_b(T)$ and $\mathfrak{M}_t(T)$ instead of $\mathfrak{M}_b(\mathsf{F}T)$ and $\mathfrak{M}_t(\mathsf{F}T)$ when T is a completely regular topological space.

The investigations of the space $\mathfrak{M}_t(T)$ in duality with $C_b(T)$ have their origin in the 1909 paper by Riesz [158] in which he proved his representation theorem for $T = [0,1]$. In the 100 years since, $\mathfrak{M}_t(T)$ and other subspaces of $\mathfrak{M}_b(T)$ in duality with $C_b(T)$ have been explored in great detail and from many angles, of which I am able to mention only few in this brief summary.

Alexandroff [2] created the general theory of representation of functionals on $C_b(T)$ for an arbitrary completely regular space T (and on more general spaces of functions) by finitely additive measures on T. He also defined several important classes of functionals in $\mathfrak{M}_b(T)$, gave their characterizations in terms of representing measures and proved basic properties of the $C_b(T)$-weak topology on $\mathfrak{M}_b(T)$ and its subspaces.

The $C_b(T)$-weak topology on $\mathfrak{M}_t(T)$ provides a natural setting for limit theorems in probability theory. A number of results about the $C_b(T)$-weak topology were obtained in a probabilistic setting, starting with Lévy [125], and later in the work of Gnedenko and Kolmogorov [83], Prokhorov [145], LeCam [121] and many others.

LeCam [121] and Varadarajan [173] consolidated early results about spaces of measures in duality with function spaces and established a framework for subsequent research. The theory then grew in depth and breath, with additional classes of functionals on $C_b(T)$ and of the representing measures, and with new concepts and methods. The comprehensive survey by Wheeler [180] describes properties of spaces of measures in duality with $C_b(T)$ and of associated strict topologies. The book by Bogachev [11, Ch. 8] includes an in-depth treatment of the $C_b(T)$-weak topology on spaces of measures, complemented by a discussion of history and many references.

In a different but related direction, the weak topology on spaces of measures may be defined on arbitrary (not necessary completely regular) topological spaces. The resulting theory is described in Schwartz [165] and Topsøe [169].

For a completely regular topological space T, certain properties of $\mathfrak{M}_b(T)$ may be usefully studied as properties of finitely additive measures on subsets of T, using the Alexandroff representation. However, the Alexandroff representation does not easily extend to $\mathfrak{M}_b(X)$ for an arbitrary uniform space X (cf. the notes in Sect. 7.6). That is why in this treatise I do not deal with the representation of functionals on $U_b(X)$ by finitely additive measures on X.

The duality $\langle \mathfrak{M}_t, C_b \rangle$ is now a basic tool in functional analysis and in probability theory. The duality $\langle \mathfrak{M}_t, U_b \rangle$ has not received as much attention, although it sometimes appears in a supporting role (see e.g. Billingsley [10]). When one is only interested in positive measures, the U_b-weak topology may be used in place of the C_b-weak topology, by Corollary 5.17. However, once we move beyond positive measures, the dualities $\langle \mathfrak{M}_t, C_b \rangle$ and $\langle \mathfrak{M}_t, U_b \rangle$ are quite different. I describe a non-traditional use of the duality $\langle \mathfrak{M}_t, U_b \rangle$ in probability theory further on in Sect. 11.2.

Molecular measures play a minor role in this chapter; they feature more prominently further on. Several references to early uses of molecular measures are given in Sect. 6.8.

The norm denoted here by $\|\cdot\|_\Delta$ is called the *Kantorovich–Rubinshtein norm* (with reference to [105]) by Bogachev [11, 8.3], who also cites a number of studies of its extensions and applications. Equivalent or similar norms were used by Arens and Eells [3] (on a subspace of $\mathrm{Mol}(X)$), Dudley [41], and Fortet and Mourier [56]. LeCam [121] considered $\mathfrak{M}_t$ and spaces of σ-additive and τ-additive measures in duality with the spaces of functions more general than C_b, with U_b as an important instance. Theorem 5.16 and Corollary 5.17 are a special case of [121, L.5]. Another version of Corollary 5.17 was proved by Billingsley [10, Th.2.1].

Later papers of LeCam [122], [124] include versions of Theorems 5.28, 5.30 and 5.36, and Corollary 5.37.

Dudley [41] proved a number of results about the metrizability of the $C_b(T)$-weak topology and uniformity on various subspaces and subsets of $\mathfrak{M}_b(T)$. Part 2 of Lemma 5.20 and Corollary 5.37 appear in [41].

Variants of Theorem 5.28 were proved by Berezanskiĭ [7, 4.7] and Fedorova [50]; in the notation that I define in Chaps. 6 and 7, they proved $\mathfrak{M}_u(X) = \mathfrak{M}_t(X)$ and $\mathfrak{M}_u(X) = \mathfrak{M}_\sigma(X)$, respectively, for every complete metric space X.

Corollaries 5.19 and 5.39 are inspired by McKennon [127, 3.9] and Granirer and Leinert [85], and by a question that A.T.-M. Lau asked me in 1978.

The duality $\langle \mathfrak{A}, \mathrm{Lip}_b(\Delta) \rangle$ in Exercise 5.33 is a useful tool in the study of the Lipschitz space $\mathrm{Lip}_b(\Delta)$. Weaver [178] describes general Lipschitz spaces and their preduals.

My original proof of Theorems 5.41 and 5.45 appeared in [136]. The proof here incorporates a simplification due to Cooper and Schachermayer [30]. Theorem 5.41 was also proved by Khurana [110, Th.1]. The second statement in Corollary 5.43 was proved by Davydov and Rotar [35, Th. 4]; it also follows from Kalton's theorem [104, 4.6] about the Schur property of certain preduals of Lipschitz spaces. A simple proof for measures on $\mathbb{R}^k$ is due to van Handel [94, B.1], who also found another proof for measures on a general metric space (private communication). Property (x) in Theorem 5.41 (and in Theorems 6.16 and 10.18 further on) is a variant of Grothedieck's double-limit condition [86].

For complete metric spaces, Theorem 5.45 strengthens the classical result of Alexandroff about C_b-weak sequential completeness.

Corollary 5.42 is due to Prokhorov [145, Th. 1.12] for complete separable metric spaces; it has led to many generalizations, discussed by Bogachev [11] and Wheeler [180].

Chapter 6
Uniform Measures

When X is a complete metric space, two properties of a functional $\mathfrak{m} \in \mathfrak{M}_b(X)$ are equivalent, by Lemma 5.2 and Theorem 5.28:

(A) $\mathfrak{m}$ is represented by a tight Borel measure on X.
(B) For every $\Delta \in \mathsf{UP}(X)$, the restriction of $\mathfrak{m}$ to $\mathsf{BLip}_b(\Delta)$ is continuous in the X-pointwise topology.

As will be seen further on, (A) and (B) are also equivalent when X is a uniformly locally compact space. The equivalence of (A) and (B) is among the reasons for the ubiquity of the space $\mathfrak{M}_t(X)$ in functional analysis on complete metric spaces and on locally compact groups.

For a general uniform space, (A) implies (B), but (B) does not necessarily imply (A); the two theories diverge. Property (A) defines the familiar space $\mathfrak{M}_t(X)$, which has been studied in great detail using powerful tools of topological measure theory. Property (B) leads to the main subject of this treatise—the theory of so called *uniform measures*. Although uniform measures are not necessarily represented by countably additive measures on X, they have many pleasing properties that make them useful in functional analysis on general uniform spaces.

In this chapter, I define uniform measures by property (B) and then present equivalent properties, each of which could be used as an alternative definition:

(1) (Theorem 6.6) Approximation by molecular measures.
(2) (Theorem 6.10) Tight image measures for every uniformly continuous mapping to a complete metric space.
(3) (Corollary 6.22) Tight image measures for a UCUD set of mappings inducing the uniformity of the underlying space.
(4) (Theorem 6.40) Regularity of integrals for vector-valued mappings.

In subsequent chapters, I add to this list of equivalent properties in Theorems 7.14 and 9.11.

J. Pachl, *Uniform Spaces and Measures*, Fields Institute Monographs 30,
DOI 10.1007/978-1-4614-5058-0_7,
© Springer Science+Business Media New York 2013

In view of the characterization in (2), the results proved in Chap. 5 for $\mathfrak{M}_t(X)$ on a complete metric space X hold more generally for $\mathfrak{M}_u(X)$ on an arbitrary uniform space X; this is explained in Sect. 6.3. In Sect. 6.5, those results are used to construct the completion and the compactification of X embedded in $\mathfrak{M}_b(X)$.

6.1 Space of Uniform Measures

Definition 6.1. Let X be a uniform space. A linear functional $\mathfrak{m}$ on $U_b(X)$ is a *uniform measure on X* iff for every $\Delta \in UP(X)$ the restriction of $\mathfrak{m}$ to $BLip_b(\Delta)$ is continuous in the X-pointwise topology. Let $\mathfrak{M}_u(X)$ denote the space of uniform measures on X. ∎

The term *uniform measure* is established in the literature. However, the reader should keep in mind that a uniform measure is not a measure. The elements of $\mathfrak{M}_u(X)$ are functionals on $U_b(X)$, not measures in the sense of the definition in Sect. P.5. Moreover, the functionals in $\mathfrak{M}_u(X)$ are not necessarily represented by measures on X. Relationships between uniform measures and measures are clarified in Chap. 7.

This inconsistency of terminology is not without precedent in mathematical literature (for example, an implicit function is not a function, and a conditional probability is not a probability). For those readers who are bothered by it, there is a simple remedy: Mentally substitute the term *u-smooth functional* for *uniform measure* everywhere in the text.

By Lemma 5.25, in order to prove that a linear functional $\mathfrak{m}$ on $U_b(X)$ is in $\mathfrak{M}_u(X)$, it is enough to prove that for every $\Delta \in UP(X)$ the restriction $\mathfrak{m}$ to $BLip_b(\Delta)^+$ is X-pointwise continuous at 0. I use this simplification in several proofs that follow in this and later chapters.

Lemma 6.2. *Let X be a uniform space, and let $\mathfrak{A}$ be a set of linear functionals on $U_b(X)$ such that for every $\Delta \in UP(X)$, the set $\mathfrak{A}$ is equicontinuous on $BLip_b(\Delta)$ in the X-pointwise topology. Then $\mathfrak{A} \subseteq \mathfrak{M}_b(X)$ and $\mathfrak{A}$ is bounded in the $\|\cdot\|$ norm.*

Proof. It suffices to prove that $\lim_j \sup_{\mathfrak{m} \in \mathfrak{A}} |\mathfrak{m}(f_j)| = 0$ for any sequence $\{f_j\}_j$ in $U_b(X)$ such that $\|f_j\|_X \leq 1$ for all j and $\lim_j \|f_j\|_X = 0$. Let $\{f_j\}_j$ be such a sequence, and

$$\Delta(x,y) := \sup_j |f_j(x) - f_j(y)|$$

for $x, y \in X$. Then $\Delta \in UP(X)$ and $f_j \in BLip_b(\Delta)$ for all j. Since $\mathfrak{A}$ is X-pointwise equicontinuous on $BLip_b(\Delta)$, it follows that $\lim_j \sup_{\mathfrak{m} \in \mathfrak{A}} |\mathfrak{m}(f_j)| = 0$. □

Corollary 6.3. *Let X be a uniform space.*

1. $\mathfrak{M}_u(X) \subseteq \mathfrak{M}_b(X)$.
2. *If X is precompact then $\mathfrak{M}_u(X) = \mathfrak{M}_b(X)$.*

Proof. Part 1 follows from Lemma 6.2. For every $\Delta \in \mathsf{UP}(X)$, on $\mathsf{BLip}_b(\Delta)$ the X-pointwise topology coincides with the topology of uniform convergence on precompact subsets of X. That proves Part 2. □

Clearly, $\mathfrak{M}_u(X)$ is a $\|\cdot\|$ closed vector subspace of $\mathfrak{M}_b(X)$. Along with the obvious inclusion $\mathsf{Mol}(X) \subseteq \mathfrak{M}_t(X)$, Lemma 5.2 and Corollary 6.3 yield

$$\mathsf{Mol}(X) \subseteq \mathfrak{M}_t(X) \subseteq \mathfrak{M}_u(X) \subseteq \mathfrak{M}_b(X).$$

In Chap. 5 we considered three topologies on the space $\mathfrak{M}_b(X)$, when X is a metric space with metric Δ: the $\mathsf{U}_b(X)$-weak topology and those given by the norms $\|\cdot\|$ and $\|\cdot\|_\Delta$. In the more general setting of an arbitrary uniform space X, the norm $\|\cdot\|$ and the $\mathsf{U}_b(X)$-weak topology on $\mathfrak{M}_b(X)$ are still available. The $\|\cdot\|_\Delta$ norm topology is replaced by the UEB topology, defined as follows.

Definition 6.4. Let X be any uniform space. The UEB(X) *uniformity*, or simply the UEB *uniformity*, on subsets of $\mathfrak{M}_b(X)$ is the uniformity induced by the pseudometrics of the seminorms $\|\cdot\|_\Delta$ where $\Delta \in \mathsf{UP}(X)$. The UEB$(X)$ *topology*, or simply the UEB *topology*, is the topology of the UEB uniformity. ∎

By Definition 1.19 and Lemma 1.20, the UEB uniformity and the UEB topology are the $\mathfrak{S}$-uniformity and the $\mathfrak{S}$-topology, where $\mathfrak{S}$ is the set of UEB subsets of $\mathsf{U}_b(X)$.

Lemma 6.5. *Let X be a uniform space. Every* UEB *continuous linear functional on* $\mathfrak{M}_u(X)$ *is of the form* $\mathfrak{m} \mapsto \mathfrak{m}(f)$, $\mathfrak{m} \in \mathfrak{M}_u(X)$, *for some* $f \in \mathsf{U}_b(X)$.

Thus the UEB dual of $\mathfrak{M}_u(X)$ naturally identifies with $\mathsf{U}_b(X)$.

Proof. On every UEB set the X-pointwise topology and the $\mathfrak{M}_u(X)$-weak topology coincide. Apply the Mackey–Arens Theorem P.8 with $E = \mathfrak{M}_u(X)$, $F = \mathsf{U}_b(X)$ and $\mathfrak{S} =$ the set of the sets $r\mathsf{BLip}_b(\Delta)$ where $\Delta \in \mathsf{UP}(X)$ and $r \in \mathbb{R}^+$. □

Now we come to the first characterization, or alternative definition, of uniform measures: Uniform measures are exactly those functionals on $\mathsf{U}_b(X)$ that are approximated by molecular measures uniformly on UEB(X) subsets of $\mathsf{U}_b(X)$.

Theorem 6.6. *Let X be any uniform space,* $\mathfrak{A} := \{\mathfrak{m} \in \mathfrak{M}_u(X)^+ \mid \|\mathfrak{m}\| = 1\}$, *and* $\mathfrak{B} := \{\mathfrak{m} \in \mathfrak{M}_u(X) \mid \|\mathfrak{m}\| \leq 1\}$.

1. *The space $\mathfrak{M}_u(X)$ is* UEB *complete.*
2. *The set* $\mathsf{Mol}(X) \cap \mathfrak{A}$ *is* UEB *dense in* $\mathfrak{A}$.
3. *The set* $\mathsf{Mol}(X) \cap \mathfrak{B}$ *is* UEB *dense in* $\mathfrak{B}$.

Proof. 1. Take any UEB Cauchy net $\{\mathfrak{m}_\gamma\}_\gamma$ in $\mathfrak{M}_u(X)$. The net $\{\mathfrak{m}_\gamma(f)\}_\gamma$ converges for every $f \in \mathsf{U}_b(X)$; $\mathfrak{m}(f) := \lim_\gamma \mathfrak{m}_\gamma(f)$ defines a linear functional $\mathfrak{m}$ on $\mathsf{U}_b(X)$, and $\lim_\gamma \|\mathfrak{m}_\gamma - \mathfrak{m}\|_\Delta = 0$ for every $\Delta \in \mathsf{UP}(X)$. Hence $\mathfrak{m}$ is X-pointwise continuous on $\mathsf{BLip}_b(\Delta)$, and $\mathfrak{m} \in \mathfrak{M}_u(X)$.

2. If $f \in U_b(X)$ and there is $r \in \mathbb{R}$ such that $\mathfrak{n}(f) \leq r$ for all $\mathfrak{n} \in \mathrm{Mol}(X) \cap \mathfrak{A}$, then $f(x) \leq r$ for every $x \in X$; and if also $\mathfrak{m} \in \mathfrak{A}$, then $\mathfrak{m}(f) \leq r$. The set $\mathrm{Mol}(X) \cap \mathfrak{A}$ is convex, so is its UEB closure, and Part 2 follows from Theorem P.5 and Lemma 6.5.

3. Similarly, if $f \in U_b(X)$ and $\mathfrak{n}(f) \leq r$ for all $\mathfrak{n} \in \mathrm{Mol}(X) \cap \mathfrak{B}$ and if $\mathfrak{m} \in \mathfrak{B}$, then $|\mathfrak{m}(f)| \leq \|f\|_X \leq r$. The set $\mathrm{Mol}(X) \cap \mathfrak{B}$ is convex, so is its UEB closure, and again part 3 follows from Theorem P.5 and Lemma 6.5. $\square$

By Theorem 6.6, $\mathfrak{M}_u(X)$ is the UEB closure of $\mathrm{Mol}(X)$ in the space $\mathfrak{M}_b(X)$. On the other hand, in the $U_b(X)$-weak topology the space $\mathrm{Mol}(X)$ is dense in $\mathfrak{M}_b(X)$, by Corollary P.6.

6.2 Image Under a Uniformly Continuous Mapping

The assignment of $\mathfrak{M}_b(X)$ to X is functorial in the sense that every uniformly continuous mapping $\varphi \colon X \to Y$ naturally extends to $\mathfrak{M}_b(\varphi) \colon \mathfrak{M}_b(X) \to \mathfrak{M}_b(Y)$.

Definition 6.7. Let X and Y be two uniform spaces. For every uniformly continuous mapping $\varphi \colon X \to Y$, the mapping $\mathfrak{M}_b(\varphi) \colon \mathfrak{M}_b(X) \to \mathfrak{M}_b(Y)$ is defined by

$$\mathfrak{M}_b(\varphi)(\mathfrak{m})(g) := \mathfrak{m}(g \circ \varphi) \quad \text{for } g \in U_b(Y).$$

The functional $\mathfrak{M}_b(\varphi)(\mathfrak{m}) \in \mathfrak{M}_b(Y)$ is called the *image of* $\mathfrak{m}$ *under* φ. It will often be written simply as $\varphi(\mathfrak{m})$ instead of $\mathfrak{M}_b(\varphi)(\mathfrak{m})$.   ∎

Note that $\mathfrak{M}_b(\varphi)(\partial_X(x)) = \partial_Y(\varphi(x))$ for $x \in X$. When X and Y are identified with $\partial_X(X) \subseteq \mathfrak{M}_b(X)$ and $\partial_Y(Y) \subseteq \mathfrak{M}_b(Y)$, the restriction of $\mathfrak{M}_b(\varphi)$ to X is φ.

Lemma 6.8. *Let X and Y be two uniform spaces and φ a uniformly continuous mapping from X to Y.*

1. *$\mathfrak{M}_b(\varphi)$ is a positive linear mapping of norm ≤ 1 from the Banach lattice $\mathfrak{M}_b(X)$ to the Banach lattice $\mathfrak{M}_b(Y)$.*
2. *$\mathfrak{M}_b(\varphi) \colon \mathfrak{M}_b(X) \to \mathfrak{M}_b(Y)$ is continuous when both spaces $\mathfrak{M}_b(X)$ and $\mathfrak{M}_b(Y)$ are equipped with the UEB topology.*
3. *$\mathfrak{M}_b(\varphi) \colon \mathfrak{M}_b(X) \to \mathfrak{M}_b(Y)$ is continuous from $\mathfrak{M}_b(X)$ with the $U_b(X)$-weak topology to $\mathfrak{M}_b(Y)$ with the $U_b(Y)$-weak topology.*
4. *If $\mathfrak{m} \in \mathrm{Mol}(X)$, then $\mathfrak{M}_b(\varphi)(\mathfrak{m}) \in \mathrm{Mol}(Y)$.*
5. *If $\mathfrak{m} \in \mathfrak{M}_t(X)$, then $\mathfrak{M}_b(\varphi)(\mathfrak{m}) \in \mathfrak{M}_t(Y)$.*
6. *If $\mathfrak{m} \in \mathfrak{M}_u(X)$, then $\mathfrak{M}_b(\varphi)(\mathfrak{m}) \in \mathfrak{M}_u(Y)$.*

Proof. Parts 1 and 3 hold because the mapping $g \mapsto g \circ \varphi$ from $U_b(Y)$ to $U_b(X)$ is linear and positive, and $\|g \circ \varphi\|_X \leq \|g\|_Y$ for $g \in U_b(Y)$.

Parts 2 and 6: For every $\mathrm{UEB}(Y)$ set $\mathscr{F}$, the set $\{g \circ \varphi \mid g \in \mathscr{F}\}$ is $\mathrm{UEB}(X)$.

Part 4: If $\mathfrak{m} = \sum_{x \in D} r(x)\partial_X(x)$, then $\varphi(\mathfrak{m}) = \sum_{x \in D} r(x)\partial_Y(\varphi(x))$.

Part 5: If $g \in U_b(Y)$ and $K \subseteq X$, then $\|g \circ \varphi\|_X \leq \|g\|_Y$ and $\|g \circ \varphi\|_K = \|g\|_{\varphi(K)}$. If $m \in \mathfrak{M}_t(X)$, then for every $\varepsilon > 0$ there are a compact set $K \subseteq X$ and $\theta > 0$ such that

$$\forall f \in U_b(X) \, [\, \|f\|_X \leq 1 \text{ and } \|f\|_K < \theta \Rightarrow |m(f)| < \varepsilon \,]$$

$$\forall g \in U_b(Y) \, [\, \|g\|_Y \leq 1 \text{ and } \|g\|_{\varphi(K)} < \theta \Rightarrow |m(g \circ \varphi)| < \varepsilon \,]$$

which means that $\varphi(m) \in \mathfrak{M}_t(Y)$. $\square$

Lemma 6.9. *Let Y be a uniform space, $X \Subset Y$, and let $\iota : X \to Y$ be the inclusion mapping.*

1. *$\mathfrak{M}_b(\iota)$ is an isometric embedding of $\mathfrak{M}_b(X)$ into $\mathfrak{M}_b(Y)$ when both spaces $\mathfrak{M}_b(X)$ and $\mathfrak{M}_b(Y)$ are equipped with their $\|\cdot\|$ norms.*
2. *$\mathfrak{M}_b(\iota)$ is a homeomorphism of $\mathfrak{M}_b(X)$ onto the subspace $\mathfrak{M}_b(\iota)(\mathfrak{M}_b(X))$ of $\mathfrak{M}_b(Y)$ when both $\mathfrak{M}_b(X)$ and $\mathfrak{M}_b(Y)$ are equipped with their UEB topologies.*
3. *$\mathfrak{M}_b(\iota)$ is a homeomorphism of $\mathfrak{M}_b(X)$ onto the subspace $\mathfrak{M}_b(\iota)(\mathfrak{M}_b(X))$ of $\mathfrak{M}_b(Y)$ when $\mathfrak{M}_b(X)$ and $\mathfrak{M}_b(Y)$ are equipped with their U_b-weak topologies.*
4. *$\mathfrak{M}_b(\iota)(\mathfrak{M}_u(X)) = \mathfrak{M}_u(Y) \cap \mathfrak{M}_b(\iota)(\mathfrak{M}_b(X))$.*
5. *If X is dense in Y, then $\mathfrak{M}_b(\iota)(\mathfrak{M}_b(X)) = \mathfrak{M}_b(Y)$.*

Proof. 1. By Theorem 2.22, every $g \in U_b(X)$ extends to a function $\widetilde{g} \in U_b(Y)$, and clearly $\widetilde{g}$ may be chosen so that $\|\widetilde{g}\|_Y = \|g\|_X$. Hence $\mathfrak{M}_b(\iota)$ is a $\|\cdot\|$ isometry onto $\mathfrak{M}_b(\iota)(\mathfrak{M}_b(X)) \subseteq \mathfrak{M}_b(Y)$.

2. By Theorem 2.20, the bounded pseudometrics in $UP(X)$ are exactly the restrictions to X of bounded pseudometrics in $UP(Y)$. Hence the sets $\mathrm{BLip}_b(\Delta)$ where $\Delta \in UP(X)$ are exactly the restrictions to X of the sets $\mathrm{BLip}_b(\Delta')$ where $\Delta' \in UP(Y)$. Hence $\mathfrak{M}_b(\iota)$ is a UEB homeomorphism onto $\mathfrak{M}_b(\iota)(\mathfrak{M}_b(X)) \subseteq \mathfrak{M}_b(Y)$.

3. As noted in the proof of Part 1, the functions in $U_b(X)$ are exactly the restrictions to X of functions in $U_b(Y)$. Hence $\mathfrak{M}_b(\iota)$ is a U_b-weak homeomorphism onto $\mathfrak{M}_b(\iota)(\mathfrak{M}_b(X)) \subseteq \mathfrak{M}_b(Y)$.

4. Take any $m \in \mathfrak{M}_b(X)$ and its image $m' := \mathfrak{M}_b(\iota)(m) \in \mathfrak{M}_b(Y)$. By Lemma 6.8, if $m \in \mathfrak{M}_u(X)$, then $m' \in \mathfrak{M}_u(Y)$. For the converse, assume $m' \in \mathfrak{M}_u(Y)$ and take any $\Delta \in UP(X)$. As noted in the proof of Part 2, there is a pseudometric $\Delta' \in UP(Y)$ with $\mathrm{BLip}_b(\Delta) = \{g \circ \iota \mid g \in \mathrm{BLip}_b(\Delta')\}$. The set $\mathrm{BLip}_b(\Delta')$ is Y-pointwise compact, the mapping $g \mapsto g \circ \iota$ is Y-pointwise to X-pointwise continuous, and m' is Y-pointwise continuous on $\mathrm{BLip}_b(\Delta')$; hence m is X-pointwise continuous on $\mathrm{BLip}_b(\Delta)$.

5. If X is dense in Y, then $g \mapsto g \circ \iota$ is a bijection from $U_b(Y)$ onto $U_b(X)$. Hence $\mathfrak{M}_b(\iota)(\mathfrak{M}_b(X)) = \mathfrak{M}_b(Y)$. $\square$

Theorem 6.10. *Let X be a uniform space and $m \in \mathfrak{M}_b(X)$.*

1. *A net $\{m_\gamma\}_\gamma$ in $\mathfrak{M}_b(X)$ converges to m in the UEB topology if and only if $\lim_\gamma \|\varphi(m_\gamma) - \varphi(m)\|_{\Delta'} = 0$ for every uniformly continuous mapping φ from X to a complete metric space Y with metric Δ'. It is also sufficient to restrict the condition to the mappings φ for which $\varphi(X)$ is dense in Y.*

2. $\mathfrak{m} \in \mathfrak{M}_u(X)$ *if and only if* $\varphi(\mathfrak{m}) \in \mathfrak{M}_t(Y)$ *for every uniformly continuous mapping* φ *from* X *to a complete metric space* Y. *It is also sufficient to restrict the condition to the mappings* φ *for which* $\varphi(X)$ *is dense in* Y.

Proof. For any pseudometric $\Delta \in \mathsf{UP}(X)$, consider the associated metric space X/Δ with metric $\Delta^\bullet$ and the canonical surjection $\chi_\Delta : X \to X/\Delta$, as defined in Sect. P.2. By Corollary 5.34, X/Δ is dense in the complete metric space $\widehat{X/\Delta}$ with metric $\widehat{\Delta^\bullet}$ such that $\widehat{\Delta^\bullet}{\upharpoonright}(X/\Delta) = \Delta^\bullet$. Let $\imath : X/\Delta \hookrightarrow \widehat{X/\Delta}$ be the inclusion mapping. Then $\overleftarrow{\imath \circ \chi_\Delta} \, \widehat{\Delta^\bullet} = \Delta$ and $\mathsf{BLip}_b(\Delta) = \{ g \circ \imath \circ \chi_\Delta \mid g \in \mathsf{BLip}_b(\widehat{\Delta^\bullet}) \}$ by Lemma P.17.

For Part 1, first note that if $\lim_\gamma \mathfrak{m}_\gamma = \mathfrak{m}$ in the UEB topology on $\mathfrak{M}_b(X)$, then $\lim_\gamma \varphi(\mathfrak{m}_\gamma) = \varphi(\mathfrak{m})$ in the UEB topology on $\mathfrak{M}_b(Y)$, by Part 2 of Lemma 6.8.

To prove the converse, assume that $\lim_\gamma \| \varphi(\mathfrak{m}_\gamma) - \varphi(\mathfrak{m}) \|_{\Delta'} = 0$ for every uniformly continuous mapping φ from X to a complete metric space Y with metric Δ' such that $\varphi(X)$ is dense in Y. If $\Delta \in \mathsf{UP}(X)$, then

$$\lim_\gamma \| \imath \circ \chi_\Delta(\mathfrak{m}_\gamma) - \imath \circ \chi_\Delta(\mathfrak{m}) \|_{\widehat{\Delta^\bullet}} = 0,$$

and therefore $\lim_\gamma \mathfrak{m}_\gamma = \mathfrak{m}$ uniformly on $\mathsf{BLip}_b(\Delta)$.

For Part 2, first note that if $\mathfrak{m} \in \mathfrak{M}_u(X)$, $\varphi : X \to Y$ and Y is complete metric, then $\varphi(\mathfrak{m}) \in \mathfrak{M}_u(Y) = \mathfrak{M}_t(Y)$ by Lemma 6.8 and Theorem 5.28.

To prove the converse, assume that $\varphi(\mathfrak{m}) \in \mathfrak{M}_t(Y)$ for every uniformly continuous mapping φ from X to a complete metric space Y such that $\varphi(X)$ is dense in Y, and take any $\Delta \in \mathsf{UP}(X)$. Then $\imath(\chi_\Delta(\mathfrak{m})) \in \mathfrak{M}_t(\widehat{X/\Delta}) = \mathfrak{M}_u(\widehat{X/\Delta})$, and $\chi_\Delta(\mathfrak{m}) \in \mathfrak{M}_u(X/\Delta)$ by Part 4 in Lemma 6.9. Therefore the restriction of $\chi_\Delta(\mathfrak{m})$ to $\mathsf{BLip}_b(\Delta^\bullet)$ is continuous in the X/Δ-pointwise topology, and the restriction of $\mathfrak{m}$ to $\mathsf{BLip}_b(\Delta)$ is continuous in the X-pointwise topology. □

Corollary 6.11. *For any uniform space* X, *the space* $\mathfrak{M}_u(X)$ *is a band in the Banach lattice* $\mathfrak{M}_b(X)$.

Proof. Apply Theorem 6.10 with Corollary 5.4. □

6.3 Topologies on the Space of Uniform Measures

With Theorem 6.10, it is now straightforward to obtain results about uniform measures from the corresponding results about tight measures in Chap. 5.

Theorem 6.12. *Let* X *be a uniform space and* $\mathfrak{m} \in \mathfrak{M}_u(X)^+$. *If* $\{\mathfrak{m}_\gamma\}_\gamma$ *is a net in* $\mathfrak{M}_b(X)^+$ *and* $\lim_\gamma \mathfrak{m}_\gamma(f) = \mathfrak{m}(f)$ *for every* $f \in \mathsf{U}_b(X)$, *then* $\lim_\gamma \mathfrak{m}_\gamma = \mathfrak{m}$ *in the* UEB *topology.*

Proof. Apply Theorem 5.36 and both parts of Theorem 6.10. □

Corollary 6.13. *For any uniform space* X, *the* $\mathsf{U}_b(X)$-*weak topology and the* UEB *topology coincide on* $\mathfrak{M}_u(X)^+$. □

Theorem 6.14. *Let X be a uniform space and $\mathfrak{m} \in \mathfrak{M}_u(X)$. If $\{\mathfrak{m}_\gamma\}_\gamma$ is a net in $\mathfrak{M}_b(X)$ such that $\lim_\gamma \|\mathfrak{m}_\gamma\| \leq \|\mathfrak{m}\|$ and $\lim_\gamma \mathfrak{m}_\gamma(f) = \mathfrak{m}(f)$ for every $f \in U_b(X)$, then $\lim_\gamma \mathfrak{m}_\gamma = \mathfrak{m}$ in the UEB topology.*

Proof. Apply Lemma P.22 and Theorem 6.12. □

Corollary 6.15. *For any uniform space X and $r \in \mathbb{R}^+$, the $U_b(X)$-weak topology and the UEB topology coincide on the set $\{\mathfrak{m} \in \mathfrak{M}_u(X) \mid \|\mathfrak{m}\| = r\}$.* □

Theorem 6.16. *Let X be a uniform space. These properties of a set $\mathfrak{A} \subseteq \mathfrak{M}_u(X)$ are equivalent:*

(i) $\mathfrak{A}$ is X-pointwise equicontinuous on $\mathrm{BLip}_b(\Delta)$ for every $\Delta \in \mathrm{UP}(X)$.
(ii) $\mathfrak{A}$ is precompact in the UEB uniformity.
(iii) $\mathfrak{A}$ is relatively UEB compact in $\mathfrak{M}_u(X)$.
(iv) $\mathfrak{A}$ is relatively UEB countably compact in $\mathfrak{M}_u(X)$.
(v) $\mathfrak{A}$ is relatively $U_b(X)$-weakly compact in $\mathfrak{M}_u(X)$.
(vi) $\mathfrak{A}$ is relatively $U_b(X)$-weakly countably compact in $\mathfrak{M}_u(X)$.
(vii) $\mathfrak{A}$ is $\|\cdot\|$ bounded and if $\{\mathfrak{m}_j\}_j$ is a sequence in $\mathfrak{A}$, $\{f_i\}_i$ is a UEB sequence in $U_b(X)$ and the two double limits $\lim_i \lim_j \mathfrak{m}_j(f_i)$ and $\lim_j \lim_i \mathfrak{m}_j(f_i)$ exist, then they are equal.

Proof. To prove (i)⇒(ii), assume $\mathfrak{A}$ is X-pointwise equicontinuous on $\mathrm{BLip}_b(\Delta)$ for every $\Delta \in \mathrm{UP}(X)$. By Lemma 6.2, the set $\mathfrak{A}$ is $\|\cdot\|$ bounded. By the Ascoli Theorem P.14, for every $\Delta \in \mathrm{UP}(X)$, the set of restrictions of the functionals in $\mathfrak{A}$ to $\mathrm{BLip}_b(\Delta)$ is $\|\cdot\|_\Delta$ precompact. By Theorem 2.9 with $\alpha = 0$, $\mathfrak{A}$ is UEB precompact.

(ii)⇒(iii) because $\mathfrak{M}_u(X)$ is complete (Theorem 6.6). Clearly (iii)⇒(iv)⇒(vi) and (iii)⇒(v)⇒(vi). The equivalence of (v) and (vii) follows from Theorem P.13 along with Lemma 6.5 and Part 1 of Theorem 6.6.

The proof of (vi)⇒(i) mimics the construction in the proof of Theorem 6.10: Assume that $\mathfrak{A} \subseteq \mathfrak{M}_u(X)$ is relatively $U_b(X)$-weakly countably compact in $\mathfrak{M}_u(X)$, and take any $\Delta \in \mathrm{UP}(X)$. By Corollary 5.34, X/Δ is dense in the complete metric space $\widehat{X/\Delta}$ with metric $\widehat{\Delta^\bullet}$ such that $\widehat{\Delta^\bullet} \restriction (X/\Delta) = \Delta^\bullet$. Let $\iota : X/\Delta \hookrightarrow \widehat{X/\Delta}$ be the inclusion mapping. By Part 3 of Lemma 6.8, the subset $\iota(\chi_\Delta(\mathfrak{A}))$ of the space $\mathfrak{M}_u(\widehat{X/\Delta}) = \mathfrak{M}_t(\widehat{X/\Delta})$ is relatively $U_b(\widehat{X/\Delta})$-weakly countably compact in $\mathfrak{M}_t(\widehat{X/\Delta})$. By Theorem 5.41, the set $\iota(\chi_\Delta(\mathfrak{A}))$ is $\widehat{X/\Delta}$-pointwise equicontinuous on $\mathrm{BLip}_b(\widehat{\Delta^\bullet})$. Since X/Δ is dense in $\widehat{X/\Delta}$, the set $\chi_\Delta(\mathfrak{A})$ is X/Δ-pointwise equicontinuous on $\mathrm{BLip}_b(\Delta^\bullet)$. By Lemma P.17, the set $\mathfrak{A}$ is X-pointwise equicontinuous on $\mathrm{BLip}_b(\Delta)$. □

Exercise 6.17. Find a uniform space X and a set $\mathfrak{A} \subseteq \mathfrak{M}_u(X)$ that is relatively $U_b(X)$-weakly compact but not relatively $U_b(X)$-weakly sequentially compact. ∎

Corollary 6.18. *For any uniform space X, the $U_b(X)$-weak topology and the UEB topology coincide on every relatively $U_b(X)$-weakly countably compact subset of $\mathfrak{M}_u(X)$. In particular, the $U_b(X)$-weak topology and the UEB topology on $\mathfrak{M}_u(X)$ have the same convergent sequences.* □

Theorem 6.19. *For any uniform space X, the space $\mathfrak{M}_u(X)$ is $\mathsf{U}_b(X)$-weakly sequentially complete.*

Proof. The argument is similar to that in the proof of Theorem 5.45. Take any $\mathsf{U}_b(X)$-weakly Cauchy sequence $\{\mathfrak{m}_j\}_j$ in $\mathfrak{M}_u(X)$. To prove that $\{\mathfrak{m}_j\}_j$ converges, by Theorem 6.6, it is enough to prove that $\{\mathfrak{m}_j\}_j$ is UEB Cauchy.

Suppose that $\{\mathfrak{m}_j\}_j$ is not UEB Cauchy. Thus there is $\Delta \in \mathsf{UP}(X)$ such that for every $j \in \omega$ there are $k(j), k'(j) \in \omega$, $k(j), k'(j) \geq j$, for which $\|\mathfrak{m}_{k(j)} - \mathfrak{m}_{k'(j)}\|_\Delta \geq 1$. Let $\mathfrak{n}_j := \mathfrak{m}_{k(j)} - \mathfrak{m}_{k'(j)}$. As $\{\mathfrak{m}_j\}_j$ is $\mathsf{U}_b(X)$-weakly Cauchy, the sequence $\{\mathfrak{n}_j\}_j$ $\mathsf{U}_b(X)$-weakly converges to 0. On the other hand, $\{\mathfrak{n}_j\}_j$ does not converge to 0 in the UEB topology, which contradicts Corollary 6.18. $\square$

6.4 Uniform Measures on Induced Spaces

By Theorem 6.10, a functional $\mathfrak{m} \in \mathfrak{M}_b(X)$ is in $\mathfrak{M}_u(X)$ whenever its image under *every* uniformly continuous mapping to a complete metric space is in $\mathfrak{M}_u$. However, frequently we are given merely the images of $\mathfrak{m}$ under some, not all, uniformly continuous mappings. It is then important to know whether $\mathfrak{m}$ is necessarily in $\mathfrak{M}_u(X)$ whenever the image measures $\varphi(\mathfrak{m})$ are in $\mathfrak{M}_u$ for some representative set of mappings φ; in particular, for a set of mappings that induces the uniformity of X. And, as in Theorem 6.10, it is important to know whether a net in $\mathfrak{M}_b(X)$ converges in the UEB topology whenever the image measures converge.

Theorem 6.20. *Let X be a uniform space whose uniformity is induced by a UCUD set Φ of mappings $\varphi \colon X \to X_\varphi$ to uniform spaces X_φ. Let $\mathfrak{A}$ be a $\|\cdot\|$-bounded subset of $\mathfrak{M}_b(X)$; let $\{\mathfrak{m}_\gamma\}_\gamma$ be a net in $\mathfrak{A}$, and $\mathfrak{m} \in \mathfrak{A}$.*

1. *If the net $\{\varphi(\mathfrak{m}_\gamma)\}_\gamma$ converges to $\varphi(\mathfrak{m})$ in the $\mathsf{UEB}(X_\varphi)$ topology for every $\varphi \in \Phi$, then the net $\{\mathfrak{m}_\gamma\}_\gamma$ converges to $\mathfrak{m}$ in the $\mathsf{UEB}(X)$ topology.*
2. *If $\varphi(\mathfrak{A}) \subseteq \mathfrak{M}_u(X_\varphi)$ and $\varphi(\mathfrak{A})$ is $\mathsf{UEB}(X_\varphi)$ relatively compact in $\mathfrak{M}_u(X_\varphi)$ for every $\varphi \in \Phi$, then $\mathfrak{A} \subseteq \mathfrak{M}_u(X)$ and $\mathfrak{A}$ is $\mathsf{UEB}(X)$ relatively compact in $\mathfrak{M}_u(X)$.*
3. *If $\varphi(\mathfrak{m}) \in \mathfrak{M}_u(X_\varphi)$ for every $\varphi \in \Phi$, then $\mathfrak{m} \in \mathfrak{M}_u(X)$.*

The proof comes after a useful approximation lemma.

Lemma 6.21. *Let X be a uniform space whose uniformity is induced by a UCUD set Φ of mappings $\varphi \colon X \to X_\varphi$ to uniform spaces X_φ. For every $\Delta \in \mathsf{UP}(X)$ and $\varepsilon > 0$, there exist $\varphi \in \Phi$ and $\Delta' \in \mathsf{UP}(X_\varphi)$ such that*

$$\forall f \in \mathsf{BLip}_b(\Delta)^+ \; \exists f' \in \mathsf{BLip}_b(\Delta')^+ \; [\, f - \varepsilon \leq f' \circ \varphi \leq f \,].$$

Proof. Take any $\Delta \in \mathsf{UP}(X)$ and $\varepsilon > 0$. By Corollary 2.8 there exist $\varphi \in \Phi$ and $\Delta' \in \mathsf{UP}(X_\varphi)$ such that $1 \wedge \Delta < \overleftarrow{\varphi}\Delta' + \varepsilon$.

For any $f \in \mathsf{BLip}_b(\Delta)^+$, define the function f' by

$$f'(x') := \sup_{y \in X} \left(f(y) - \varepsilon - \Delta'(x', \varphi(y)) \right)^+ \quad \text{for} \;\; x' \in X_\varphi.$$

For every $y \in X$, the function $x' \mapsto (f(y) - \varepsilon - \Delta'(x', \varphi(y)))^+$ is in $\mathsf{BLip_b}(\Delta')^+$, hence $f' \in \mathsf{BLip_b}(\Delta')^+$. For $x \in X$, we have on one hand

$$f(x) - \varepsilon \le (f(x) - \varepsilon - \Delta'(\varphi(x), \varphi(x)))^+ \le f'(\varphi(x))$$

and on the other hand

$$f(y) - f(x) \le 1 \wedge \Delta(x,y) < \overset{\leftarrow}{\varphi} \Delta'(x,y) + \varepsilon = \Delta'(\varphi(x), \varphi(y)) + \varepsilon$$

for every $y \in X$, which yields $(f(y) - \varepsilon - \Delta'(\varphi(x), \varphi(y)))^+ \le f(x)$, and therefore $f'(\varphi(x)) \le f(x)$.   □

Proof of Theorem 6.20. Replace each X_φ by $\varphi(X)$ with its subspace uniformity, so that $\varphi(X) = X_\varphi$; this causes no loss of generality, in view of Part 4 in Lemma 6.9. Let $r \in \mathbb{R}^+$ be such that $\|\mathfrak{m}\| \le r$ for all $\mathfrak{m} \in \mathfrak{A}$.

Take any $\Delta \in \mathsf{UP}(X)$ and $\varepsilon > 0$. By Lemma 6.21, there are $\varphi \in \Phi$ and $\Delta' \in \mathsf{UP}(X_\varphi)$ such that

$$\forall f \in \mathsf{BLip_b}(\Delta)^+ \; \exists f' \in \mathsf{BLip_b}(\Delta')^+ \; [\, f - \varepsilon \le f' \circ \varphi \le f \,].$$

Proof of Part 1: Since the net $\{\varphi(\mathfrak{m}_\gamma)\}_\gamma$ converges to $\varphi(\mathfrak{m})$ in the $\mathsf{UEB}(X_\varphi)$ topology, there is γ_0 such that $|\varphi(\mathfrak{m}_\gamma)(f') - \varphi(\mathfrak{m})(f')| < \varepsilon$ for all $f' \in \mathsf{BLip_b}(\Delta')^+$ and $\gamma \ge \gamma_0$. For any $f \in \mathsf{BLip_b}(\Delta)^+$, let f' be such that $f - \varepsilon \le f' \circ \varphi \le f$, so that

$$|\mathfrak{m}_\gamma(f) - \mathfrak{m}(f)| \le |\mathfrak{m}_\gamma(f - f' \circ \varphi)| + |\mathfrak{m}_\gamma(f' \circ \varphi) - \mathfrak{m}(f' \circ \varphi)| + |\mathfrak{m}(f' \circ \varphi - f)|$$

$$\le \varepsilon \|\mathfrak{m}_\gamma\| + \varepsilon + \varepsilon \|\mathfrak{m}\| \le (1 + 2r)\varepsilon$$

for $\gamma \ge \gamma_0$.

Proof of Part 2: Assume $\varphi(\mathfrak{A})$ is $\mathsf{UEB}(X_\varphi)$ relatively compact in $\mathfrak{M}_u(X_\varphi)$. Then $\varphi(\mathfrak{A})$ is X_φ-pointwise equicontinuous on $\mathsf{BLip_b}(\Delta')$ by Theorem 6.16, so that there are a finite set $D' \subseteq X_\varphi$ and $\theta > 0$ such that

$$\forall \mathfrak{m} \in \mathfrak{A} \; \forall f' \in \mathsf{BLip_b}(\Delta') \; [\, \|f'\|_{D'} < \theta \Rightarrow |\varphi(\mathfrak{m})(f')| < \varepsilon \,].$$

Since $\varphi(X) = X_\varphi$, there is a finite set $D \subseteq X$ such that $\varphi(D) = D'$. Take any $\mathfrak{m} \in \mathfrak{A}$ and any $f \in \mathsf{BLip_b}(\Delta)^+$ such that $\|f\|_D < \theta$. There is $f' \in \mathsf{BLip_b}(\Delta')^+$ for which $f - \varepsilon \le f' \circ \varphi \le f$. Then $\|f'\|_{D'} = \|f' \circ \varphi\|_D \le \|f\|_D < \theta$, and

$$|\mathfrak{m}(f)| \le |\mathfrak{m}(f' \circ \varphi)| + |\mathfrak{m}(f - f' \circ \varphi)| \le \varepsilon + \|\mathfrak{m}\| \varepsilon \le (1 + r)\varepsilon.$$

This proves that $\mathfrak{A}$ is X-pointwise equicontinuous on $\mathsf{BLip_b}(\Delta)^+$ at 0, hence $\mathsf{UEB}(X)$ relatively compact in $\mathfrak{M}_u(X)$ by Lemma 5.25 and Theorem 6.16.

Part 3 is a special case of Part 2.   □

Corollary 6.22. *Let X be a uniform space and $\mathfrak{m} \in \mathfrak{M}_b(X)$. Then $\mathfrak{m} \in \mathfrak{M}_u(X)$ if and only if there exists a UCUD set Φ of mappings $\varphi \colon X \to X_\varphi$ inducing the uniformity of X such that $\varphi(\mathfrak{m}) \in \mathfrak{M}_t(X_\varphi)$ for every $\varphi \in \Phi$.* □

Exercise 6.23. Find uniform spaces X_i, $i = 0, 1$, and a net $\{\mathfrak{m}_\gamma\}_\gamma$ in $\mathfrak{M}_t(X_0 \times X_1)^+$ such that $\|\mathfrak{m}_\gamma\| \leq 1$ for all γ, the nets $\{\pi_i(\mathfrak{m}_\gamma)\}_\gamma$, $i = 0, 1$, converge in the UEB topology, but the net $\{\mathfrak{m}_\gamma\}_\gamma$ does not. ∎

Exercise 6.24. Find two uniform spaces X_i, $i = 0, 1$, and $\mathfrak{m} \in \mathfrak{M}_b(X_0 \times X_1)$ such that the image of $\mathfrak{m}$ under the canonical projection onto X_i is in $\mathfrak{M}_u(X_i)$ for $i = 0, 1$ but $\mathfrak{m} \notin \mathfrak{M}_u(X_0 \times X_1)$. ∎

The preceding two exercises show that Theorem 6.20 does not hold with the UCUD property of Φ omitted. Nevertheless, Parts 2 and 3 of the theorem do hold for positive uniform measures:

Theorem 6.25. *Let X be a uniform space whose uniformity is induced by a set Φ of mappings $\varphi \colon X \to X_\varphi$ to uniform spaces X_φ.*

1. *Let $\mathfrak{A} \subseteq \mathfrak{M}_b(X)^+$ be such that $\varphi(\mathfrak{A}) \subseteq \mathfrak{M}_u(X_\varphi)$ and $\varphi(\mathfrak{A})$ is $\mathsf{UEB}(X_\varphi)$ relatively compact in $\mathfrak{M}_u(X_\varphi)$ for every $\varphi \in \Phi$. Then $\mathfrak{A} \subseteq \mathfrak{M}_u(X)^+$ and $\mathfrak{A}$ is $\mathsf{UEB}(X)$ relatively compact in $\mathfrak{M}_u(X)$.*
2. *If $\mathfrak{m} \in \mathfrak{M}_b(X)^+$ and $\varphi(\mathfrak{m}) \in \mathfrak{M}_u(X_\varphi)$ for every $\varphi \in \Phi$, then $\mathfrak{m} \in \mathfrak{M}_u(X)^+$.*

First I prove a special case of the theorem, in the following lemma.

Lemma 6.26. *Let the uniform space X be the product of finitely many uniform spaces X_i, $i = 0, 1, \ldots, k$, where $k \geq 1$, with the canonical projections $\pi_i \colon X \to X_i$. Let $\mathfrak{A} \subseteq \mathfrak{M}_b(X)^+$ and assume the set $\pi_i(\mathfrak{A})$ is X_i-pointwise equicontinuous on $\mathsf{BLip}_b(\Delta_i)$ for all $i = 0, 1, \ldots, k$ and all $\Delta_i \in \mathsf{UP}(X_i)$. Then $\mathfrak{A}$ is X-pointwise equicontinuous on $\mathsf{BLip}_b(\Delta)$ for all $\Delta \in \mathsf{UP}(X)$. In particular, if $\mathfrak{m} \in \mathfrak{M}_b(X)^+$ is such that $\pi_i(\mathfrak{m}) \in \mathfrak{M}_u(X_i)$ for $i = 0, 1, \ldots, k$ then $\mathfrak{m} \in \mathfrak{M}_u(X)^+$.*

Proof. It suffices to prove the case of $k = 1$. The general case then follows by induction.

Let $X = X_0 \times X_1$. Take any $\Delta \in \mathsf{UP}(X)$ and $\varepsilon > 0$. By Part 2 of Theorem 2.6, there are $\Delta_i \in \mathsf{UP}(X_i)$, $i = 0, 1$, with $1 \wedge \Delta < \overleftarrow{\pi_0}\Delta_0 \vee \overleftarrow{\pi_1}\Delta_1 + \varepsilon$. For $i = 0, 1$, as $\pi_i(\mathfrak{A})$ is X_i-pointwise equicontinuous on $\mathsf{BLip}_b(\Delta_i)$, there is a finite set $D_i \subseteq X_i$ such that

$$\forall \mathfrak{m} \in \mathfrak{A} \ \forall f_i \in \mathsf{BLip}_b(\Delta_i) \ [\ \|f_i\|_{D_i} = 0 \Rightarrow |\pi_i(\mathfrak{m})(f_i)| < \varepsilon\ ].$$

In particular, $\mathfrak{m}(h_i \circ \pi_i) = \pi_i(\mathfrak{m})(h_i) < \varepsilon$ for every $\mathfrak{m} \in \mathfrak{A}$ and the functions h_i defined by

$$h_i(x) := 1 \wedge \Delta_i(x, D_i) \ \text{ for } x \in X_i.$$

By Lemma 6.2, there is $r \in \mathbb{R}^+$ such that $\mathfrak{m}(1) \leq r$ for all $\mathfrak{m} \in \mathfrak{A}$. Set $D := D_0 \times D_1$. I shall prove that

$$\forall \mathfrak{m} \in \mathfrak{A} \ \forall f \in \mathsf{BLip}_b(\Delta)^+ \ [\ \|f\|_D < \varepsilon \Rightarrow \mathfrak{m}(f) < 2(r+1)\varepsilon\ ],$$

which along with Lemma 5.25 will demonstrate that $\mathfrak{A}$ is X-pointwise equicontinuous on $\mathsf{BLip}_b(\Delta)^+$.

Take any $f \in \mathsf{BLip}_b(\Delta)^+$ such that $\|f\|_D < \varepsilon$. For $x \in X$, we have

$$f(x) < \varepsilon + 1 \wedge \Delta(x, D)$$
$$< 2\varepsilon + \left(1 \wedge \overleftarrow{\pi_0}\Delta_0(x, D)\right) \vee \left(1 \wedge \overleftarrow{\pi_1}\Delta_1(x, D)\right)$$
$$= 2\varepsilon + (1 \wedge \Delta_0(\pi_0(x), D_0)) \vee (1 \wedge \Delta_1(\pi_1(x), D_1))$$
$$\leq 2\varepsilon + h_0(\pi_0(x)) + h_1(\pi_1(x)),$$

hence $\mathfrak{m}(f) \leq 2\varepsilon\mathfrak{m}(1) + \mathfrak{m}(h_0 \circ \pi_0) + \mathfrak{m}(h_1 \circ \pi_1) \leq 2r\varepsilon + 2\varepsilon$ for every $\mathfrak{m} \in \mathfrak{A}$. ☐

Proof of Theorem 6.25. To prove Part 1, replace Φ by a set Φ' of mappings that correspond to non-empty finite subsets of Φ. The $\varphi' \in \Phi'$ corresponding to a finite set $\Psi \subseteq \Phi$ is the unique mapping $\varphi' : X \to \prod_{\varphi \in \Psi} X_\varphi$ such that $\varphi = \pi_{\Psi\varphi} \circ \varphi'$ for every $\varphi \in \Psi$, where $\pi_{\Psi\varphi}$ is the canonical projection. The set Φ' is UCUD. Since the uniformity of X is induced by Φ, it is also induced by Φ'.

Applying Theorem 6.16 along with Lemma 6.26, we get that if $\varphi(\mathfrak{A})$ is $\mathsf{UEB}(X_\varphi)$ relatively compact in $\mathfrak{M}_u(X_\varphi)$ for every $\varphi \in \Phi$, then $\varphi(\mathfrak{A})$ is also $\mathsf{UEB}(X_\varphi)$ relatively compact in $\mathfrak{M}_u(X_\varphi)$ for every φ in the UCUD set Φ'. Thus the conclusion follows from Theorem 6.20.

Part 2 is a special case of Part 1. ☐

The following variation of Lemma 6.26 will be needed in Chap. 11.

Lemma 6.27. *Let X_0 and X_1 be two uniform spaces and $X = X_0 \times X_1$ with the canonical projections $\pi_i : X \to X_i$. If $\mathfrak{m} \in \mathfrak{M}_b(X)^+$ is such that $\pi_i(\mathfrak{m}) \in \mathfrak{M}_t(X_i)$ for $i = 0, 1$, then $\mathfrak{m} \in \mathfrak{M}_t(X)$.*

Proof. The proof is similar to the proof of Lemma 6.26. Take any $\mathfrak{m} \in \mathfrak{M}_b(X)^+$ such that $\pi_i(\mathfrak{m}) \in \mathfrak{M}_t(X_i)$ for $i = 0, 1$, and any $\varepsilon > 0$. There are compact sets $K_i \subseteq X_i$, $i = 0, 1$, such that

$$\forall f_i \in \mathsf{U}_b(X_i) \ \left[0 \leq f_i \leq 1 \text{ and } \|f_i\|_{K_i} = 0 \Rightarrow \pi_i(\mathfrak{m})(f_i) < \varepsilon\right]$$

for $i = 0, 1$. Set $K := K_0 \times K_1$ and take any $f \in \mathsf{U}_b(X)$ such that $0 \leq f \leq 1$ and $\|f\|_K < \varepsilon$.

By Part 1 of Theorem 2.6, there are $\Delta_i \in \mathsf{UP}(X_i)$, $i = 0, 1$, such that if $x_i, y_i \in X_i$ and $\Delta_i(x_i, y_i) < 1$ for $i = 0, 1$, then $|f((x_0, x_1)) - f((y_0, y_1))| < \varepsilon$. It follows that $f \leq h_0 \circ \pi_0 + h_1 \circ \pi_1 + 2\varepsilon$ where the functions $h_i \in \mathsf{U}_b(X_i)$, $i = 0, 1$, are defined by $h_i(x) := 1 \wedge \Delta_i(x, K_i)$ for $x \in X_i$. As $0 \leq h_i \leq 1$ and $\|h_i\|_{K_i} = 0$, we have $\mathfrak{m}(h_i \circ \pi_i) < \varepsilon$ for $i = 0, 1$, and

$$\mathfrak{m}(f) \leq \mathfrak{m}(h_0 \circ \pi_0) + \mathfrak{m}(h_1 \circ \pi_1) + 2\|\mathfrak{m}\|\varepsilon < 2\varepsilon + 2\|\mathfrak{m}\|\varepsilon,$$

which proves that $\mathfrak{m} \in \mathfrak{M}_t(X)$. ☐

Theorems 6.20 and 6.25 tell us when $\mathfrak{m} \in \mathfrak{M}_b(X)$ is a uniform measure, given its images under $\varphi \in \Phi$. They do not tell us whether there exists $\mathfrak{m} \in \mathfrak{M}_b(X)$ with the specified images in $\mathfrak{M}_b(X_\varphi)$. A sufficient condition for the existence of such $\mathfrak{m} \in \mathfrak{M}_b(X)$ is in Theorem 6.29 below, phrased in terms of projective systems.

Definition 6.28. Let Γ be an upwards-directed set. An $\mathfrak{M}_b$-*projective system indexed by* Γ is a collection $\langle \Gamma, X_\gamma, \varphi_{\beta\gamma}, \mathfrak{m}_\gamma \rangle$ consisting of uniform spaces X_γ for $\gamma \in \Gamma$, uniformly continuous mappings $\varphi_{\beta\gamma} \colon X_\gamma \to X_\beta$ for $\beta < \gamma$ and $\mathfrak{m}_\gamma \in \mathfrak{M}_b(X_\gamma)$ for $\gamma \in \Gamma$ such that $\varphi_{\beta\gamma}(\mathfrak{m}_\gamma) = \mathfrak{m}_\beta$ for all $\beta, \gamma \in \Gamma, \beta < \gamma$.

Let S be a non-empty set. A family $\{\varphi_\gamma\}_{\gamma \in \Gamma}$ of mappings $\varphi_\gamma \colon S \to X_\gamma$ is *consistent with the* $\mathfrak{M}_b$-*projective system* $\langle \Gamma, X_\gamma, \varphi_{\beta\gamma}, \mathfrak{m}_\gamma \rangle$ iff $\varphi_\beta = \varphi_{\beta\gamma} \circ \varphi_\gamma$ for all $\beta, \gamma \in \Gamma, \beta < \gamma$.   ∎

Projective systems are usually assumed to satisfy also $\varphi_{\beta\gamma} \circ \varphi_{\gamma\delta} = \varphi_{\beta\delta}$ whenever $\beta < \gamma < \delta$. As this assumption is not needed in the following result, I do not include it in the definition.

The prototypical example of a projective system is formed by finite products of an infinite family of spaces. For an infinite set Λ and uniform spaces Y_λ, $\lambda \in \Lambda$, let Γ be the family of non-empty finite subsets of Λ ordered by inclusion; for $\gamma \in \Gamma$ let $X_\gamma := \prod_{\lambda \in \gamma} Y_\lambda$, and let $\varphi_{\beta\gamma} \colon X_\gamma \to X_\beta$ be the natural projection for $\beta < \gamma$. When $\mathfrak{m}_\gamma \in \mathfrak{M}_b(X_\gamma)$, $\gamma \in \Gamma$, are such that $\langle \Gamma, X_\gamma, \varphi_{\beta\gamma}, \mathfrak{m}_\gamma \rangle$ is an $\mathfrak{M}_b$-projective system, the functionals $\mathfrak{m}_\gamma$ are the "finite-dimensional marginals" of a putative functional on the product $\prod_{\lambda \in \Lambda} Y_\lambda$.

Theorem 6.29. *Let X be a uniform space, and let $\{\varphi_\gamma\}_{\gamma \in \Gamma}$ be a family of mappings $\varphi_\gamma \colon X \to X_\gamma$ consistent with a $\mathfrak{M}_b$-projective system $\langle \Gamma, X_\gamma, \varphi_{\beta\gamma}, \mathfrak{m}_\gamma \rangle$. Assume that*

(i) *the uniformity of X is induced by the set $\{\varphi_\gamma \mid \gamma \in \Gamma\}$;*
(ii) *for every $\gamma \in \Gamma$, the image set $\varphi_\gamma(X)$ is dense in X_γ; and*
(iii) *there is $r \in \mathbb{R}^+$ such that $\|\mathfrak{m}_\gamma\| \leq r$ for all $\gamma \in \Gamma$.*

Then there exists a unique $\mathfrak{m} \in \mathfrak{M}_b(X)$ such that $\varphi_\gamma(\mathfrak{m}) = \mathfrak{m}_\gamma$ for every $\gamma \in \Gamma$.

Proof. Let $\mathscr{F} \subseteq \mathbb{R}^X$ be the set of functions $h_\gamma \circ \varphi_\gamma$ where $h_\gamma \in U_b(X_\gamma)$, $\gamma \in \Gamma$. By assumption (i), every $\varphi_\gamma \colon X \to X_\gamma$ is uniformly continuous, and $\mathscr{F} \subseteq U_b(X)$.

If $f = h_\gamma \circ \varphi_\gamma = h'_\gamma \circ \varphi_\gamma$, $h_\gamma, h'_\gamma \in U_b(X_\gamma)$, then $h_\gamma = h'_\gamma$ by (ii). If $\beta < \gamma$, $h_\beta \in U_b(X_\beta)$ and $f = h_\beta \circ \varphi_\beta$, then the function $h_\gamma := h_\beta \circ \varphi_{\beta\gamma}$ is in $U_b(X_\gamma)$, $f = h_\gamma \circ \varphi_\gamma$, and $\mathfrak{m}_\beta(h_\beta) = \mathfrak{m}_\gamma(h_\gamma)$. Since the index set Γ is directed, it follows that $\mathscr{F}$ is a vector subspace of $U_b(X)$ and that there is a unique linear functional $\mathfrak{m}'$ on $\mathscr{F}$ such that $\mathfrak{m}'(h_\gamma \circ \varphi_\gamma) = \mathfrak{m}_\gamma(h_\gamma)$ whenever $h_\gamma \in U_b(X_\gamma)$.

By assumption (iii), the norm of $\mathfrak{m}'$ on $\mathscr{F}$ is $\leq r$. By the Hahn–Banach theorem P.9, there is $\mathfrak{m} \in \mathfrak{M}_b(X)$ that extends $\mathfrak{m}'$. Clearly $\varphi_\gamma(\mathfrak{m}) = \mathfrak{m}_\gamma$ for $\gamma \in \Gamma$.

By Lemma 6.21, the set $\mathscr{F}$ is $\|\cdot\|_X$ dense in $U_b(X)$; therefore the $\mathfrak{m} \in \mathfrak{M}_b(X)$ extending $\mathfrak{m}'$ is unique.   □

Exercise 6.30. Show that Theorem 6.29 does not hold when assumption (iii) is omitted.   ∎

6.5 Completion and Compactification

The completion and the uniform compactification of a uniform space X are closely related to the spaces $\mathfrak{M}_u(X)$ and $\mathfrak{M}_b(X)$. On one hand, I show in this section that the completion and the compactification of X are naturally embedded in $\mathfrak{M}_b(X)$. On the other hand, as will be seen in Sect. 7.2, the compactification is a valuable tool in the study of functionals in $\mathfrak{M}_b(X)$ and $\mathfrak{M}_u(X)$.

Definition 6.31. When X and Y are uniform spaces, Y is said to be a *completion* of X iff $X \Subset Y$, X is dense in Y and Y is complete. ∎

Theorem 6.32. *If Y_0 and Y_1 are two completions of a uniform space X, then there is a uniform isomorphism $\varphi \colon Y_0 \to Y_1$ such that $\varphi(x) = x$ for all $x \in X$.*

Proof. Apply Theorem 2.24 to the inclusion mappings $X \hookrightarrow Y_0$ and $X \hookrightarrow Y_1$. □

Definition 6.33. Let X be a uniform space. A compact uniform space Y is a *uniform compactification* of X (also known as a *Samuel compactification* of X) iff the set X is a dense subset of Y, the inclusion mapping from X to Y is uniformly continuous, and Y has the following universal property: For every uniformly continuous mapping φ from X to a compact uniform space K, there is a continuous mapping $\varphi' \colon Y \to K$ that extends φ. ∎

A uniform space is not necessarily a uniform subspace of its compactification, but we will see further on that it is a topological subspace. The continuous extension φ' of φ in the definition is obviously unique.

Theorem 6.32 has the following parallel, stating that a uniform compactification is unique up to a homeomorphism. Recall (Corollary 1.7) that every homeomorphism of compact uniform spaces is a uniform isomorphism.

Theorem 6.34. *If Y_0 and Y_1 are two uniform compactifications of a uniform space X, then there is a homeomorphism $\varphi \colon Y_0 \to Y_1$ such that $\varphi(x) = x$ for all $x \in X$.*

Proof. Apply the universal property in Definition 6.33 to the uniformly continuous inclusion mappings $X \hookrightarrow Y_0$ and $X \hookrightarrow Y_1$. □

The next theorem is the main result of this section—the construction of a completion and a uniform compactification of X embedded in $\mathfrak{M}_b(X)$. In the theorem, X^u, $\widehat{X}$ and $\widehat{p}X$ are three uniform spaces whose point sets are subsets of $\mathfrak{M}_b(X)$:

- X^u is the set $\partial_X(X)$ with the $\mathrm{UEB}(X)$ uniformity;
- $\widehat{X}$ is the $\mathrm{UEB}(X)$ closure of $\partial_X(X)$ in $\mathfrak{M}_u(X)$ with the $\mathrm{UEB}(X)$ uniformity;
- $\widehat{p}X$ is the $\mathrm{U}_b(X)$-weak closure of $\partial_X(X)$ in $\mathfrak{M}_b(X)$ with the $\mathrm{U}_b(X)$-weak uniformity.

Clearly $X^u \Subset \widehat{X}$ and $\widehat{X} \subseteq \widehat{p}X$.

Exercise 6.35. Show that $\widehat{p}X = \widehat{pX}$ for every uniform space X. ∎

Lemma 6.36. *For any uniform space X, the mapping ∂_X is a uniform isomorphism from X onto X^{u}.*

Proof. Follows from Lemma 5.12. □

Theorem 6.37. *Let X be an arbitrary uniform space.*

1. *$\widehat{X}$ is a completion of X^{u}.*
2. *$\widehat{\mathsf{p}}X$ is a uniform compactification of X^{u}.*
3. *$\widehat{X} = \widehat{\mathsf{p}}X \cap \mathfrak{M}_{\mathsf{u}}(X)$, where $=$ means equality of topological spaces.*

Note that $\widehat{\mathsf{p}}X \cap \mathfrak{M}_{\mathsf{u}}(X)$ (as a subspace of $\widehat{\mathsf{p}}X$) and $\widehat{X}$ are not equal as uniform spaces, unless X is precompact.

Proof. 1. By definition, X^{u} is UEB(X) dense in $\widehat{X}$. By Theorems 2.17 and 6.6, the space $\widehat{X}$ is complete.
2. By definition, X^{u} is $\mathsf{U}_{\mathsf{b}}(X)$-weakly dense in $\widehat{\mathsf{p}}X$. Since the UEB(X) uniformity on $\mathfrak{M}_{\mathsf{b}}(X)$ is finer than the $\mathsf{U}_{\mathsf{b}}(X)$-weak uniformity, the inclusion mapping $X^{\mathsf{u}} \hookrightarrow \widehat{\mathsf{p}}X$ is uniformly continuous. The space $\widehat{\mathsf{p}}X$ is $\mathsf{U}_{\mathsf{b}}(X)$-weakly compact because it is a $\mathsf{U}_{\mathsf{b}}(X)$-weakly closed subset of the $\|\cdot\|$ unit ball in $\mathfrak{M}_{\mathsf{b}}(X)$.

   To prove the universal property in Definition 6.33, take any uniformly continuous mapping φ from X^{u} to a compact space K. When $\mathfrak{M}_{\mathsf{b}}(K)$ has the $\mathsf{U}_{\mathsf{b}}(K)$-weak topology, the mapping $\partial_K \colon K \to \mathfrak{M}_{\mathsf{b}}(K)$ is continuous; hence it is a homeomorphism from K onto $\partial_K(K)$.

   Since $\varphi \circ \partial_X \colon X \to K$ is uniformly continuous, the corresponding mapping $\mathfrak{M}_{\mathsf{b}}(\varphi \circ \partial_X) \colon \mathfrak{M}_{\mathsf{b}}(X) \to \mathfrak{M}_{\mathsf{b}}(K)$ is continuous in the $\mathsf{U}_{\mathsf{b}}(X)$-weak and $\mathsf{U}_{\mathsf{b}}(K)$-weak topologies by Lemma 6.8. Let φ' be the restriction of $\mathfrak{M}_{\mathsf{b}}(\varphi \circ \partial_X)$ to $\widehat{\mathsf{p}}X$.

   The set $K' := \varphi'(\widehat{\mathsf{p}}X) \subseteq \mathfrak{M}_{\mathsf{b}}(K)$ is $\mathsf{U}_{\mathsf{b}}(K)$-weakly compact, and $\varphi'(X^{\mathsf{u}})$ is dense in K'. Therefore $\partial_K(K) \supseteq K'$, and the mapping $\partial_K^{-1} \circ \varphi' \colon \widehat{\mathsf{p}}X \to K$ is a continuous extension of φ.
3. By Corollary 6.13, the $\mathsf{U}_{\mathsf{b}}(X)$-weak topology and the UEB topology coincide on $\mathfrak{M}_{\mathsf{u}}(X)^{+}$. Thus the UEB($X$) closure of $\partial_X(X)$ in $\mathfrak{M}_{\mathsf{u}}(X)$ is the same as the $\mathsf{U}_{\mathsf{b}}(X)$-weak closure, and the two topologies coincide on the set $\widehat{X}$. □

By Lemma 6.36, X may be identified with X^{u} by means of ∂_X, thus making $\widehat{X}$ a completion of X, and $\widehat{\mathsf{p}}X$ a uniform compactification of X.

Definition 6.38. Identify X with X^{u} by means of ∂_X, so that $X \Subset \widehat{X} \subseteq \widehat{\mathsf{p}}X$. The space $\widehat{X}$ will be called *the completion of X*, and the space $\widehat{\mathsf{p}}X$ *the uniform compactification of X*. ∎

As a rule, when we work with a completion or a compactification, the specific nature of the points added to X is not important. However, in Chap. 9 it will be convenient to use the spaces $\widehat{X}$ and $\widehat{\mathsf{p}}X$ constructed here, so that $\widehat{X} \subseteq \widehat{\mathsf{p}}X \subseteq \mathfrak{M}_{\mathsf{b}}(X)$.

For the sake of brevity, from now on I no longer make the distinction between X and X^{u}. It will be clear from context when X is identified with X^{u} by means of ∂_X.

By Theorem 2.24, every function $f \in U(X)$ extends to a unique function in $U(\widehat{X})$, which will be denoted $\widehat{f}$. When $f \in U_b(X)$, the range of f is included in a compact interval $[r, r'] \subseteq \mathbb{R}$, hence f extends to a unique function in $U_b(\widehat{p}X)$, which will be denoted $\overline{f}$.

6.6   Vector-Valued Integrals

A functional $m \in \mathfrak{M}_b(X)$ assigns the value $m(f)$ to every scalar-valued mapping $f \in U_b(X)$. In this section, I describe the role played by uniform measures when we attempt to assign values in a similar way also to vector-valued mappings, that is, mappings from X to locally convex spaces.

If $\varphi \colon X \to E$ is a uniformly continuous mapping from a uniform space X to a locally convex space E with its additive uniformity and if the set $\varphi(X)$ is bounded in E, then the function $w \circ \varphi$ is in $U_b(X)$ for every $w \in E^*$, and $m(w \circ \varphi)$ is defined for every linear functional m on the space $U_b(X)$.

Definition 6.39.   Let X be a uniform space, m a linear functional on the space $U_b(X)$ and E a locally convex space. When φ is a uniformly continuous mapping from X to E with its additive uniformity and the range $\varphi(X)$ is bounded in E, define the linear functional $m^E(\varphi)$ on E^* by $m^E(\varphi)(w) := m(w \circ \varphi)$ for $w \in E^*$.   ∎

When m is represented by a measure μ, the value $m^E(\varphi)$ is a "weak integral" of φ with respect to μ. It is then important to know whether $m^E(\varphi)$ identifies with an element of E. More generally, the same question makes sense for any functional m on $U_b(X)$, regardless of any representing measure (which may or may not exist). The following answer is another characterization of uniform measures:

Theorem 6.40.   *Let X be any uniform space. These two properties of a linear functional m on $U_b(X)$ are equivalent:*

(i) $m \in \mathfrak{M}_u(X)$.
(ii)   For every complete locally convex space E and for every uniformly continuous mapping $\varphi \colon X \to E$ whose range $\varphi(X)$ is bounded in E, the functional $m^E(\varphi)$ is E-weakly continuous on E^* (and therefore identifies with an element of E).

Proof.   Assume $m \in \mathfrak{M}_u(X)$, and take any complete locally convex space E and any uniformly continuous mapping $\varphi \colon X \to E$ with bounded range. If a set $B \subseteq E^*$ is equicontinuous on E, then $\mathscr{F} := \{w \circ \varphi \mid w \in B\} \subseteq U_b(X)$ is a $UEB(X)$ set, the restriction of m to $\mathscr{F}$ is X-pointwise continuous, and thus the restriction of $m^E(\varphi)$ to B is E-weakly continuous. Hence $m^E(\varphi)$ is E-weakly continuous by Theorem P.12. That proves (i)⇒(ii).

To prove the converse, let E be the space $\mathfrak{M}_u(X)$ with the UEB topology, and $\varphi := \partial_X$. The space E is complete by Theorem 6.6, φ is uniformly continuous by Lemma 6.36 and clearly $\varphi(X)$ is bounded in E.

Every $f \in U_b(X)$ defines $w_f \in E^*$ by $w_f(\mathfrak{n}) := \mathfrak{n}(f)$ for $\mathfrak{n} \in E$. Now if $\mathfrak{m}$ is a linear functional on $U_b(X)$ with property (ii), then there is $\mathfrak{n} \in E = \mathfrak{M}_u(X)$ such that $\mathfrak{m}^E(\varphi)(w) = w(\mathfrak{n})$ for every $w \in E^*$, and in particular

$$\mathfrak{m}(f) = \mathfrak{m}(w_f \circ \varphi) = \mathfrak{m}^E(\varphi)(w_f) = w_f(\mathfrak{n}) = \mathfrak{n}(f),$$

for every $f \in U_b(X)$, so that $\mathfrak{m} = \mathfrak{n} \in \mathfrak{M}_u(X)$. $\square$

Exercise 6.41. Give a direct proof of the implication (i)$\Rightarrow$(ii) in Theorem 6.40, without appealing to Theorem P.12.   ∎

Exercise 6.42. Prove that the two conditions in Theorem 6.40 are also equivalent to condition (ii) in which E is restricted to be a Banach space.   ∎

6.7 Vector-Valued Uniform Measures

Another manifestation of Theorem 6.16, briefly described here, is a regularity property of vector-valued uniform measures.

Definition 6.43. Let X be a uniform space and E a locally convex space. A linear mapping $\overrightarrow{\mathfrak{m}} \colon U_b(X) \to E$ is an *E-valued uniform measure on X* iff, for every $\Delta \in UP(X)$, the restriction of $\overrightarrow{\mathfrak{m}}$ to $BLip_b(\Delta)$ is continuous in the X-pointwise topology. Let $\mathfrak{M}_u(X, E)$ denote the space of E-valued uniform measures on X.   ∎

Evidently $\mathfrak{M}_u(X, E)$ with the $U_b(X)$-pointwise operations is a vector space.

Theorem 6.44. *Let X be a uniform space and E a locally convex space. Two properties of a linear mapping $\overrightarrow{\mathfrak{m}} \colon U_b(X) \to E$ are equivalent:*

(i) $\overrightarrow{\mathfrak{m}} \in \mathfrak{M}_u(X, E)$.
(ii) $w \circ \overrightarrow{\mathfrak{m}} \in \mathfrak{M}_u(X)$ *for every* $w \in E^*$.

Thus E-valued uniform measures are the same for all locally convex topologies on E having the same dual E^*.

Proof. The implication (i)$\Rightarrow$(ii) is obvious. To prove the converse, take any linear mapping $\overrightarrow{\mathfrak{m}} \colon U_b(X) \to E$ with property (ii), any $\Delta \in UP(X)$ and any continuous seminorm α on E. The set $B := \{w \in E^* \mid |w(y)| \le \alpha(y)$ for all $y \in E\}$ is E-weakly compact, and the mapping $w \mapsto w \circ \overrightarrow{\mathfrak{m}}$ is continuous from E^* with the E-weak topology to $\mathfrak{M}_u(X)$ with the $U_b(X)$-weak topology. Hence $\mathfrak{A} := \{w \circ \overrightarrow{\mathfrak{m}} \mid w \in B\}$ is $U_b(X)$-weakly compact. Thus $\mathfrak{A}$ is X-pointwise equicontinuous on $BLip_b(\Delta)$ by Theorem 6.16, and $\overrightarrow{\mathfrak{m}}$ is X-pointwise continuous on $BLip_b(\Delta)$ as a mapping to E with α by Corollary P.10. $\square$

6.8   Notes for Chap. 6

The theory of uniform measures is an outcome of several lines of research which, although initiated independently and motivated by different questions, converged to the same core concepts. In these notes, I replace the wide variety of terms and notations in the cited sources by their equivalents used throughout this treatise.

Molecular measures were utilized by Arens and Eells [3], Katětov [107], [108], Pták [147], [148] and Raĭkov [153], [154], who studied the embedding of a topological, uniform or metric space X into the space of finite linear combinations of the elements of X.

Raĭkov's approach was further developed by Berezanskiĭ [7] and Fedorova [50], who investigated integral representation of functionals on spaces $U(X)$ and $U_b(X)$ for a uniform space X, and in particular of the functionals approximated by molecular measures. Along with other spaces of functionals, they defined the space $\mathfrak{M}_u(X)$ and derived its basic properties.

In a separate effort, Tomášek [168] applied Katětov's approach in the study of spaces of molecular measures (Λ-*structures* in Katětov's terminology) on uniform spaces.

Csiszár [32] and LeCam [122] identified a defining property of uniform measures in the course of their work on convolution of measures on topological vector spaces, and more generally on topological groups. LeCam proved results about projective limits, thus showing that uniform measures arise naturally when one is concerned with projective limits of Radon (tight) measures. Although LeCam's paper [122] remained unpublished, his approach was further developed by Caby [20], [21].

Extending the work of LeCam, Frolík [63], [64] derived additional properties of uniform measures using tools of the theory of uniform spaces and placed $\mathfrak{M}_u(X)$ in the context of topological measure theory. He also guided and inspired research on relationships between measures and uniform spaces in the group formed around his Prague seminar on uniform spaces [67], [70], [73].

From a different direction, in the work of Deaibes [36], the space $\mathfrak{M}_u(X)$ and the duality $\langle \mathfrak{M}_u(X), U_b(X) \rangle$ are an important instance in the general theory of compactological spaces and their duals created by Buchwalter [15] and Waelbroeck [177]. Deaibes [38], [39] also used the compactology framework to define other classes of functionals on $U_b(X)$ and $U(X)$. In a related setting, the duality $\langle \mathfrak{M}_u(X), U_b(X) \rangle$ is an illustrative example in the theory of Saks spaces described by Cooper [29, II.7] and in the approach to spaces of measures via strict topologies surveyed by Wheeler [180, Sect. 15].

An important source of inspiration for the general theory of uniform measures was the special case $\mathfrak{M}_u(\mathsf{F}T)$ where T is a completely regular topological space. As I explain in Sect. 8.4, $\mathfrak{M}_u(\mathsf{F}T)$ is the space of (functionals represented by) separable measures on T. Kirk [113] described $\mathfrak{M}_u(\mathsf{F}T)$ as a completion of the space of molecular measures. More references to related work in topological measure theory are in Sect. 8.7.

The principal concepts and results in Sect. 6.1 are due to Berezanskiĭ [7], Fedorova [50] and LeCam [122], [124]. Fedorova (loc. cit.) derived Theorem 6.6 from Grothendieck's completion theorem [164, IV.6.2]. Theorem 6.10 and Theorem 6.12 with its Corollary 6.13 are due to LeCam [122], [124].

The derivation of Theorems 6.16 and 6.19 from the corresponding results for tight measures on complete metric spaces (Theorems 5.41 and 5.45) appears in [30] and [136]. LeCam [122], [124] pointed out the equivalence of properties (v) and (vii) in Theorem 6.16.

The main results about projective limits in Sect. 6.4 are due to LeCam [122] and Zahradník [181]. Uniform measures on products were also considered by Berezanskiĭ [7] and Fedorova [50]. In a somewhat imprecise abbreviated form, Part 3 in Theorem 6.20 states that the projective limit of uniform measures is a uniform measure. In contrast, the projective limit of tight Borel measures need not be tight; in fact, by Corollary 6.22 every uniform measure is a projective limit of tight Borel measures. Conditions for the tightness of projectively defined measures are covered in detail by Schwartz [165, I.10], with references to related work. Caby [22] investigated the tightness of projective limits of uniform measures.

By Part 1 of Theorem 6.37, every uniform space has a completion. This is a key classical result about uniform spaces, appearing in the original memoir of Weil [179, Th. II]. Various proofs are for instance in [12, II.3.7], [31, 6.3.c], [81, 15.9], [100, II.16] and [109, 6.28]. Buchwalter and Pupier [18], Fedorova [49] and Tomášek [168, s.6] describe the completion $\widehat{X}$ as a subset of $\mathfrak{M}_u(X)$ or $\mathfrak{M}_b(X)$. Parallels between the completion and the uniform compactification are discussed in Isbell [100].

The characterization of uniform measures by the regularity of weak integrals of vector-valued mappings in Theorem 6.40 and Exercise 6.42 is due to Frolík [63]. Property (ii) in Theorem 6.40 may be interpreted as a certain universal property of the functor $\mathfrak{M}_u$ from the category of uniform spaces to a category of vector spaces with additional structure (see Exercise 10.24); details and references to related work are in [29], [30]. I describe a similar universal property for the related space $\mathfrak{M}_F(X)$ in Sect. 10.2.

Theorem 6.44 is a uniform-measure version of the Orlicz–Pettis theorem; for more traditional versions, see Kalton [103]. Vector-valued uniform measures were investigated by Aguayo–Garrido [1] and Khurana [110], [111].

Chapter 7
Uniform Measures as Measures

In this chapter I discuss the representation of functionals in $\mathfrak{M}_b(X)$ by measures on X and on the uniform compactification $\hat{p}X$. For a general uniform space X, uniform measures on X are represented by certain tight Borel measures on $\hat{p}X$ (Theorem 7.14), but not necessarily by measures on X itself (Examples 7.18 and 7.19). It is thus natural to ask what properties of X ensure that every uniform measure on X is represented by a measure on X. By the theory covered so far, this is the case when X is a complete metric space, because then $\mathfrak{M}_u(X) = \mathfrak{M}_t(X)$. The results in this chapter give answers for other classes of uniform spaces:

1. If X is uniformly locally compact, then $\mathfrak{M}_u(X) = \mathfrak{M}_t(X)$ (Theorem 7.20).
2. If X is inversion-closed, then every $\mathfrak{m} \in \mathfrak{M}_u(X)$ is represented by a measure on X (Theorem 7.21).
3. If X is supercomplete, then every $\mathfrak{m} \in \mathfrak{M}_u(X)$ is represented by a τ-additive Borel measure on X (Theorem 7.22).

Theorem 7.25 answers the converse question: Every functional represented by a measure on X is in $\mathfrak{M}_u(X)$ if and only if the cardinality of every uniformly discrete subset of X is measure-free.

The last section of this chapter is another application of the representation of uniform measures by measures on the compactification: For every uniform space there is a finer and in some respects more tractable uniform space with the same uniform measures.

7.1 Functionals Represented by Measures

So far we have dealt with three subspaces of $\mathfrak{M}_b(X)$ for a uniform space X, namely, $\mathrm{Mol}(X)$, $\mathfrak{M}_t(X)$ and $\mathfrak{M}_u(X)$. Generalizing slightly the smoothness (or regularity) properties traditionally studied in topological measure theory, I now add two more spaces.

J. Pachl, *Uniform Spaces and Measures*, Fields Institute Monographs 30,
DOI 10.1007/978-1-4614-5058-0_8,
© Springer Science+Business Media New York 2013

Definition 7.1. Let X be a uniform space.

1. A linear functional m on $U_b(X)$ is τ-*smooth* iff $\lim_\gamma m(f_\gamma) = 0$ for every net $\{f_\gamma\}_\gamma$ of functions in $U_b(X)$ such that $f_\gamma \searrow 0$. Let $\mathfrak{M}_\tau(X)$ denote the space of τ-smooth functionals on $U_b(X)$.
2. A linear functional m on $U_b(X)$ is σ-*smooth* iff $\lim_\gamma m(f_j) = 0$ for every sequence $\{f_j\}_j$ of functions in $U_b(X)$ such that $f_j \searrow 0$. Let $\mathfrak{M}_\sigma(X)$ denote the space of σ-smooth functionals on $U_b(X)$. ∎

Clearly $\mathfrak{M}_\tau(X)$ and $\mathfrak{M}_\sigma(X)$ are $\|\cdot\|$ closed subspaces of $\mathfrak{M}_b(X)$ for every uniform space X, and

$$\mathsf{Mol}(X) \subseteq \mathfrak{M}_t(X) \subseteq \mathfrak{M}_\tau(X) \subseteq \mathfrak{M}_u(X) \subseteq \mathfrak{M}_b(X)$$

$$\mathfrak{M}_\tau(X) \subseteq \mathfrak{M}_\sigma(X) \subseteq \mathfrak{M}_b(X)$$

If X is compact, then $\mathfrak{M}_t(X) = \mathfrak{M}_b(X)$, and therefore

$$\mathfrak{M}_t(X) = \mathfrak{M}_\tau(X) = \mathfrak{M}_\sigma(X) = \mathfrak{M}_u(X) = \mathfrak{M}_b(X).$$

Farther on in this chapter, I discuss other conditions for various inclusions among $\mathfrak{M}_t(X)$, $\mathfrak{M}_\tau(X)$, $\mathfrak{M}_\sigma(X)$ and $\mathfrak{M}_u(X)$.

Exercise 7.2. Let X be any uniform space. Show that $\mathfrak{M}_u(X) = \mathfrak{M}_b(X)$ if and only if X is precompact. ∎

With the notation in Sect. P.2, $\sigma(U(X))$ is the smallest σ-algebra for which all uniformly continuous functions on the uniform space X are measurable. Obviously $\sigma(U(X)) = \sigma(U_b(X))$.

Definition 7.3. For a uniform space X, a *Baire measure on X* is a measure on the σ-algebra $\sigma(U(X))$ of subsets of X. ∎

Evidently $\sigma(U(X))$ is the smallest σ-algebra containing all zero sets in X, and it is also the smallest σ-algebra containing all cozero sets in X. Sometimes $\sigma(U(X))$ is called the *Baire σ-algebra on X* (not to be confused with the σ-algebra of sets having the Baire property).

By Theorem P.25 and Lemma 4.5, every Baire measure μ on X is inner regular with respect to zero sets, in the sense that

$$|\mu|(A) = \sup\{ |\mu|(Z) \mid Z \subseteq A \text{ and } Z \text{ is a zero set in } X\}$$

for every $A \in \sigma(U(X))$.

General theorems in Sect. P.5 now immediately produce representation of functionals in $\mathfrak{M}_\sigma(X)$ and $\mathfrak{M}_\tau(X)$.

Theorem 7.4. *Let X be any uniform space. The following two properties of a functional $\mathfrak{m} \in \mathfrak{M}_b(X)$ are equivalent:*

(i) $\mathfrak{m}$ is σ-smooth.
(ii) There is a Baire measure $\mu: \sigma(\mathsf{U}(X)) \to \mathbb{R}$ such that $\mathfrak{m}(f) = \int f \, d\mu$ for every $f \in \mathsf{U}_b(X)$.

If (ii) holds, then there is a unique such measure μ, and moreover, $\mathfrak{m}^+(f) = \int f \, d\mu^+$, $\mathfrak{m}^-(f) = \int f \, d\mu^-$ and $|\mathfrak{m}|(f) = \int f \, d|\mu|$ for $f \in \mathsf{U}_b(X)$, and $\|\mathfrak{m}\| = \|\mu\|_{\mathsf{TV}}$.

Proof. Apply Theorem P.24.   □

Corollary 7.5. *For every uniform space X, the space $\mathfrak{M}_\sigma(X)$ is a band in the Banach lattice $\mathfrak{M}_b(X)$.*

Proof. Apply Theorem 7.4 with Theorem P.21.   □

Theorem 7.6. *Let X be any uniform space. The following two properties of a functional $\mathfrak{m} \in \mathfrak{M}_b(X)$ are equivalent:*

(i) $\mathfrak{m}$ is τ-smooth.
(ii) There is a τ-additive Borel measure $\mu: \mathsf{Bo}(X) \to \mathbb{R}$ such that $\mathfrak{m}(f) = \int f \, d\mu$ for every $f \in \mathsf{U}_b(X)$.

If (ii) holds, then there is a unique such τ-additive Borel measure μ, and moreover, $\mathfrak{m}^+(f) = \int f \, d\mu^+$, $\mathfrak{m}^-(f) = \int f \, d\mu^-$ and $|\mathfrak{m}|(f) = \int f \, d|\mu|$ for $f \in \mathsf{U}_b(X)$, and $\|\mathfrak{m}\| = \|\mu\|_{\mathsf{TV}}$.

Proof. By Corollary 1.22, the topology of X is the coarsest topology for which every function in $\mathsf{U}_b(X)$ is continuous. Apply Theorem P.28.   □

Corollary 7.7. *For every uniform space X, the space $\mathfrak{M}_\tau(X)$ is a band in the Banach lattice $\mathfrak{M}_b(X)$.*

Proof. Apply Theorem 7.6 with Theorem P.21.   □

The next result generalizes Corollary 5.5.

Corollary 7.8. *Let X be a uniform space. For every $\mathfrak{m} \in \mathfrak{M}_\tau(X)$ there exists a unique $\widetilde{\mathfrak{m}} \in \mathfrak{M}_\tau(\mathsf{F}X)$ such that $\mathfrak{m}(f) = \widetilde{\mathfrak{m}}(f)$ for all $f \in \mathsf{U}_b(X)$. Moreover, $\mathfrak{m}^+(f) = \widetilde{\mathfrak{m}}^+(f)$, $\mathfrak{m}^-(f) = \widetilde{\mathfrak{m}}^-(f)$ and $|\mathfrak{m}|(f) = |\widetilde{\mathfrak{m}}|(f)$ for $f \in \mathsf{U}_b(X)$, and $\|\mathfrak{m}\| = \|\widetilde{\mathfrak{m}}\|$.*

Proof. Since the topology of X is the same as that of $\mathsf{F}X$, Borel sets and τ-additive Borel measures for X are the same as those for $\mathsf{F}X$.   □

The comment after Corollary 5.5 in Sect. 5.1 now applies with the space $\mathfrak{M}_\tau$ in place of $\mathfrak{M}_t$: If X and Y are two uniform spaces with the same set of points and compatible with the same topology, then $\mathsf{F}X = \mathsf{F}Y$, and by Corollary 7.8 there is a norm-preserving Banach-lattice isomorphism between $\mathfrak{M}_\tau(X)$ and $\mathfrak{M}_\tau(Y)$. The isomorphism is produced by extending each $\mathfrak{m} \in \mathfrak{M}_\tau(X)$ to a unique $\widetilde{\mathfrak{m}} \in \mathfrak{M}_\tau(\mathsf{F}X)$ and then restricting $\widetilde{\mathfrak{m}}$ to $\mathsf{U}_b(Y)$.

Thus the space $\mathfrak{M}_\tau(X)$ depends only on the topology of X. However, the same is not true for $\mathfrak{M}_\sigma(X)$. There are a uniform space X and $m \in \mathfrak{M}_\sigma(X)$ for which there is no $\tilde{m} \in \mathfrak{M}_\sigma(\mathsf{F}X)$ with $m(f) = \tilde{m}(f)$ for all $f \in U_b(X)$ (see Exercise 8.12).

7.2   Measures on the Compactification

Let X be a uniform space. Recall from Sect. 6.5 that every $f \in U_b(X)$ extends uniquely to a function $\overline{f} \in U_b(\hat{p}X) = C_b(\hat{p}X)$. If μ is a tight Borel measure on $\hat{p}X$, then the mapping $f \mapsto \int \overline{f}\,d\mu$, $f \in U_b(X)$, is a functional in $\mathfrak{M}_b(X)$. By the Riesz representation theorem P.31, every $m \in \mathfrak{M}_b(X)$ is of this form for a unique tight Borel measure μ on $\hat{p}X$, which will be denoted by ρ_m. Thus $m(f) = \int \overline{f}\,d\rho_m$ for $f \in U_b(X)$.

The results in this section characterize the spaces $\mathfrak{M}_t(X)$, $\mathfrak{M}_\tau(X)$, $\mathfrak{M}_\sigma(X)$ and $\mathfrak{M}_u(X)$ by the different ways in which the representing measures ρ_m on $\hat{p}X$ are concentrated on the set $X \subseteq \hat{p}X$.

Theorem 7.9. *Let X be a uniform space. A functional $m \in \mathfrak{M}_b(X)$ belongs to $\mathfrak{M}_t(X)$ if and only if there is $B \in Bo(\hat{p}X)$ such that $B \subseteq X$ and $|\rho_m|(\hat{p}X \setminus B) = 0$.*

Proof. Since $m \in \mathfrak{M}_t(X)$ if and only if $|m| \in \mathfrak{M}_t(X)$, assume without loss of generality that $m \geq 0$. By Theorem 5.3, $m \in \mathfrak{M}_t(X)$ if and only if there is a tight Borel measure μ on X such that $m(f) = \int f\,d\mu$ for $f \in U_b(X)$.

To prove that the condition in the theorem is necessary, let μ be such tight Borel measure on X. Define a tight Borel measure μ' on $\hat{p}X$ by $\mu'(A) := \mu(A \cap X)$ for $A \in Bo(\hat{p}X)$. Then $m(f) = \int \overline{f}\,d\mu'$ for $f \in U_b(X)$, and $\mu' = \rho_m$ by the uniqueness of ρ_m. Since μ is tight on X, there are compact sets $K_j \subseteq X$, $j = 1,2,\ldots$, such that $\mu(X \setminus K_j) < 1/j$ for each j. For $B := \bigcup_j K_j$ we have

$$\rho_m(\hat{p}X \setminus B) = \mu'(\hat{p}X \setminus B) = \mu(X \setminus B) = 0.$$

To prove that the condition is sufficient, let $B \in Bo(\hat{p}X)$ be such that $B \subseteq X$ and $\rho_m(\hat{p}X \setminus B) = 0$. Then $B \in Bo(X)$, and $\mu(A) := \rho_m(A \cap B)$, $A \in Bo(X)$, defines a measure $\mu: Bo(X) \to \mathbb{R}$. Since ρ_m is tight Borel on $\hat{p}X$, it follows that μ is tight Borel on X, and

$$m(f) = \int \overline{f}\,d\rho_m = \int_B f\,d\rho_m = \int f\,d\mu.$$

Hence $m \in \mathfrak{M}_t(X)$. $\square$

Theorem 7.10. *Let X be a uniform space. A functional $m \in \mathfrak{M}_b(X)$ belongs to $\mathfrak{M}_\tau(X)$ if and only if $|\rho_m|(K) = 0$ for every compact set $K \subseteq \hat{p}X$ such that $K \cap X = \emptyset$.*

Proof. Since $m \in \mathfrak{M}_\tau(X)$ if and only if $|m| \in \mathfrak{M}_\tau(X)$, assume without loss of generality that $m \geq 0$.

To prove the condition is necessary, take any $m \in \mathfrak{M}_\tau(X)^+$ and any compact set $K \subseteq \hat{p}X \setminus X$. Since every point in X is separated from K by a function in $C_b(\hat{p}X)$, there is a net of functions $g_\gamma \in C_b(\hat{p}X)^+$ such that $g_\gamma \searrow g$, $g(x) = 0$ for every $x \in X$ and $g(x) = 1$ for every $x \in K$. Let f_γ be the restriction of g_γ to X, so that $f_\gamma \in U_b(X)$ and $g_\gamma = \overline{f_\gamma}$. Then $f_\gamma \searrow 0$, therefore $\lim_\gamma m(f_\gamma) = 0$, and

$$\rho_m(K) \leq \lim_\gamma \int g_\gamma d\rho_m = \lim_\gamma m(f_\gamma) = 0.$$

To prove the condition is sufficient, assume that $\rho_m(K) = 0$ for every compact set $K \subseteq \hat{p}X \setminus X$. Take any net $\{f_\gamma\}_\gamma$ in $U_b(X)$ such that $1 \geq f_\gamma \searrow 0$, and any $\varepsilon > 0$. Let $K_\gamma := \{x \in \hat{p}X \mid \overline{f_\gamma} \geq \varepsilon\}$ and $K := \bigcap_\gamma K_\gamma$. The measure ρ_m is τ-additive, $K_\gamma \searrow K$, and $K \subseteq \hat{p}X \setminus X$. Thus $\lim_\gamma \rho_m(K_\gamma) = \rho_m(K) = 0$,

$$\lim_\gamma m(f_\gamma) = \lim_\gamma \int \overline{f_\gamma} d\rho_m \leq \varepsilon \rho_m(\hat{p}X) + \lim_\gamma \rho_m(K_\gamma) = \varepsilon \|\rho_m\|_{TV},$$

and $\lim_\gamma m(f_\gamma) = 0$.   □

Corollary 7.11. *Let X be a uniform space. If the set X is measurable for every tight Borel measure on $\hat{p}X$ (in particular, if $X \in \mathrm{Bo}(\hat{p}X)$), then $\mathfrak{M}_\tau(X) = \mathfrak{M}_t(X)$.*

Proof. If $m \in \mathfrak{M}_\tau(X)$ and X is $|\rho_m|$-measurable, then $|\rho_m|(\hat{p}X \setminus X) = 0$ by Theorem 7.10. Hence there is a set $B \in \mathrm{Bo}(X)$ such that $|\rho_m|(X \setminus B) = 0$. Therefore $m \in \mathfrak{M}_t(X)$ by Theorem 7.9.   □

Theorem 7.12. *Let X be a uniform space. A functional $m \in \mathfrak{M}_b(X)$ belongs to $\mathfrak{M}_\sigma(X)$ if and only if $|\rho_m|(Z) = 0$ for every zero set Z in $\hat{p}X$ such that $Z \cap X = \emptyset$.*

Proof. This is similar to the proof of Theorem 7.10. Since $m \in \mathfrak{M}_\sigma(X)$ if and only if $|m| \in \mathfrak{M}_\sigma(X)$, assume without loss of generality that $m \geq 0$.

To prove the condition is necessary, take any $m \in \mathfrak{M}_\sigma(X)^+$ and any zero set $Z = g^{-1}(0) \subseteq \hat{p}X \setminus X$ where $g \in C_b(\hat{p}X)$, $0 \leq g \leq 1$. Let $g_j(x) := (1 - jg(x))^+$ for $x \in \hat{p}(X)$, $j \in \omega$, and let f_j be the restriction of g_j to X, so that $f_j \in U_b(X)$ and $g_j = \overline{f_j}$. Then $f_j \searrow 0$; therefore $\lim_j m(f_j) = 0$, and

$$\rho_m(Z) \leq \lim_j \int g_j d\rho_m = \lim_j m(f_j) = 0.$$

To prove the condition is sufficient, assume that $\rho_m(Z) = 0$ for every zero set Z in $\hat{p}X$ such that $Z \cap X = \emptyset$. Take any sequence $\{f_j\}_j$ in $U_b(X)$ such that $1 \geq f_j \searrow 0$, and any $\varepsilon > 0$. Let $Z_j := \{x \in \hat{p}X \mid \overline{f_j} \geq \varepsilon\}$ and $Z := \bigcap_j Z_j$. Every Z_j is a zero set in $\hat{p}X$, so is Z by Lemma 4.5, and $Z \cap X = \emptyset$. Thus $\lim_j \rho_m(Z_j) = \rho_m(Z) = 0$,

$$\lim_j m(f_j) = \lim_j \int \overline{f_j} d\rho_m \leq \varepsilon \rho_m(\hat{p}X) + \lim_j \rho_m(Z_j) = \varepsilon \|\rho_m\|_{TV},$$

and $\lim_j m(f_j) = 0$.   □

Definition 7.13. For a uniform space X and a pseudometric Δ on the set X, define subsets $Œ(\Delta, X)$ and $Œ^\infty(\Delta, X)$ of $\hat{p}X$ by

$$Œ(\Delta, X) := \bigcup_{x \in X} \text{int } \overline{\bigodot[x, \Delta]}$$

$$Œ^\infty(\Delta, X) := \bigcap_{j \in \omega} Œ(j\Delta, X)$$

where $\bigodot[x, \Delta] := \{y \in X \mid \Delta(x, y) < 1\}$ and the interior and closure are taken in $\hat{p}X$. The sets $Œ(\Delta, X)$ and $Œ^\infty(\Delta, X)$ are, respectively, the *open envelope* and the *reduced envelope of X in $\hat{p}X$ relative to* Δ.

Usually the underlying uniform space X is understood from the context, and then $Œ(\Delta, X)$ and $Œ^\infty(\Delta, X)$ are written simply as $Œ(\Delta)$ and $Œ^\infty(\Delta)$.   ∎

If $\Delta \in UP(X)$, then $\overline{\bigodot[x, (1+\varepsilon)\Delta]} \subseteq \text{int} \overline{\bigodot[x, \Delta]}$ for all $x \in X$ and $\varepsilon > 0$. In fact, for the function $f_x \in U_b(X)^+$ defined by $f_x(y) := 1 \wedge \Delta(x, y)$, $y \in X$, we have

$$\overline{\bigodot[x, (1+\varepsilon)\Delta]} \subseteq \{y \in \hat{p}X \mid (1+\varepsilon)\overline{f_x}(y) \leq 1\} \subseteq \text{int} \overline{\bigodot[x, \Delta]}.$$

Therefore $X \subseteq \bigcap_{j \in \omega} \bigcup_{x \in X} \overline{\bigodot[x, j\Delta]} = Œ^\infty(\Delta)$ for every $\Delta \in UP(X)$.

Theorem 7.14. *For any uniform space X, these three properties of a functional* $\mathfrak{m} \in \mathfrak{M}_b(X)$ *are equivalent:*

(i) $\mathfrak{m} \in \mathfrak{M}_u(X)$.
(ii) $|\rho_{\mathfrak{m}}|(\hat{p}X \setminus Œ(\Delta)) = 0$ *for every* $\Delta \in UP(X)$.
(iii) $|\rho_{\mathfrak{m}}|(\hat{p}X \setminus Œ^\infty(\Delta)) = 0$ *for every* $\Delta \in UP(X)$.

Proof. Since $\mathfrak{m} \in \mathfrak{M}_u(X)$ if and only if $|\mathfrak{m}| \in \mathfrak{M}_u(X)$, it is enough to prove the theorem assuming $\mathfrak{m} \geq 0$.

Properties (ii) and (iii) are equivalent because $j\Delta \in UP(X)$ for every $\Delta \in UP(X)$.

To prove the implication (i)⇒(ii), take any $\mathfrak{m} \in \mathfrak{M}_u(X)^+$ and $\Delta \in UP(X)$. Define a net $\{f_D\}_D$ of functions $f_D \in \text{BLip}_b(\Delta)^+$ indexed by non-empty finite subsets D of X (ordered by inclusion): $f_D(x) := 1 \wedge \Delta(x, D)$ for $x \in X$. Then $\lim_D f_D = 0$ in the X-pointwise topology; hence $\lim_D \mathfrak{m}(f_D) = 0$.

By continuity, the extensions $\overline{f_D} \in C_b(\hat{p}X)$ of f_D satisfy $\overline{f_D} = \min_{x \in D} \overline{f_{\{x\}}}$, and $\overline{f_D} \searrow g$ for a function g on $\hat{p}X$. By Theorem P.26,

$$\int g \, d\rho_{\mathfrak{m}} = \lim_D \int \overline{f_D} \, d\rho_{\mathfrak{m}} = \lim_D \mathfrak{m}(f_D) = 0.$$

To verify $\rho_{\mathfrak{m}}(\hat{p}X \setminus Œ(\Delta)) = 0$, I now show that $g(y) \geq 1$ for every $y \in \hat{p}X \setminus Œ(\Delta)$. Take any $y \in \hat{p}X$ such that $g(y) < 1$. Then $\overline{f_D}(y) < 1$ for some D; hence $\overline{f_{\{x\}}}(y) < 1$ for some $x \in X$. It follows that $y \in \text{int } \overline{\bigodot[x, \Delta]}$, and $y \in Œ(\Delta)$.

To prove (ii)⇒(i), assume that $\rho_{\mathfrak{m}}(\hat{p}X \setminus Œ(\Delta)) = 0$ for every $\Delta \in UP(X)$. Take any $\Delta' \in UP(X)$ and any net $\{f_\gamma\}_\gamma$ of functions in $\text{BLip}_b(\Delta')^+$ such that

$\lim_{\gamma} f_{\gamma} = 0$ in the X-pointwise topology. Set $h_{\gamma}(x) := \sup\{f_{\beta}(x) \mid \beta \geq \gamma\}$ for $x \in X$. The functions h_{γ} are in $\mathsf{BLip}_b(\Delta')^+$ and $h_{\gamma} \searrow 0$. By continuity, the extensions $\overline{h_{\gamma}} \in C_b(\hat{p}X)$ of h_{γ} satisfy $\overline{h_{\beta}} \geq \overline{h_{\gamma}}$ for $\beta \leq \gamma$. Let g be the pointwise limit of the net $\{\overline{h_{\gamma}}\}_{\gamma}$ on $\hat{p}X$. By Theorem P.26,

$$0 \leq \lim_{\gamma} \mathsf{m}(f_{\gamma}) \leq \lim_{\gamma} \mathsf{m}(h_{\gamma}) = \lim_{\gamma} \int \overline{h_{\gamma}} \, d\rho_{\mathsf{m}} = \int g \, d\rho_{\mathsf{m}}.$$

Next, I show that $\int g \, d\rho_{\mathsf{m}} = 0$, thus concluding the proof of (ii)$\Rightarrow$(i).

Fix $\varepsilon > 0$ and take any $y \in \mathsf{Œ}(\Delta'/\varepsilon)$, so that $y \in \overline{\odot}[x_0, \Delta'/\varepsilon]$ for some $x_0 \in X$. There is γ such that $h_{\gamma}(x_0) < \varepsilon$; for every $x \in \odot[x_0, \Delta'/\varepsilon]$, we have

$$h_{\gamma}(x) \leq h_{\gamma}(x_0) + \Delta'(x_0, x) < 2\varepsilon,$$

and $g(y) \leq \overline{h_{\gamma}}(y) \leq 2\varepsilon$ because $\overline{h_{\gamma}}$ is continuous. Therefore

$$\{y \in \hat{p}X \mid g(y) > 2\varepsilon\} \subseteq \hat{p}X \setminus \mathsf{Œ}(\Delta'/\varepsilon)$$

and $\rho_{\mathsf{m}}(\{y \in \hat{p}X \mid g(y) > 2\varepsilon\}) = 0$ by the assumption. Hence $\int g \, d\rho_{\mathsf{m}} = 0$. □

Corollary 7.15. *1. For any uniform space X, the point set of the completion $\hat{X} \subseteq \hat{p}X$ is equal to $\bigcap\{\mathsf{Œ}(\Delta) \mid \Delta \in \mathsf{UP}(X)\}$.*
2. For any metric space X with metric Δ, the point set of the completion $\hat{X} \subseteq \hat{p}X$ is $\mathsf{Œ}^{\infty}(\Delta)$.

Proof. For Part 1, combine Theorem 7.14 and Part 3 of Theorem 6.37.

If X is a metric space with metric Δ and $\Delta' \in \mathsf{UP}(X)$, then there is $j \in \omega$ such that $\mathsf{Œ}(j\Delta) \subseteq \mathsf{Œ}(\Delta')$. Thus Part 2 follows from Part 1. □

Exercise 7.16. Using the results in this section, prove that $\mathfrak{M}_u(X) = \mathfrak{M}_t(X)$ for every complete metric space X (Theorem 5.28). ∎

Exercise 7.17. Let X be a uniform space and $\mathsf{m} \in \mathfrak{M}_b(X)^+$. Prove that there are $\mathsf{m}_0 \in \mathfrak{M}_u(X)^+$ and $\mathsf{m}_1 \in \mathfrak{M}_b(X)^+$ such that $\mathsf{m} = \mathsf{m}_0 + \mathsf{m}_1$ and m_1 is "purely nonuniform"; that is, if $\mathsf{n} \in \mathfrak{M}_u(X)$ and $0 \leq \mathsf{n} \leq \mathsf{m}_1$, then $\mathsf{n} = 0$. ∎

7.3 Conditions for Uniform Measures to be Measures

It is easy to find a uniform space X for which $\mathfrak{M}_u(X) \not\subseteq \mathfrak{M}_{\sigma}(X)$:

Example 7.18. Let T be the closed interval $[0,1]$ with the usual compact uniformity, let μ be the Lebesgue measure on T, and let X be the set of rational numbers in T with the subspace uniformity. By Parts 4 and 5 of Lemma 6.9, the inclusion mapping $\iota : X \hookrightarrow T$ defines a natural bijection $\mathfrak{M}_b(\iota) : \mathfrak{M}_u(X) \to \mathfrak{M}_u(T)$; thus there is $\mathsf{m} \in \mathfrak{M}_u(X)$ such that $\mathsf{m}(g \circ \iota) = \int g \, d\mu$ for every $g \in \mathsf{U}_b(T)$.

However, the set X is countable, and every subset of X is in $\sigma(\mathsf{U}(X))$. If $\mathfrak{m}$ were represented by a Baire measure ν on X, we would have $\nu(\{x\}) > 0$ for some $x \in X$. Then $\mu(\{x\}) > 0$, which would contradict μ being the Lebesgue measure on $[0,1]$. ∎

The space X in the example is far from being complete. Since $\mathfrak{M}_u(X) = \mathfrak{M}_t(X)$ for every complete metric space X, it is tempting to imagine that $\mathfrak{M}_u(X) \subseteq \mathfrak{M}_\sigma(X)$ for every complete uniform space X. That, however, is false by virtue of the modified example.

Example 7.19. Let T be the closed interval $[0,1]$ with its usual compact topology, and Q the set of irrational numbers in T. As in Sect. 1.3, $\mathbb{R}$ is the space of real numbers with the usual metric $\Delta_{\mathbb{R}}$.

For every $q \in Q$, define the mapping $\psi_q : T \setminus \{q\} \to \mathbb{R}$ by $\psi_q(x) := 1/(q-x)$, $x \in T \setminus \{q\}$. Let X_q be the metric space on the set $T \setminus \{q\}$ with the metric $\overleftarrow{\psi_q}\Delta_{\mathbb{R}}$. The space X_q is complete because ψ_q is a uniform isomorphism between X_q and the closed set $(-\infty, 1/(q-1)] \cup [1/q, \infty)$ in $\mathbb{R}$.

$T \setminus Q$ is the set of rational numbers in $[0,1]$, and the domain of every ψ_q, $q \in Q$, includes $T \setminus Q$. Define X to be the uniform space on the set $T \setminus Q$ whose uniformity is induced by the set $\{\psi_q \mid q \in Q\}$ of mappings from $T \setminus Q$ to $\mathbb{R}$; equivalently, the uniformity of X is induced by the set $\{\overleftarrow{\psi_q}\Delta_{\mathbb{R}}{\upharpoonright}(T \setminus Q) \mid q \in Q\}$ of metrics on $T \setminus Q$. I now show that the space X is complete and $\mathfrak{M}_u(X) \not\subseteq \mathfrak{M}_\sigma(X)$.

To prove that X is complete, consider the uniform product $P := \prod_{q \in Q} X_q$ with the canonical projections $\pi_q : P \to X_q$. The space P is complete (see Exercise 2.31). The natural mapping $\iota : X \to P$, for which $\pi_q \circ \iota$ is the inclusion mapping $X \to X_q$ for all $q \in Q$, defines a uniform isomorphism between X and its image $\iota(X) \Subset P$. The set

$$\iota(X) = \{z \in P \mid \pi_q(z) = \pi_r(z) \text{ for all } q, r \in Q\}$$

is closed in P, therefore complete in the subspace uniformity (Theorem 2.17). Hence X is complete.

To find $\mathfrak{m} \in \mathfrak{M}_u(X) \setminus \mathfrak{M}_\sigma(X)$, I construct an $\mathfrak{M}_b$-projective system and use results in Sect. 6.4. Let Γ be the directed set of non-empty finite subsets of Q, partially ordered by inclusion. For every $D \in \Gamma$, define X_D to be the uniform space on the set $T \setminus D$ whose uniformity is induced by the set $\{\overleftarrow{\psi_q}\Delta_{\mathbb{R}}{\upharpoonright}(T \setminus D) \mid q \in D\}$ of metrics on $T \setminus D$. Note that $X_D \hookrightarrow T$ is a topological embedding, that is, as a topological space X_D is a subspace of T. The inclusion mappings $\varphi_{DD'} : X_{D'} \to X_D$ for $D, D' \in \Gamma$, $D \subseteq D'$, are uniformly continuous.

Let again μ be the Lebesgue measure on $\mathrm{Bo}(T)$. If $D \in \Gamma$ and $f \in \mathsf{U}_b(X_D)$, then f is continuous on $T \setminus D$. The expression $\mathfrak{m}_D(f) := \int_{T \setminus D} f \, d\mu$, $f \in \mathsf{U}_b(X_D)$, defines $\mathfrak{m}_D \in \mathfrak{M}_t(X_D)$. For every $D \in \Gamma$, let $\varphi_D : X \to X_D$ be the inclusion mapping. Then $\langle \Gamma, X_D, \varphi_{DD'}, \mathfrak{m}_D \rangle$ is an $\mathfrak{M}_b$-projective system, and the family $\{\varphi_D\}_{D \in \Gamma}$ is consistent with it.

By Theorem 6.29, there exists a unique $\mathfrak{m} \in \mathfrak{M}_b(X)$ such that $\varphi_D(\mathfrak{m}) = \mathfrak{m}_D$ for all $D \in \Gamma$, and $\mathfrak{m} \in \mathfrak{M}_u(X)$ by Theorem 6.20.

As in the previous example, it is easy to see that $\mathfrak{m}$ cannot be represented by a measure on X because the countable set $T \setminus Q$ has Lebesgue measure 0.   ∎

Now we come to sufficient conditions for every $\mathfrak{m} \in \mathfrak{M}_u(X)$ to be in $\mathfrak{M}_\sigma(X)$, or even in $\mathfrak{M}_\tau(X)$ or $\mathfrak{M}_t(X)$.

Theorem 7.20. *Let X be a uniform space. If X is uniformly locally compact, then $\mathfrak{M}_u(X) = \mathfrak{M}_t(X)$.*

Proof. This follows from Theorem 7.22 below, but there is also a simple direct proof: Let $\Delta \in \mathrm{UP}(X)$ be a pseudometric such that for every $x \in X$ the closure $K(x)$ of $\odot[x, \Delta]$ in X is compact. Then $Œ(\Delta) = X$ because every $K(x)$ is closed in $\hat{p}X$. If $\mathfrak{m} \in \mathfrak{M}_u(X)$, then $|\rho_{\mathfrak{m}}|(\hat{p}X \setminus X) = 0$ by Theorem 7.14, and $\mathfrak{m} \in \mathfrak{M}_t(X)$ by Theorem 7.9.   □

Theorem 7.21. *If X is an inversion-closed uniform space, then $\mathfrak{M}_u(X) \subseteq \mathfrak{M}_\sigma(X)$.*

Proof. If X is inversion-closed, $\{f_j\}_j$ is a sequence of functions in $U_b(X)$ and $f_j \searrow 0$, then the set $\{f_j \mid j \in \omega\}$ is uniformly equicontinuous by Theorem 4.7. The inclusion $\mathfrak{M}_u(X) \subseteq \mathfrak{M}_\sigma(X)$ follows from the definition of $\mathfrak{M}_u(X)$ and $\mathfrak{M}_\sigma(X)$.   □

The sufficient condition in Theorem 7.21 is far from being necessary. For example, $\mathfrak{M}_u(\mathbb{R}) = \mathfrak{M}_\sigma(\mathbb{R})$, but $\mathbb{R}$ is not inversion-closed. Nevertheless, the theorem does have a partial converse, Theorem 7.37 in Sect. 7.5.

Theorem 7.22. *If X is a supercomplete uniform space, then $\mathfrak{M}_u(X) = \mathfrak{M}_\tau(X)$.*

Proof. Take any supercomplete space X, $\mathfrak{m} \in \mathfrak{M}_u(X)^+$, and any compact set $K \subseteq \hat{p}X$ such that $\rho_{\mathfrak{m}}(K) > 0$. In light of Theorem 7.10, the conclusion will follow once I prove $K \cap X \neq \emptyset$.

For $\Delta \in \mathrm{UP}(X)$, write

$$A(\Delta) := \{x \in X \mid \rho_{\mathfrak{m}}(K \cap \mathrm{int}\,\overline{\odot[x, \Delta]}) > 0\}$$

where again the interior and closure are taken in $\hat{p}X$. Let $B(\Delta)$ be the closure of $A(\Delta)$ in X.

I claim that $A(\Delta) \neq \emptyset$ for all $\Delta \in \mathrm{UP}(X)$, and that the net $\{B(\Delta)\}_{\Delta \in \mathrm{UP}(X)}$, indexed by the upwards-directed set $\mathrm{UP}(X)$ ordered by $\leq$, is Cauchy in the hyperspace HX. To prove the claim, take any $\Delta_0 \leq \Delta_1 \in \mathrm{UP}(X)$ such that $A(\Delta_0) \neq \emptyset$, and any $x_0 \in A(\Delta_0)$. As $\mathfrak{m} \in \mathfrak{M}_u(X)^+$, Theorem 7.14 implies

$$\rho_{\mathfrak{m}}\left(Œ(\Delta_1) \cap K \cap \mathrm{int}\,\overline{\odot[x_0, \Delta_0]}\right) = \rho_{\mathfrak{m}}\left(K \cap \mathrm{int}\,\overline{\odot[x_0, \Delta_0]}\right) > 0.$$

Since the measure $\rho_{\mathfrak{m}}$ is τ-additive on $\hat{p}X$, there is $x_1 \in X$ such that

$$\rho_{\mathfrak{m}}\left(\mathrm{int}\,\overline{\odot[x_1, \Delta_1]} \cap K \cap \mathrm{int}\,\overline{\odot[x_0, \Delta_0]}\right) > 0.$$

Thus $x_1 \in A(\Delta_1)$ and $\Delta_0(x_0, x_1) \leq 2$, so that $\sup\{\Delta_0(x, A(\Delta_1)) \mid x \in A(\Delta_0)\} \leq 2$. As $A(\Delta_0) \supseteq A(\Delta_1)$, it follows that $\Delta_0^{\mathsf{H}}(B(\Delta_0), B(\Delta_1)) \leq 2$. For the pseudometric Δ_0 that is identically 0, we have $A(\Delta_0) = X \neq \emptyset$; hence $A(\Delta) \neq \emptyset$ for all $\Delta \in \mathsf{UP}(X)$, concluding the proof of the claim.

As the space $\mathsf{H}X$ is complete, the net $\{B(\Delta)\}_{\Delta \in \mathsf{UP}(X)}$ converges in $\mathsf{H}X$, and by Corollary 4.14 there is $x \in \bigcap_\Delta B(\Delta)$. Then $x \in \bigcap_\Delta A(\Delta)$ because $B(2\Delta) \subseteq A(\Delta)$ for every $\Delta \in \mathsf{UP}(X)$; hence $x \in K$ and $K \cap X \neq \emptyset$. $\square$

Exercise 7.23. Let X be any uniform space. Prove that if $\mathfrak{M}_{\mathsf{u}}(X) = \mathfrak{M}_\tau(X)$, then X is complete.   ∎

Exercise 7.24. Let X be any uniform space. Prove that $\mathfrak{M}_\tau(X) = \mathfrak{M}_{\mathsf{b}}(X)$ if and only if X is compact.   ∎

7.4 Condition for Measures to be Uniform Measures

In this section I characterize the uniform spaces for which $\mathfrak{M}_\sigma(X) \subseteq \mathfrak{M}_{\mathsf{u}}(X)$.

Theorem 7.25. *Two properties of a uniform space X are equivalent:*

(i) $\mathfrak{M}_\sigma(X) \subseteq \mathfrak{M}_{\mathsf{u}}(X)$.
(ii) The cardinality of every uniformly discrete subset of X is measure-free.

Proof. To prove that (i) implies (ii), assume that X has a uniformly discrete subset Y whose cardinality is not measure-free. There is $\Delta \in \mathsf{UP}(X)$ such that $\Delta(x, y) \geq 1$ for $x, y \in Y$, $x \neq y$, and there is a measure $v \geq 0$ on the σ-algebra of all subsets of Y such that $v(\{x\}) = 0$ for every $x \in Y$ and $v(Y) = 1$. Then $\mathsf{m}(f) := \int_Y f \, dv$, $f \in \mathsf{U}_{\mathsf{b}}(X)$, defines a functional $\mathsf{m} \in \mathfrak{M}_\sigma(X)$.

Let Γ be the directed set of non-empty finite subsets of Y, partially ordered by inclusion. For $D \in \Gamma$, let $f_D(x) := 1 \wedge \Delta(x, D)$, $x \in X$. Then $f_D \in \mathsf{BLip}_{\mathsf{b}}(\Delta)$ for every $D \in \Gamma$, and $f_D \searrow f$ where $f \in \mathsf{BLip}_{\mathsf{b}}(\Delta)$. Since $f_D(x) = 1$ for $x \in Y \setminus D$, we have $\mathsf{m}(f_D) = 1$ for every $D \in \Gamma$. On the other hand, $f(x) = 0$ for $x \in Y$, and $\mathsf{m}(f) = 0$. Thus $\mathsf{m} \notin \mathfrak{M}_{\mathsf{u}}(X)$.

To prove (ii)⟹(i), assume (ii) and take any $\mathsf{m} \in \mathfrak{M}_\sigma(X)^+$ and $\Delta \in \mathsf{UP}(X)$. The weight of the associated metric space X/Δ (Sect. P.2) is measure-free because, by Corollary 1.9, for every $j \in \omega$ there is a set $Y_j \subseteq X$ whose cardinality $|Y_j|$ is measure-free and $X = \bigcup\{\bigodot[x, j\Delta] \mid x \in Y_j\}$; then $\{\bigodot[x^\bullet, i\Delta^\bullet] \mid x \in \bigcup_j Y_j, i \in \omega\}$ is a base of the topology of X/Δ.

By Theorem 7.4, there is a Baire measure μ on X such that $\mathsf{m}(f) = \int f \, d\mu$ for $f \in \mathsf{U}_{\mathsf{b}}(X)$. Apply Theorem P.34 with the canonical surjection $\chi_\Delta : X \to X/\Delta$ to get a closed separable set $S \subseteq X/\Delta$ such that $\mu(X \setminus \chi_\Delta^{-1}(S)) = 0$.

Now take any net of functions $f_\gamma \in \mathsf{BLip}_{\mathsf{b}}(\Delta)^+$ such that $\lim_\gamma f_\gamma(x) = 0$ for every $x \in X$. Let $g_\gamma(x) := \sup\{f_\beta(x) \mid \beta \geq \gamma\}$ for $x \in X$. Then $g_\gamma \in \mathsf{BLip}_{\mathsf{b}}(\Delta)$ and $g_\gamma \searrow 0$. The set $C := \chi_\Delta^{-1}(S)$ is the closure of a countable set $C_0 \subseteq X$ in the topology of the pseudometric Δ. There is an increasing sequence of indices $\gamma(i)$, $i \in \omega$, such that $\lim_i g_{\gamma(i)}(x) = 0$ for every $x \in C_0$, hence also for every $x \in C$, and

$$\lim_{\gamma} \mathfrak{m}(f_\gamma) \leq \lim_{\gamma} \mathfrak{m}(g_\gamma) \leq \lim_i \mathfrak{m}(g_{\gamma(i)}) = \lim_i \left(\int_C g_{\gamma(i)} \, d\mu + \int_{X \setminus C} g_{\gamma(i)} \, d\mu \right) = 0$$

which along with Lemma 5.25 proves that $\mathfrak{m} \in \mathfrak{M}_u(X)$. $\square$

The theorem shows that, in a certain sense, uniform measures generalize Baire measures on uniform spaces: It is consistent with the ZFC set theory to assume that, for every uniform space X, every functional on $U_b(X)$ represented by a Baire measure is a uniform measure. In the next chapter I show that uniform measures generalize (σ-additive) measures also in another sense; namely, several familiar spaces of measures may be identified with $\mathfrak{M}_u(X)$ for a suitably chosen uniform space X.

Exercise 7.26. Let X be an inversion-closed uniform space in which the cardinality of every uniformly discrete subset is measure-free. Prove that if X is complete, then so is cX.   ∎

7.5 Measure-Fine Uniform Spaces

For every uniform space, Theorem 7.14 yields a finer space with the same uniform measures, in the sense made precise in this section.

Recall from Sect. 1.4 that a pseudometric Δ is in $\mathsf{UP}(\mathsf{F}X)$ if and only if the function $\backslash_y \Delta(x,y)$ is continuous on X for every $x \in X$.

Lemma 7.27. *Let X be a uniform space and $\Delta_0, \Delta_1 \in \mathsf{UP}(\mathsf{F}X)$. Then*

$$\text{Œ}(3\Delta_0, X) \cap \text{Œ}(3\Delta_1, X) \subseteq \text{Œ}(\Delta_0 \vee \Delta_1, X).$$

Proof. Take any $z \in \text{Œ}(3\Delta_0) \cap \text{Œ}(3\Delta_1)$. There are points $x_0, x_1 \in X$ such that the set $V := \text{int} \odot [x_0, 3\Delta_0] \cap \text{int} \odot [x_1, 3\Delta_1]$ is an open neighbourhood of z in $\widehat{\mathsf{p}}X$. As X is dense in $\widehat{\mathsf{p}}X$, it follows that $z \in \text{int} \overline{X \cap V}$. Fix a point $x \in X \cap V$.

From the continuity of Δ_i, $i = 0, 1$, we get

$$X \cap \text{int} \, \overline{\odot [x_i, 3\Delta_i]} \subseteq X \cap \overline{\odot [x_i, 3\Delta_i]} \subseteq \odot [x_i, 2\Delta_i] \subseteq \odot [x, \Delta_i]$$

$$X \cap V \subseteq \odot [x, \Delta_0] \cap \odot [x, \Delta_1] = \odot [x, \Delta_0 \vee \Delta_1]$$

and thus $z \in \text{int} \, \overline{\odot [x, \Delta_0 \vee \Delta_1]} \subseteq \text{Œ}(\Delta_0 \vee \Delta_1)$. $\square$

Theorem 7.28. *Let X be a uniform space, and let $\mathscr{U}$ be the set of all pseudometrics $\Delta \in \mathsf{UP}(\mathsf{F}X)$ such that $\rho_\mathfrak{m}(\widehat{\mathsf{p}}X \setminus \text{Œ}^\infty(\Delta, X)) = 0$ for all $\mathfrak{m} \in \mathfrak{M}_u(X)^+$. Then $\mathscr{U}$ is a uniform structure on X and $\mathscr{U} \supseteq \mathsf{UP}(X)$.*

Proof. Take any $\Delta_0, \Delta_1 \in \mathcal{U}$, $m \in \mathfrak{M}_u(X)^+$ and $j \in \omega$, and let $\Delta := \Delta_0 \vee \Delta_1$. Then $\text{Œ}(3j\Delta_0) \cap \text{Œ}(3j\Delta_1) \subseteq \text{Œ}(j\Delta)$ by Lemma 7.27. Thus $\rho_m(\widehat{p}X \setminus \text{Œ}(j\Delta)) = 0$ for every $j \in \omega$, and $\rho_m(\widehat{p}X \setminus \text{Œ}^\circ(\Delta)) = 0$. Therefore $\mathcal{U}$ has property (U1) in Definition 1.1.

To prove property (U2), take any pseudometric Δ such that $\Delta \ll \mathcal{U}$ and any $m \in \mathfrak{M}_u(X)^+$ and $j \in \omega$. There is $\Delta' \in \mathcal{U}$ such that $\odot[x, \Delta'] \subseteq \odot[x, j\Delta]$ for all $x \in X$. Hence $\text{Œ}(\Delta') \subseteq \text{Œ}(j\Delta)$, and $\rho_m(\widehat{p}X \setminus \text{Œ}(j\Delta)) = 0$. It follows that $\Delta \in \mathcal{U}$.

Property (U3) follows from the inclusion $\mathcal{U} \supseteq \text{UP}(X)$, which holds by virtue of Theorem 7.14. □

Definition 7.29. When X is a uniform space, the *measure-fine uniformity of X* is the uniformity $\mathcal{U}$ in Theorem 7.28. The *measure-fine uniform space of X*, denoted by $\text{M}X$, is the set X with the measure-fine uniformity of X.

A uniform space X is *measure-fine* iff $X = \text{M}X$. ∎

Evidently $\text{M}X$ is finer than X and coarser than $\text{F}X$.

Exercise 7.30. Prove that if X is a uniform space such that $\mathfrak{M}_u(X) = \mathfrak{M}_t(X)$, then $\text{M}X = \text{F}X$. ∎

As in Sect. 1.3, $\Delta_\mathbb{R}$ is the usual metric on $\mathbb{R}$ so that $\overleftarrow{f}\Delta_\mathbb{R}(x,y) = |f(x) - f(y)|$ for any real-valued function f on X and $x, y \in X$.

Lemma 7.31. *Let X be a uniform space, $f \in C_b(X)$, $z \in \widehat{p}X$ and $j \in \omega$, $j \geq 1$. Let Δ be the pseudometric $\overleftarrow{f}\Delta_\mathbb{R}$ on X.*

1. *If $z \in \text{Œ}(j\Delta, X)$, then there exist $g, g' \in C_b(\widehat{p}X)$ such that $g(x) \leq f(x) \leq g'(x)$ for every $x \in X$ and $g'(z) - g(z) = 2/j$.*
2. *If there exist $g, g' \in C_b(\widehat{p}X)$ such that $g(x) \leq f(x) \leq g'(x)$ for every $x \in X$ and $g'(z) - g(z) < 1/3j$, then $z \in \text{Œ}(j\Delta, X)$.*

Proof. To prove Part 1, take any $z \in \text{Œ}(j\Delta)$. Then there is $x_0 \in X$ such that the set $V := \text{int} \odot[x_0, j\Delta]$ is a neighbourhood of z in $\widehat{p}X$. With $r := \|f\|_X + 1/j$, there are functions $g, g' \in C_b(\widehat{p}X)$ such that

$$-r \leq g \leq f(x_0) - 1/j, \qquad g(z) = f(x_0) - 1/j, \qquad g \text{ is } -r \text{ on } \widehat{p}X \setminus V,$$
$$r \geq g' \geq f(x_0) + 1/j, \qquad g'(z) = f(x_0) + 1/j, \qquad g' \text{ is } r \text{ on } \widehat{p}X \setminus V.$$

Since f is continuous,

$$f(x_0) - 1/j \leq f(x) \leq f(x_0) + 1/j$$

for every $x \in X \cap \overline{\odot[x_0, j\Delta]}$. Thus $g \leq f \leq g'$ on X.

To prove Part 2, take $g, g' \in C_b(\widehat{p}X)$ such that $g(x) \leq f(x) \leq g'(x)$ for all $x \in X$ and $g'(z) - g(z) < 1/3j$. The set

$$V := \{y \in \widehat{p}X \mid |g(y) - g(z)| < 1/3j \text{ and } |g'(y) - g'(z)| < 1/3j\}$$

is open in $\widehat{p}X$. Fix $x_0 \in X \cap V$. For all $x \in X \cap V$, we have

$$g(z) - 1/3j < g(x) \le f(x) \le g'(x) < g'(z) + 1/3j < g(z) + 2/3j;$$

hence $|f(x) - f(x_0)| < 1/j$, and it follows that $X \cap V \subseteq \bigodot[x_0, j\Delta]$. Therefore z is in $V \subseteq \mathrm{int}\bigodot[x_0, j\Delta] \subseteq \mathfrak{E}(j\Delta)$. □

Exercise 7.32. Let X be a uniform space and $\varphi: X \to Z$ a continuous mapping to a complete metric space Z with metric Δ. Prove that $\mathfrak{E}^\infty(\overleftarrow{\varphi}\Delta)$ is the largest subset A of $\widehat{p}X$ for which there is a continuous mapping $\widetilde{\varphi}: A \to Z$ extending φ. ∎

The next theorem states that a bounded function is in $U_b(MX)$ if and only if it is "Riemann-integrable" with respect to every uniform measure on X.

Theorem 7.33. *Let X be a uniform space. Two properties of a real-valued function f on X are equivalent:*

(i) $f \in U_b(MX)$.
(ii) For every $\mathfrak{m} \in \mathfrak{M}_u(X)^+$ *and for every* $\varepsilon > 0$, *there are* $h, h' \in U_b(X)$ *such that* $h \le f \le h'$ *and* $\mathfrak{m}(h' - h) < \varepsilon$.

Proof. Set $\Delta := \overleftarrow{f}\Delta_{\mathbb{R}}$. Thus $f \in U_b(MX)$ if and only if $\rho_\mathfrak{m}(\widehat{p}X \setminus \mathfrak{E}(j\Delta)) = 0$ for all $\mathfrak{m} \in \mathfrak{M}_u(X)^+$ and $j \in \omega$, by the definition of MX.

To prove (i)⇒(ii), assume $f \in U_b(MX)$ and take any $\mathfrak{m} \in \mathfrak{M}_u(X)^+$ and $\varepsilon > 0$. Choose $j \in \omega$ such that $2\|\mathfrak{m}\|/j < \varepsilon$. By part 1 of Lemma 7.31, there are nets $\{g_\gamma\}_\gamma$ and $\{g'_\gamma\}_\gamma$ in $C_b(\widehat{p}X)$ such that $g_\gamma(x) \le f(x) \le g'_\gamma(x)$ for all γ and all $x \in X$, $g_\gamma \nearrow g$, $g'_\gamma \searrow g'$ where g and g' are functions on $\widehat{p}X$, and $g'(z) - g(z) \le 2/j$ for all $z \in \mathfrak{E}(j\Delta)$. By Theorem P.26,

$$\lim_\gamma \int (g'_\gamma - g_\gamma) \, d\rho_\mathfrak{m} = \int (g' - g) \, d\rho_\mathfrak{m} = \int_{\mathfrak{E}(j\Delta)} (g' - g) \, d\rho_\mathfrak{m} \le 2\|\mathfrak{m}\|/j < \varepsilon.$$

Thus $\int (g'_\gamma - g_\gamma) \, d\rho_\mathfrak{m} < \varepsilon$ for some γ, and (ii) holds when h and h' are the restrictions of such g_γ and g'_γ to X.

To prove the converse, assume property (ii). Take any $\mathfrak{m} \in \mathfrak{M}_u(X)^+$, $j \in \omega$ and $\varepsilon > 0$. Let $h, h' \in U_b(X)$ be such that $h \le f \le h'$ and $\mathfrak{m}(h' - h) < \varepsilon/3j$. By Part 2 of Lemma 7.31, if $z \in \widehat{p}X \setminus \mathfrak{E}(j\Delta)$, then $\overline{h'}(z) - \overline{h}(z) \ge 1/3j$. Thus

$$\rho_\mathfrak{m}(\widehat{p}X \setminus \mathfrak{E}(j\Delta)) \le 3j \int (\overline{h'} - \overline{h}) \, d\rho_\mathfrak{m} = 3j\,\mathfrak{m}(h' - h) < \varepsilon,$$

which proves $\rho_\mathfrak{m}(\widehat{p}X \setminus \mathfrak{E}(j\Delta)) = 0$ and $f \in U_b(MX)$. □

Exercise 7.34. Let X be any uniform space. Prove that the $UEB(X)$ topology, the $UEB(MX)$ topology and the $U_b(X)$-weak and $U_b(MX)$-weak topologies coincide on the positive cone $\mathfrak{M}_u(MX)^+$. Prove that the four topologies coincide on $\|\cdot\|$ spheres in $\mathfrak{M}_u(MX)$. ∎

The identity mapping $\iota\colon \mathsf{M}X \to X$ is uniformly continuous and extends to $\mathfrak{M}_b(\iota)\colon \mathfrak{M}_b(\mathsf{M}X) \to \mathfrak{M}_b(X)$ as in Definition 6.7. The image $\iota(\mathfrak{m}) = \mathfrak{M}_b(\iota)(\mathfrak{m})$ of $\mathfrak{m}\in\mathfrak{M}_b(\mathsf{M}X)$ is simply the restriction of $\mathfrak{m}$ to the space $\mathsf{U}_b(X) \subseteq \mathsf{U}_b(\mathsf{M}X)$, and $\iota(\mathfrak{m})\in\mathfrak{M}_u(X)$ for all $\mathfrak{m}\in\mathfrak{M}_u(\mathsf{M}X)$ by Lemma 6.8. By the next theorem, every $\mathfrak{m}\in\mathfrak{M}_u(\mathsf{M}X)$ is uniquely determined by $\iota(\mathfrak{m})$, and every uniform measure on X is of the form $\iota(\mathfrak{m})$ for some $\mathfrak{m}\in\mathfrak{M}_u(\mathsf{M}X)$. In that sense, uniform measures on $\mathsf{M}X$ and on X are the same.

Theorem 7.35. *Let X be a uniform space and $\iota\colon \mathsf{M}X \to X$ the identity mapping.*

1. *If $\mathfrak{m}\in\mathfrak{M}_b(\mathsf{M}X)$ and $\iota(\mathfrak{m})\in\mathfrak{M}_u(X)$, then $\mathfrak{m}\in\mathfrak{M}_u(\mathsf{M}X)$.*
2. *The mapping $\mathfrak{m}\mapsto \iota(\mathfrak{m})$ is a bijection from $\mathfrak{M}_u(\mathsf{M}X)$ onto $\mathfrak{M}_u(X)$.*
3. $\mathsf{MM}X = \mathsf{M}X$.

The main step in the proof is the following lemma.

Lemma 7.36. *Let X be a uniform space, $\Delta\in\mathsf{UP}(\mathsf{F}X)$, let $\iota\colon \mathsf{M}X \to X$ be the identity mapping and $\mathfrak{m}\in\mathfrak{M}_b(\mathsf{M}X)^+$. Then*

$$\rho_\mathfrak{m}(\widehat{\mathsf{p}}\mathsf{M}X \setminus \text{\OE}^\infty(\Delta,\mathsf{M}X)) \leq \rho_{\iota(\mathfrak{m})}(\widehat{\mathsf{p}}X \setminus \text{\OE}^\infty(\Delta,X)).$$

Proof. Let $\varphi\colon \widehat{\mathsf{p}}\mathsf{M}X \to \widehat{\mathsf{p}}X$ denote the restriction of the mapping $\mathfrak{M}_b(\iota)$ to the set $\widehat{\mathsf{p}}\mathsf{M}X \subseteq \mathfrak{M}_b(\mathsf{M}X)$. Then $\rho_{\iota(\mathfrak{m})}$ is equal to $\varphi(\rho_\mathfrak{m})$, the image of $\rho_\mathfrak{m}$ under φ as defined in Sect. P.5. Thus it is enough to prove $\text{\OE}^\infty(\Delta,\mathsf{M}X) \supseteq \varphi^{-1}(\text{\OE}^\infty(\Delta,X))$.

In this proof, I write int_M and $\overline{}^\mathsf{M}$ for the interior and closure operations in $\widehat{\mathsf{p}}\mathsf{M}X$, and int and $\overline{}$ for the operations in $\widehat{\mathsf{p}}X$.

Take any $z\in\varphi^{-1}(\text{\OE}^\infty(\Delta,X))$ and any $j\in\omega$. Then $\varphi(z)\in\text{\OE}(2j\Delta,X)$; therefore there is $x\in X$ for which $\varphi(z)\in\text{int}\,\odot[x,2j\Delta]$. As the mapping φ is continuous,

$$V := \varphi^{-1}\left(\text{int}\,\overline{\odot[x,2j\Delta]}\right)$$

is an open subset of $\widehat{\mathsf{p}}\mathsf{M}X$, and $z\in\text{int}_\mathsf{M}\overline{X\cap V}^\mathsf{M}$ because X is dense in $\widehat{\mathsf{p}}\mathsf{M}X$. From the continuity of Δ we get

$$X\cap V \subseteq X\cap\overline{\odot[x,2j\Delta]} \subseteq \odot[x,j\Delta];$$

hence $z\in\text{int}_\mathsf{M}\overline{\odot[x,j\Delta]}^\mathsf{M} \subseteq \text{\OE}(j\Delta,\mathsf{M}X)$. That proves $z\in\text{\OE}^\infty(\Delta,\mathsf{M}X)$. $\square$

Proof of Theorem 7.35. To prove Part 1, take any $\mathfrak{m}\in\mathfrak{M}_b(\mathsf{M}X)$ with $\iota(\mathfrak{m})\in\mathfrak{M}_u(X)$, and any $\Delta\in\mathsf{UP}(\mathsf{M}X)$. From Theorem 7.14 and Lemma 7.36 we get

$$|\rho_\mathfrak{m}|(\widehat{\mathsf{p}}\mathsf{M}X \setminus \text{\OE}^\infty(\Delta,\mathsf{M}X)) \leq |\rho_{\iota(\mathfrak{m})}|(\widehat{\mathsf{p}}X \setminus \text{\OE}^\infty(\Delta,X)) = 0,$$

and another application of Theorem 7.14 gives $\mathfrak{m}\in\mathfrak{M}_u(\mathsf{M}X)$.

The mapping $\mathfrak{m}\mapsto \iota(\mathfrak{m})$ is injective on $\mathfrak{M}_u(\mathsf{M}X)$ by Theorem 7.33. To prove that it is surjective, in view of Part 1 it is enough to show that every uniform measure in

$\mathfrak{M}_u(X)$ is of the form $\iota(\mathfrak{m})$ for some $\mathfrak{m} \in \mathfrak{M}_b(MX)$. But such $\mathfrak{m}$ is easily obtained from the Hahn–Banach theorem P.9; alternatively, an explicit extension of a uniform measure on X to a $\|\cdot\|$ bounded functional on $U_b(MX)$ is just as easily obtained from Theorem 7.33.

To prove Part 3, take any $\Delta \in UP(MMX)$, $\mathfrak{m} \in \mathfrak{M}_u(X)^+$, $j \in \omega$ and $\varepsilon > 0$. By Part 2, there is $\mathfrak{m}' \in \mathfrak{M}_u(MMX)$ whose restriction to $U_b(X)$ is $\mathfrak{m}$.

Define a net $\{f_D\}_D$ of functions $f_D \in BLip_b(j\Delta)^+$ indexed by non-empty finite subsets D of X (ordered by inclusion): $f_D(x) := 1 \wedge j\Delta(x,D)$ for $x \in X$. Then $\lim_D f_D = 0$ in the X-pointwise topology; hence $\lim_D \mathfrak{m}'(f_D) = 0$, and there is D for which $\mathfrak{m}'(f_D) < \varepsilon$. By Theorem 7.33, there are functions $h_x, h'_x \in U_b(X)$ for $x \in D$ such that $h_x \leq f_{\{x\}} \leq h'_x$ and $\sum_{x \in D} \mathfrak{m}(h'_x - h_x) < \varepsilon$. Let $h := \min_{x \in D} h_x$ and $h' := \min_{x \in D} h'_x$. Then $h \leq f_D \leq h'$ and $\mathfrak{m}(h' - h) < \varepsilon$, and thus $\mathfrak{m}(h') < 2\varepsilon$.

I claim that if $z \in \hat{p}X \setminus \text{Œ}(j\Delta, X)$, then $\overline{h'}(z) \geq 1$. In fact, if $z \in \hat{p}X$ and $\overline{h'}(z) < 1$, then there is $x \in D$ such that $\overline{h'_x}(z) < 1$. Thus $\overline{h'_x}$ is < 1 on an open neighbourhood V of z in $\hat{p}X$. But $V \cap X \subseteq \text{Ⓞ}[x, j\Delta]$ because $f_{\{x\}} \leq h'_x$, and therefore

$$z \in V \subseteq \text{int}\,\overline{V \cap X} \subseteq \text{int}\,\overline{\text{Ⓞ}[x, j\Delta]} \subseteq \text{Œ}(j\Delta, X)$$

(where the interior and closure are taken in $\hat{p}X$), as claimed. From the claim, we get

$$\rho_\mathfrak{m}(\hat{p}X \setminus \text{Œ}(j\Delta, X)) \leq \int h'\, d\rho_\mathfrak{m} = \mathfrak{m}(h') < 2\varepsilon,$$

which proves that $\Delta \in UP(MX)$.   □

The following theorem is a partial converse to Theorem 7.21.

Theorem 7.37. *Let X be a measure-fine uniform space and $\mathfrak{M}_u(X) \subseteq \mathfrak{M}_\sigma(X)$. Then X is inversion-closed.*

Proof. Take any sequence of functions $f_j \in U_b(X)$, $j \in \omega$, such that $f_j \searrow 0$. Write $\Delta(x,x') := \sup_j |f_j(x) - f_j(x')|$ for $x, x' \in X$. Then $\Delta \in UP(FX)$ because for every $x \in X$ and every $\varepsilon > 0$ there is a neighbourhood W of x in X such that $\|f_j\|_W < \varepsilon$ for almost all j.

Since $f_j \searrow 0$ on X, there is a function g on $\hat{p}X$ for which $\overline{f_j} \searrow g$. I claim that $coz(g) \cup \text{Œ}^\infty(\Delta) = \hat{p}X$.

To prove the claim, take any $z \in \hat{p}X$ such that $g(z) = 0$ and any $k \in \omega$, $k \geq 1$. Then $\overline{f_j}(z) < 1/3k$ for some $j \in \omega$. The set

$$V := \{y \in \hat{p}X \mid |\overline{f_i}(y) - \overline{f_i}(z)| < 1/3k \text{ for all } i \leq j\}$$

is open in $\hat{p}X$; hence $z \in \text{int}\,\overline{X \cap V}$. If $y \in V$ and $i \geq j$ then $\overline{f_i}(y) \leq \overline{f_j}(y) < 2/3k$. Therefore $\Delta(x,x') \leq 2/3k$ for all $x, x' \in X \cap V$; hence $X \cap V \subseteq \text{Ⓞ}[x_0, k\Delta]$ for any $x_0 \in X \cap V$, and

$$z \in \text{int}\,\overline{X \cap V} \subseteq \text{int}\,\overline{\text{Ⓞ}[x_0, k\Delta]} \subseteq \text{Œ}^\infty(\Delta),$$

which proves the claim.

Now take any $m \in \mathfrak{M}_u(X)^+$. Then $m \in \mathfrak{M}_\sigma(X)^+$; hence $\rho_m(Z) = 0$ for every zero set $Z \subseteq \hat{p}X \setminus X$ by Theorem 7.12. The set $\mathrm{coz}(g)$ is a countable union of zero sets $\bigcap_j \{z \in \hat{p}X \mid \overline{f_j}(z) \geq 1/i\}$, $i = 1, 2, \ldots$; therefore

$$\rho_m(\hat{p}X \setminus \text{Œ}^\infty(\Delta)) \leq \rho_m(\mathrm{coz}(g)) = 0.$$

Thus $\Delta \in \mathrm{UP}(\mathsf{M}X)$. Then $\Delta \in \mathrm{UP}(X)$ because X is measure-fine, the set $\{f_j \mid j \in \omega\}$ is uniformly equicontinuous, and X is inversion-closed by Theorem 4.7.   □

Corollary 7.38. *These properties of a uniform space X are equivalent:*

(i) $\mathfrak{M}_u(X) \subseteq \mathfrak{M}_\sigma(X)$.
(ii) $\mathfrak{M}_u(\mathsf{M}X) \subseteq \mathfrak{M}_\sigma(\mathsf{M}X)$.
(iii) $\mathsf{M}X$ *is inversion-closed.*

Proof. The implication (ii)⇒(iii) follows from Theorem 7.37, (iii)⇒(ii) from Theorem 7.21 and (ii)⇒(i) from Part 2 of Theorem 7.35.

Assume (i), and take any $m \in \mathfrak{M}_u(\mathsf{M}X)^+$, functions $f_j \in U_b(\mathsf{M}X)$, $j \in \omega$, such that $f_j \searrow 0$, and $\varepsilon > 0$. By Theorems 7.33 and 7.35, there are $h_j, h'_j \in U_b(X)$ for $j \in \omega$ such that $h_j \leq f_j \leq h'_j$ and $m(h'_j - h_j) < \varepsilon/2^{j+1}$. Define functions $g_j := \min_{i \leq j} h_i$ and $g'_j := \min_{i \leq j} h'_i$ for $j \in \omega$. Then $g_j \leq f_j \leq g'_j$ and $m(g'_j - g_j) < \varepsilon$. Hence

$$\lim_j m(f_j) \leq \lim_j m(g'_j) \leq \lim_j m(g_j) + \varepsilon = \varepsilon$$

because $g_j \searrow 0$ and $\lim_j m(g_j) = 0$ by (i). That proves (i)⇒(ii).   □

7.6 Notes for Chap. 7

Alexandroff [2] proved that, on certain spaces of bounded functions with the usual sup norm, every norm-bounded linear functional is represented as integral with respect to a finitely additive measure. A representing finitely additive measure exists under fairly general assumptions, and it is unique when the function space satisfies a certain normality condition.

The function space properties required in Alexandroff's construction may be described in the language of uniform structures: Let X be a uniform space and $\mathscr{B}$ the algebra of subsets of X generated by the cozero sets in X. If X is an Alexandroff space (Sect. 4.1), then every functional in $\mathfrak{M}_b(X)$ is represented by a unique finitely additive measure μ on $\mathscr{B}$ that is inner regular with respect to zero sets, in the sense that

$$|\mu|(A) = \sup\{|\mu|(Z) \mid Z \subseteq A \text{ and } Z \text{ is a zero set}\}$$

for all $A \in \mathscr{B}$.

Now if X is a general uniform space, it can be shown that there is an Alexandroff space X' on the same set of points such that X and X' have the same bounded cozero-continuous functions. The algebra $\mathscr{B}$ for X is the same as for X', $\mathsf{U_b}(X) \subseteq \mathsf{U_b}(X')$, and by the Hahn–Banach theorem P.9, every functional in $\mathfrak{M}_b(X)$ extends to a functional in $\mathfrak{M}_b(X')$. Thus every functional in $\mathfrak{M}_b(X)$ is represented by a finitely additive measure μ on $\mathscr{B}$; however, in general, μ need not be unique, which makes such a representation less useful.

Alexandroff (loc. cit.) also described σ-smooth and τ-smooth functionals (not under those names), and their integral representation, and representation by measures on a certain compactification of the underlying space. When Alexandroff's definitions are expressed in the language of uniform structures, the results in Sect. 7.1, when restricted to the class of Alexandroff spaces, follow from those in [2].

Theorems 7.9, 7.10 and 7.12 were proved by LeCam [121], who considered σ-smooth, τ-smooth and tight functionals on function spaces that include $\mathsf{U_b}(X)$ for an arbitrary uniform space X and described representation by measures on the uniform compactification. Theorem 7.14 is due to Frolík [64], [69].

The decomposition in Exercise 7.17 is a special case of a general result about projections in Dedekind-complete Riesz spaces [59, 353I]. The general result yields also analogous decompositions for $\mathfrak{M}_t(X)$, $\mathfrak{M}_\tau(X)$, $\mathfrak{M}_\sigma(X)$ and other bands in the Riesz space $\mathfrak{M}_b(X)$.

Example 7.19 was constructed by Zahradník [181]. Theorem 7.20 was proved by Berezanskiĭ [7], and a weaker version by Fedorova [50]. Theorem 7.22 is essentially due to Fedorova [52], who, however, stated it in terms of σ-smooth (not τ-smooth) functionals.

Theorem 7.25 is a generalization of a theorem in topological measure theory to which I return in Sect. 8.4. Another result in topological measure theory demonstrates that the converse of Theorem 7.22 is false (see Sect. 8.7).

I formulate Exercise 7.26 using measure-free cardinals so that Theorem 7.25 can be applied. However, in this application the theorem is used only for multiplicative functionals in $\mathfrak{M}_\sigma(X)$, and those are represented by $\{0,1\}$-valued measures. Hence the same approach with a modified version of Theorem 7.25 yields the following result of Rice [155, 2.8]: If X is a complete inversion-closed uniform space in which the cardinality of every uniformly discrete subset is non-measurable (i.e. not measurable by a $\{0,1\}$-valued measure), then cX is complete. The case $X = \mathsf{F}T$, where T is a completely regular topological space, is the Shirota theorem [81, 15.20]; its generalization for locally fine uniform spaces was proved by Isbell [100, VII.18].

The concept of a measure-fine space and the results in Sect. 7.5 are due to Frolík [69], [74], who proved also additional properties of $\mathsf{M}X$ and related M to other functors on the category of uniform spaces.

Research Problem 2.   Find an intrinsic characterization of those uniform spaces X for which $\mathfrak{M}_u(X) = \mathfrak{M}_\tau(X)$, or at least narrow the gap between the sufficient condition in Theorem 7.22 (supercompleteness) and the necessary condition in Exercise 7.23 (completeness).

Frolík's theory of measure-fine spaces (Sect. 7.5) offers a framework and tools for attacking this problem. Another promising approach using strict topologies was developed by Fedorova [54] (but beware: parts of Theorem 1 in [54] are disproved by Example 7.19).

For completely regular topological spaces T, the property $\mathfrak{M}_u(\mathsf{F}T) = \mathfrak{M}_\tau(\mathsf{F}T)$ along with closely related measure-compactness (Sects. 8.3 and 8.4) have been extensively studied in topological measure theory [11, 7.14] [60, Ch.43] [180]. ∎

Chapter 8
Instances of Uniform Measures

In this chapter I show how to obtain several familiar spaces of measures and measure-like functionals as $\mathfrak{M}_u(X)$ for suitably chosen uniform spaces X, and how to derive properties of spaces of measures from the results in Chap. 6. Although many of the theorems derived here are well known, it is noteworthy that they are all obtained as special cases of general theorems about uniform measures. This brings out parallels between various spaces of measures and indicates to what extent uniform measures are their common generalization.

For each instance of $\mathfrak{M}_u(X)$ in this chapter, I follow the same basic steps: First, I find a uniform space X such that $\mathfrak{M}_u(X)$ is naturally isomorphic to the given space of measures, and identify an interesting class of $UEB(X)$ subsets of $U_b(X)$. The rest is then a straightforward application of the results in Chap. 6.

8.1 Sequentially Uniform Measures

For any uniform space X, define $\mathfrak{M}_{u\sigma}(X)$ to be the space of those linear functionals on $U_b(X)$ that are sequentially X-pointwise continuous on the $UEB(X)$ sets. In other words, $\mathfrak{M}_{u\sigma}(X)$ is the space of the functionals $\mathfrak{m} \in \mathfrak{M}_b(X)$ such that if $\Delta \in UP(X)$ and if $\{f_j\}_j$ is a sequence in $BLip_b(\Delta)$ and $\lim_j f_j(x) = 0$ for all $x \in X$, then $\lim_j \mathfrak{m}(f_j) = 0$.

Obviously $\mathfrak{M}_\sigma(X) \subseteq \mathfrak{M}_{u\sigma}(X)$ and $\mathfrak{M}_u(X) \subseteq \mathfrak{M}_{u\sigma}(X)$.

Lemma 8.1. *Let X be a uniform space, $\Delta \in UP(X)$, and let $\{f_j\}_j$ be a sequence in $BLip_b(\Delta)$ converging X-pointwise to 0. Then the set $\{f_j \mid j \in \omega\}$ is $UEB(p_1X)$.*

Proof. Define the pseudometric Δ' on X by

$$\Delta'(x,y) := \sup_j |f_j(x) - f_j(y)| \quad \text{for} \quad x,y \in X.$$

Then $\Delta' \in UP(p_1X)$ and $f_j \in BLip_b(\Delta')$ for $j \in \omega$. □

J. Pachl, *Uniform Spaces and Measures*, Fields Institute Monographs 30,
DOI 10.1007/978-1-4614-5058-0_9,
© Springer Science+Business Media New York 2013

Theorem 8.2. $\mathfrak{M}_{u\sigma}(X) = \mathfrak{M}_u(p_1 X)$ *for every uniform space X.*

Proof. Take any $\Delta \in UP(p_1 X)$. Then X has a countable Δ-dense subset; hence the X-pointwise topology on $BLip_b(\Delta)$ is metrizable. If $\mathfrak{m} \in \mathfrak{M}_{u\sigma}(X)$, then $\mathfrak{m}$ is sequentially continuous on $BLip_b(\Delta)$, hence continuous on $BLip_b(\Delta)$. That proves $\mathfrak{M}_{u\sigma}(X) \subseteq \mathfrak{M}_u(p_1 X)$.

The opposite inclusion follows from Lemma 8.1.   □

Corollary 8.3. $\mathfrak{M}_\sigma(X) \subseteq \mathfrak{M}_u(p_1 X)$ *for every uniform space X.*   □

By Theorem 8.2, spaces $\mathfrak{M}_{u\sigma}(X)$ have all the properties of general $\mathfrak{M}_u(X)$ spaces. In particular, the space $\mathfrak{M}_{u\sigma}(X)$ is a band in $\mathfrak{M}_b(X)$, and the results in Chap. 6 yield the following properties of the $U_b(X)$-weak topology on $\mathfrak{M}_{u\sigma}(X)$:

Theorem 8.4. *Let X be any uniform space.*

1. *Let* $\{\mathfrak{m}_\gamma\}_\gamma$ *be a net in* $\mathfrak{M}_{u\sigma}(X)$, $\mathfrak{m} \in \mathfrak{M}_{u\sigma}(X)$ *and* $\lim_\gamma \mathfrak{m}_\gamma(f) = \mathfrak{m}(f)$ *for every* $f \in U_b(X)$. *If* $\mathfrak{m}_\gamma \geq 0$ *or* $\|\mathfrak{m}_\gamma\| = \|\mathfrak{m}\|$ *for all* γ *or if* $\{\mathfrak{m}_\gamma\}_\gamma$ *is a sequence, then*

$$\lim_\gamma \sup_j |\mathfrak{m}_\gamma(f_j) - \mathfrak{m}(f_j)| = 0$$

   for every $\Delta \in UP(X)$ *and every sequence of functions* $f_j \in BLip_b(\Delta)$ *that converges X-pointwise to 0.*

2. *A set* $\mathfrak{A}$ *is relatively compact in the space* $\mathfrak{M}_{u\sigma}(X)$ *with the* $U_b(X)$-weak topology *if and only if* $\mathfrak{A}$ *is* $\|\cdot\|$ *bounded and*

$$\lim_j \sup_{\mathfrak{m} \in \mathfrak{A}} |\mathfrak{m}(f_j)| = 0$$

   for every $\Delta \in UP(X)$ *and every sequence of functions* $f_j \in BLip_b(\Delta)$ *that converges X-pointwise to 0.*

3. *The space* $\mathfrak{M}_{u\sigma}(X)$ *is* $U_b(X)$-*weakly sequentially complete.*

Proof. Clearly $U_b(X) = U_b(p_1 X)$. If $\Delta \in UP(X)$, $f_j \in BLip_b(\Delta)$ for all $j \in \omega$ and $\lim_j f_j(x) = 0$ for all $x \in X$, then $\mathscr{F} := \{f_j \mid j \in \omega\}$ is a $UEB(p_1 X)$ set by Lemma 8.1. If a net in the space $\mathfrak{M}_b(X) = \mathfrak{M}_b(p_1 X)$ converges in the $UEB(p_1 X)$ topology, then it converges uniformly on $\mathscr{F}$.

The $U_b(X)$-weak topology and the $UEB(p_1 X)$ topology agree on the positive cone $\mathfrak{M}_u(p_1 X)^+$ (Corollary 6.13), on $\|\cdot\|$ spheres in $\mathfrak{M}_u(p_1 X)$ (Corollary 6.15) and on $U_b(X)$-weakly convergent sequences in $\mathfrak{M}_u(p_1 X)$ (Corollary 6.18). Part 1 now follows from Theorem 8.2.

To prove Part 2, take any set $\mathfrak{A} \subseteq \mathfrak{M}_{u\sigma}(X)$. If $\mathfrak{A}$ is $U_b(X)$-weakly relatively compact in $\mathfrak{M}_{u\sigma}(X) = \mathfrak{M}_u(p_1 X)$, then it is $\|\cdot\|$ bounded and X-pointwise equicontinuous on $\mathscr{F}$ (Theorem 6.16); hence $\lim_j \sup_{\mathfrak{m} \in \mathfrak{A}} |\mathfrak{m}(f_j)| = 0$.

Conversely, assume that $\mathfrak{A}$ is $\|\cdot\|$ bounded and $\lim_j \sup_{\mathfrak{m} \in \mathfrak{A}} |\mathfrak{m}(f_j)| = 0$ whenever $\Delta \in UP(X)$, $f_j \in BLip_b(\Delta)$ for $j \in \omega$ and $\lim_j f_j(x) = 0$ for all $x \in X$. Then $\mathfrak{A}$ is $U_b(X)$-weakly relatively compact in $\mathfrak{M}_b(X)$, and any element $\mathfrak{n}$ of the $U_b(X)$-weak closure of $\mathfrak{A}$ in $\mathfrak{M}_b(X)$ satisfies $\lim_j |\mathfrak{n}(f_j)| \leq \lim_j \sup_{\mathfrak{m} \in \mathfrak{A}} |\mathfrak{m}(f_j)| = 0$, which means that $\mathfrak{n} \in \mathfrak{M}_{u\sigma}(X)$. Thus $\mathfrak{A}$ is relatively compact in $\mathfrak{M}_{u\sigma}(X)$ with the $U_b(X)$-weak topology.

Part 3 is a special case of Theorem 6.19.   □

By Theorem 8.2, $\mathfrak{M}_u(X) = \mathfrak{M}_{u\sigma}(X)$ whenever $X = p_1 X$. If X is a discrete space such that $|X|$ is not measure-free, then $\mathfrak{M}_u(X) \neq \mathfrak{M}_{u\sigma}(X)$ because $\mathfrak{M}_\sigma(X) \neq \mathfrak{M}_u(X)$ by Theorem 7.25. There are uniform spaces X for which $\mathfrak{M}_u(X) \neq \mathfrak{M}_{u\sigma}(X)$ even if all cardinals are measure-free, although the construction is then more difficult; see the notes in Sect. 8.7.

8.2   Measures on Abstract σ-Algebras

For any point-separating σ-algebra Σ of subsets of a non-empty set S, let X be the uniform space constructed in Example 2.3: The point set of X is S, and the uniformity $\mathsf{UP}(X) = \mathscr{U}(\Sigma)$ is induced by the mappings $\varphi \colon S \to \mathbb{N}$ such that $\varphi^{-1}(j) \in \Sigma$ for every $j \in \mathbb{N}$, where $\mathbb{N}$ is the discrete uniform space on the set ω. A real-valued function on X is uniformly continuous if and only if it is Σ-measurable, and in particular, $U_b(X) = \ell_\infty(S, \Sigma)$.

Clearly X is inversion-closed and $X = p_1 X$, so that $\mathfrak{M}_u(X) = \mathfrak{M}_\sigma(X) = \mathfrak{M}_{u\sigma}(X)$ by Theorems P.33, 7.21, 7.25 and 8.2. The space of finite linear combinations of characteristic functions I_E, $E \in \Sigma$, is $\|\cdot\|_S$ dense in $\ell_\infty(S, \Sigma)$. Define the mapping $\vartheta \colon \mathfrak{M}_b(X) \to \mathbb{R}^\Sigma$ by $\vartheta(\mathfrak{m})(E) := \mathfrak{m}(I_E)$ for $\mathfrak{m} \in \mathfrak{M}_b(X)$, $E \in \Sigma$. The space $\mathfrak{M}_u(X) = \mathfrak{M}_\sigma(X)$ naturally identifies, via the mapping ϑ, with the space of all measures on Σ.

Lemma 8.5. *Let Σ be a point-separating σ-algebra of subsets of a non-empty set S, and $f_j \in \ell_\infty(S, \Sigma)$ for $j \in \omega$. If $\lim_j f_j(x) = 0$ for all $x \in S$, then the set $\{f_j \mid j \in \omega\}$ is uniformly equicontinuous with respect to the uniformity $\mathscr{U}(\Sigma)$.*

Proof. Take any $\varepsilon > 0$. The sets $A_k := \{x \in S \mid \forall j \geq k \, [|f_j(x)| < \varepsilon]\}$, $k \in \omega$, are in Σ, and their union is S. Since the functions f_j are Σ-measurable, each A_k may be written as a countable union $A_k = \bigcup_{i \in \omega} B_{k,i}$ of sets $B_{k,i} \in \Sigma$ such that if $x, y \in B_{k,i}$ and $j < k$, then $|f_j(x) - f_j(y)| < \varepsilon$. Then also $|f_j(x) - f_j(y)| < 2\varepsilon$ for $x, y \in B_{k,i}$ and any $j \in \omega$. By renumbering the sets $B_{k,i}$, we get $E_i \in \Sigma$, $i \in \omega$, such that $S = \bigcup_i E_i$, and if $x, y \in E_i$, $i \in \omega$, then $|f_j(x) - f_j(y)| < 2\varepsilon$. Hence there is a mapping $\varphi \colon S \to \mathbb{N}$ such that $\varphi^{-1}(z) \in \Sigma$ for every $z \in \mathbb{N}$ and $|f_j(x) - f_j(y)| < 2\varepsilon$ whenever $j \in \omega$, $x, y \in S$ and $\varphi(x) = \varphi(y)$.   □

Theorem 8.6. *Let Σ be a σ-algebra of subsets of a non-empty set S.*

1. Let $\{\mu_\gamma\}_\gamma$ be a net of measures on Σ, and let μ be a measure on Σ such that $\lim_\gamma \int f \, d\mu_\gamma = \int f \, d\mu$ for every $f \in \ell_\infty(S, \Sigma)$. If $\mu_\gamma \geq 0$ or $|\mu_\gamma|(S) = |\mu|(S)$ for all γ or if $\{\mu_\gamma\}_\gamma$ is a sequence, then

$$\lim_{\gamma} \sup_{j} \left| \int f_j \, d\mu_\gamma - \int f_j \, d\mu \right| = 0$$

for every $\|\cdot\|_S$ bounded sequence $\{f_j\}_j$ in $\ell_\infty(S,\Sigma)$ that converges X-pointwise to 0.

2. These properties of a $\|\cdot\|_{\mathrm{TV}}$ bounded set M of measures on Σ are equivalent:

   (i) M is relatively compact in the space of measures on Σ in the $\ell_\infty(S,\Sigma)$-weak topology.
   (ii) If $\{f_j\}_j$ is a $\|\cdot\|_S$ bounded sequence in $\ell_\infty(S,\Sigma)$ and $\lim_j f_j(x) = 0$ for all $x \in X$, then

$$\lim_{j} \sup_{\mu \in M} \left| \int f_j \, d\mu \right| = 0.$$

   (iii) If $\{f_j\}_j$ is a sequence in $\ell_\infty(S,\Sigma)$ and $f_j \searrow 0$, then

$$\lim_{j} \sup_{\mu \in M} \left| \int f_j \, d\mu \right| = 0.$$

3. If $\{\mu_j\}_j$ is a sequence of measures on Σ and the limit $\lim_j \int f \, d\mu_j$ exists for every $f \in \ell_\infty(S,\Sigma)$, then $\mu(E) := \lim_j \mu_j(E)$ defines a measure μ on Σ.

Proof. Assume, without loss of generality, that Σ separates the points of S. Let again X be the uniform space constructed from Σ as in Example 2.3. If $\{f_j\}_j$ is a $\|\cdot\|_S$ bounded sequence in $\ell_\infty(S,\Sigma)$ such that $\lim_j f_j(x) = 0$ for all $x \in X$, then the set $\mathscr{F} := \{f_j \mid j \in \omega\}$ is UEB(X) by Lemma 8.5. Thus Part 1 and implication (i)$\Rightarrow$(ii) in Part 2 follow from the corresponding parts in Theorem 8.4.

Implication (ii)$\Rightarrow$(iii) in Part 2 is obvious. Now assume that (iii) holds, and let $\mathfrak{A} \subseteq \mathfrak{M}_u(X)$ be the set of functionals represented by the measures in M. Then $\mathfrak{A}$ is $U_b(X)$-weakly relatively compact in $\mathfrak{M}_b(X)$, and any element $\mathfrak{n}$ of the $U_b(X)$-weak closure of $\mathfrak{A}$ in $\mathfrak{M}_b(X)$ satisfies $\lim_j |\mathfrak{n}(f_j)| \le \lim_j \sup_{\mathfrak{m} \in \mathfrak{A}} |\mathfrak{m}(f_j)| = 0$, which means that $\mathfrak{n} \in \mathfrak{M}_\sigma(X) = \mathfrak{M}_u(X)$. Thus $\mathfrak{A}$ is relatively compact in $\mathfrak{M}_u(X)$ with the $U_b(X)$-weak topology, so that (i) holds.

Part 3 is a special case of Theorem 6.19. □

Exercise 8.7. Strengthen Theorem 8.6 as follows:

1. Part 1 still holds if the condition $\lim_\gamma \int f \, d\mu_\gamma = \int f \, d\mu$ for $f \in \ell_\infty(S,\Sigma)$ is replaced by $\lim_\gamma \mu_\gamma(E) = \mu(E)$ for $E \in \Sigma$.
2. In Part 2, a set of measures is relatively compact in the $\ell_\infty(S,\Sigma)$-weak topology if and only if it is relatively compact in the Σ-pointwise topology. Condition (iii) in Part 2 is equivalent to $\lim_j \sup_{\mu \in M} |\mu(E_j)| = 0$ for every sequence of sets $E_j \in \Sigma$ such that $E_j \searrow \emptyset$.
3. Part 3 still holds if the condition $\lim_j \int f \, d\mu_j = \int f \, d\mu$ for $f \in \ell_\infty(S,\Sigma)$ is replaced by $\lim_j \mu_j(E) = \mu(E)$ for $E \in \Sigma$. ∎

Exercise 8.8. Let $\vartheta \colon \mathfrak{M}_b(X) \to \mathbb{R}^\Sigma$ be the mapping defined before Lemma 8.5. Show that ϑ is a bijection from $\mathfrak{M}_b(X)$ onto the space of finitely additive measures on Σ.   ∎

8.3   Baire Measures on Completely Regular Spaces

Let T be a completely regular topological space and consider the uniform space $X := \mathsf{c}\mathsf{F}T$; in other words, X is the set T with the uniformity induced by $C(T)$.

The cardinality of every uniformly discrete subset of X is measure-free by Theorems P.33 and 2.9, and X is inversion-closed because $U(X) = C(T)$. In fact, the following comments apply more generally with any uniform space X such that $U(X) = C(T)$ and the cardinality of every uniformly discrete subset of X is measure-free. For example, we may take $X := \mathsf{p}_\alpha\mathsf{F}T$ as long as every cardinal $< \aleph_\alpha$ is measure-free.

As X is inversion-closed, $\mathfrak{M}_{u\sigma}(X) \subseteq \mathfrak{M}_\sigma(X)$ by Theorem 4.7. As the cardinality of every uniformly discrete subset of X is measure-free, $\mathfrak{M}_\sigma(X) \subseteq \mathfrak{M}_u(X)$ by Theorem 7.25. Thus

$$\mathfrak{M}_u(X) = \mathfrak{M}_{u\sigma}(X) = \mathfrak{M}_\sigma(X) = \mathfrak{M}_\sigma(\mathsf{F}T),$$

and the space $\mathfrak{M}_u(X)$ naturally identifies with the space of Baire measures on $\mathsf{F}T$ (known as *Baire measures on T* in topological measure theory)—that is, measures on the σ-algebra $\sigma(C(T))$. In the identification, the $U_b(X)$-weak topology on $\mathfrak{M}_u(X)$ becomes the $C_b(T)$-weak topology on the space of Baire measures on $\mathsf{F}T$ that arises from the duality $\langle \mu, f \rangle = \int f \, d\mu$ for a measure μ and $f \in C_b(T)$.

The uniform compactification of X is the Čech–Stone compactification βT. Recall from Sect. 7.2 that every $\|\cdot\|_T$ continuous linear functional $\mathfrak{m}$ on the space $U_b(X) = C_b(T)$ is represented by a unique tight Borel measure $\rho_\mathfrak{m}$ on βT.

Theorem 8.9. *Let T be a completely regular space. A $\|\cdot\|_T$ continuous linear functional $\mathfrak{m}$ on the space $C_b(T)$ is σ-smooth if and only if $|\rho_\mathfrak{m}|(Z) = 0$ for every zero set $Z \subseteq \beta T$ such that $Z \cap T = \emptyset$.*

Proof. This is a special case of Theorem 7.12.   □

A completely regular space T is called *measure-compact* iff $\mathfrak{M}_\sigma(\mathsf{F}T) = \mathfrak{M}_\tau(\mathsf{F}T)$. This is an important concept in topological measure theory. By the result in Exercise 7.23, every measure-compact space is realcompact.

Exercise 8.10. The *Sorgenfrey line* is a topological space on the set $\mathbb{R}$. A base of its topology is the set of half-open intervals $[a,b) = \{x \in \mathbb{R} \mid a \leq x < b\}$, $a,b \in \mathbb{R}$. The *Sorgenfrey plane* is the topological product of two copies of the Sorgenfrey line.

Prove that the Sorgenfrey plane is realcompact but not measure-compact.   ∎

Theorem 8.11. *Let T be any completely regular topological space.*

1. *Let $\{\mu_\gamma\}_\gamma$ be a net of measures on the σ-algebra $\sigma(C(T))$, and μ another measure on $\sigma(C(T))$ such that $\lim_\gamma \int f \, d\mu_\gamma = \int f \, d\mu$ for every $f \in C_b(T)$. If $\mu_\gamma \geq 0$ or $|\mu_\gamma|(T) = |\mu|(T)$ for all γ or if $\{\mu_\gamma\}_\gamma$ is a sequence, then*

$$\lim_\gamma \sup_j \left| \int f_j \, d\mu_\gamma - \int f_j \, d\mu \right| = 0$$

   for every sequence of functions $f_j \in C_b(T)$ such that $f_j \searrow 0$.
2. *A set M is relatively compact in the space of measures on $\sigma(C(T))$ with the $C_b(T)$-weak topology if and only if M is $\|\cdot\|_{\mathrm{TV}}$ bounded and*

$$\lim_j \sup_{\mu \in M} \left| \int f_j \, d\mu \right| = 0$$

   for every sequence of functions $f_j \in C_b(T)$ such that $f_j \searrow 0$.
3. *If $\{\mu_j\}_j$ is a sequence of measures on $\sigma(C(T))$ and $\mathrm{m}(f) := \lim_j \int f \, d\mu_j$ exists for every $f \in C_b(T)$, then there is a measure μ on the σ-algebra $\sigma(C(T))$ such that $\mathrm{m}(f) = \int f \, d\mu$ for every $f \in C_b(T)$.*

Proof. This is similar to the proof of Theorem 8.6. If $f_j \in C_b(T)$ for $j \in \omega$ and $f_j \searrow 0$, then the set $\mathscr{F} := \{f_j \mid j \in \omega\}$ is $\mathsf{UEB}(\mathsf{cF}T)$ by Theorem 4.7. Thus Part 1 and the necessity of the condition in Part 2 follow from the corresponding parts in Theorem 8.4.

To prove that the condition in Part 2 is also sufficient, let $\mathfrak{A} \subseteq \mathfrak{M}_u(\mathsf{cF}T)$ be the set of functionals represented by the measures in M, and assume that $\mathfrak{A}$ is $\|\cdot\|$ bounded and

$$\lim_j \sup_{\mathrm{m} \in \mathfrak{A}} |\mathrm{m}(f_j)| = 0$$

for every sequence of functions $f_j \in C_b(T)$ such that $f_j \searrow 0$. Then $\mathfrak{A}$ is $\mathsf{U}_b(\mathsf{cF}T)$-weakly relatively compact in $\mathfrak{M}_b(\mathsf{cF}T)$, and any element n of the $\mathsf{U}_b(\mathsf{cF}T)$-weak closure of $\mathfrak{A}$ in $\mathfrak{M}_b(\mathsf{cF}T)$ satisfies $\lim_j |\mathrm{n}(f_j)| \leq \lim_j \sup_{\mathrm{m} \in \mathfrak{A}} |\mathrm{m}(f_j)| = 0$, which means that $\mathrm{n} \in \mathfrak{M}_\sigma(\mathsf{cF}T) = \mathfrak{M}_u(\mathsf{cF}T)$. Thus $\mathfrak{A}$ is relatively compact in $\mathfrak{M}_u(\mathsf{cF}T)$ with the $\mathsf{U}_b(\mathsf{cF}T)$-weak topology.

Part 3 is a special case of Theorem 6.19. $\square$

To what extent may Theorem 8.11 be generalized to $\mathfrak{M}_\sigma(X)$ for an arbitrary uniform space X? First, observe that to define $\mathfrak{M}_\sigma(X)$, it is not enough to know the topology of X. Although the spaces $\mathfrak{M}_t(X)$ and $\mathfrak{M}_\tau(X)$ are determined, up to a natural isomorphism, by the topology of X (see Corollaries 5.5 and 7.8 and the comments after them), the same does not hold for the space $\mathfrak{M}_\sigma(X)$:

Exercise 8.12. Find a uniform space X and $\mathrm{m} \in \mathfrak{M}_\sigma(X)$ for which there is no $\tilde{\mathrm{m}} \in \mathfrak{M}_\sigma(\mathsf{F}X)$ such that $\mathrm{m}(f) = \tilde{\mathrm{m}}(f)$ for all $f \in \mathsf{U}_b(X)$.   ∎

Now let X be an arbitrary uniform space. By combining Theorem 7.4 and the construction in Sect. 8.2, we get a uniform space Y on the set X such that Y is finer than X and the identity mapping $Y \to X$ yields a natural isomorphism of the Banach lattices $\mathfrak{M}_u(Y)$ and $\mathfrak{M}_\sigma(X)$. However, there need not exist such Y for which $U_b(X) = U_b(Y)$ and $\mathfrak{M}_\sigma(X) = \mathfrak{M}_u(Y)$:

Exercise 8.13. As in Example 7.18, consider the compact space $[0,1]$, and let X be the uniform subspace of rational numbers in $[0,1]$. Show that the space $\mathfrak{M}_\sigma(X)$ is not $U_b(X)$-weakly sequentially complete. Thus there is no space Y on the set X such that $U_b(X) = U_b(Y)$ and $\mathfrak{M}_\sigma(X) = \mathfrak{M}_u(Y)$.   ∎

8.4   Separable Measures

Let T be a completely regular topological space. A measure μ on the σ-algebra $\sigma(C(T))$ is called *separable* iff for every continuous mapping $\varphi : T \to Z$ to a metrizable topological space Z there is a closed separable set $Z_0 \subseteq Z$ such that $|\mu|(T \setminus \varphi^{-1}(Z_0)) = 0$.

Theorem 8.14. *If T is a paracompact topological space, then $\mathfrak{M}_u(\mathsf{F}T) = \mathfrak{M}_\tau(\mathsf{F}T)$.*

Proof. Apply Theorems 4.20 and 7.22.   □

By the next theorem, the space of separable measures on $\sigma(C(T))$ naturally identifies with $\mathfrak{M}_u(\mathsf{F}T)$.

Theorem 8.15. *Let T be a completely regular topological space, $\mathfrak{m} \in \mathfrak{M}_\sigma(\mathsf{F}T)$, and let μ be the unique measure on $\sigma(C(T))$ such that $\mathfrak{m}(f) = \int f \, d\mu$ for all $f \in C_b(T)$. Then the measure μ is separable if and only if $\mathfrak{m} \in \mathfrak{M}_u(\mathsf{F}T)$.*

Proof. Assume that μ is separable. Take any metric space Z and any uniformly continuous mapping $\varphi : \mathsf{F}T \to Z$. There is a closed separable $Z_0 \Subset Z$ such that $|\mu|(T \setminus \varphi^{-1}(Z_0)) = 0$. Define a measure μ_0 on $\sigma(U(Z_0))$ and $\mathfrak{m}_0 \in \mathfrak{M}_\sigma(Z_0)$ by

$$\mu_0(E) := \mu(\varphi^{-1}(E)) \quad \text{for } E \in \sigma(U(Z_0)),$$

$$\mathfrak{m}_0(f) := \int f \, d\mu_0 \quad \text{for } f \in U_b(Z_0).$$

Let $\imath : Z_0 \hookrightarrow Z$ be the inclusion mapping. Since Z_0 is separable, $\mathfrak{m}_0 \in \mathfrak{M}_u(Z_0)$ by Theorem 7.25, and $\varphi(\mathfrak{m}) = \imath(\mathfrak{m}_0) \in \mathfrak{M}_u(Z)$. This proves that the image of $\mathfrak{m}$ under every uniformly continuous mapping from $\mathsf{F}T$ to a metric space Z is in $\mathfrak{M}_u(Z)$, and $\mathfrak{m} \in \mathfrak{M}_u(\mathsf{F}T)$ by Part 3 of Theorem 6.20.

To prove the converse, assume that $\mathfrak{m} \in \mathfrak{M}_u(\mathsf{F}T)$, and take any continuous mapping $\varphi : T \to Z$ to a metric space Z. Then φ is uniformly continuous from $\mathsf{F}T$ to $\mathsf{F}Z$ (Theorem 1.26), and $\varphi(\mathfrak{m}) \in \mathfrak{M}_\tau(\mathsf{F}Z)$ by Theorem 8.14. Let μ_Z be the τ-additive measure on $Bo(Z)$ that represents $\varphi(m)$ (Theorem 7.6). The support Z_0 of μ_Z is separable by Theorem P.27, and $|\mu|(T \setminus \varphi^{-1}(Z_0)) = 0$.   □

Theorem 8.16. *Two properties of a completely regular topological space T are equivalent:*

(i) *Every measure on the σ-algebra $\sigma(C(T))$ is separable.*
(ii) *If Δ is a continuous pseudometric on T and S is a Δ-discrete subset of T, then the cardinality of S is measure-free.*

Thus the statement that every measure on $\sigma(C(T))$ is separable is consistent with the ZFC set theory.

Proof. A pseudometric Δ on T is continuous if and only if $\Delta \in UP(FT)$. Apply Theorems 7.25 and 8.15.   □

In view of Theorem 8.15, the results in Chap. 6 when specialized to the uniform space FT become statements about separable measures on T:

Theorem 8.17. *Let T be any completely regular topological space.*

1. *Let $\{\mu_\gamma\}_\gamma$ be a net of separable measures on T, and μ a separable measure on T such that $\lim_\gamma \int f \, d\mu_\gamma = \int f \, d\mu$ for all $f \in C_b(T)$. If $\mu_\gamma \geq 0$ or $|\mu_\gamma|(T) = |\mu|(T)$ for all γ or if $\{\mu_\gamma\}_\gamma$ is a sequence, then*

$$\lim_\gamma \sup_{f \in \mathscr{F}} \left| \int f \, d\mu_\gamma - \int f \, d\mu \right| = 0$$

   for every $\|\cdot\|_T$ bounded equicontinuous set $\mathscr{F} \subseteq C_b(T)$.
2. *A set M is relatively compact in the space of separable measures on T with the $C_b(T)$-weak topology if and only if M is $\|\cdot\|_{TV}$ bounded and*

$$\lim_\gamma \sup_{\mu \in M} \left| \int f_\gamma \, d\mu \right| = 0$$

   for every $\|\cdot\|_T$ bounded equicontinuous net of functions $f_\gamma \in C_b(T)$ that converges T-pointwise to 0.
3. *If $\{\mu_j\}_j$ is a sequence of separable measures on T and $m(f) := \lim_j \int f \, d\mu_j$ exists for every $f \in C_b(T)$, then there is a separable measure μ on T such that $m(f) = \int f \, d\mu$ for every $f \in C_b(T)$.*

Proof. A subset of $C_b(T)$ is $\|\cdot\|_T$ bounded and equicontinuous on T if and only if it is $UEB(FT)$. Part 1 follows from Corollaries 6.13, 6.15 and 6.18, Part 2 from Theorem 6.16 and Part 3 from Theorem 6.19.   □

8.5   Tight Measures on Locally Compact Groups

Results in prior chapters yield theorems about tight Borel measures on a locally compact group G in duality with the space $U_b(rG) = LUC(G)$.

Theorem 8.18. *Let G be a locally compact group. The following properties of $\mathfrak{m} \in \mathfrak{M}_b(rG)$ are equivalent:*

(i) $\mathfrak{m} \in \mathfrak{M}_t(rG)$.
(ii) $\mathfrak{m} \in \mathfrak{M}_u(rG)$.
(iii) $\mathfrak{m}$ *is G-pointwise continuous on $\overline{orb}(f)$ for every $f \in U_b(rG)$.*

Proof. (i) $\Leftrightarrow$ (ii) by Theorem 7.20. By Lemma 3.19 the set $\overline{orb}(f)$ is $UEB(rG)$ for every $f \in U_b(rG)$; hence (ii)$\Rightarrow$(iii).

To prove (iii)$\Rightarrow$(ii), it is enough to consider the case where G is not compact. Then G is ambitable by Part 4 of Theorem 3.35, so that every $UEB(rG)$ set is contained in $\overline{orb}(f)$ for some $f \in U_b(rG)$.   □

Theorem 8.19. *Let G be a locally compact group.*

1. *Let $\{\mu_\gamma\}_\gamma$ be a net of tight Borel measures on G, and μ another tight Borel measure on G such that $\lim_\gamma \int f \, d\mu_\gamma = \int f \, d\mu$ for every $f \in U_b(rG)$. If $\mu_\gamma \geq 0$ or $|\mu_\gamma|(G) = |\mu|(G)$ for all γ or if $\{\mu_\gamma\}_\gamma$ is a sequence, then*

$$\limsup_\gamma \left\{ \left| \int f \, d\mu_\gamma - \int f \, d\mu \right| \ \middle| \ f \in \overline{orb}(g) \right\} = 0$$

   for every $g \in U_b(rG)$.
2. *A set M is relatively compact in the space of tight Borel measures on G with the $U_b(rG)$-weak topology if and only if M is $\|\cdot\|_{TV}$-bounded, and*

$$\limsup_{\gamma \ \mu \in M} \left| \int f_\gamma \, d\mu \right| = 0$$

   for every $g \in U_b(rG)$ and for every net of functions $f_\gamma \in \overline{orb}(g)$ that converges G-pointwise to 0.
3. *If $\{\mu_j\}_j$ is a sequence of tight Borel measures on G and $\mathfrak{m}(f) := \lim_j \int f \, d\mu_j$ exists for every $f \in U_b(rG)$, then there is a tight Borel measure μ on G such that $\mathfrak{m}(f) = \int f \, d\mu$ for every $f \in U_b(rG)$.*

Proof. By Lemma 3.19, the orbit closure $\overline{orb}(g)$ is $UEB(rG)$ for every $g \in U_b(rG)$, and by Part 4 of Theorem 3.35, either G is compact or every $UEB(rG)$ set is contained in an orbit closure. Part 1 follows from Corollaries 6.13, 6.15 and 6.18, Part 2 from Theorem 6.16 and Part 3 from Theorem 6.19.   □

8.6 Cylindrical Measures

A cylindrical measure is a particular projective system of tight Borel measures on finite-dimensional vector spaces.

Definition 8.20. Let E be a locally convex space. Let Γ_E denote the directed set of all closed vector subspaces of E of finite codimension ordered by $\supseteq$. For $\gamma \in \Gamma_E$, E/γ is the finite-dimensional quotient space with its usual uniform structure and $\varphi_\gamma \colon E \to E/\gamma$ is the quotient mapping. In addition, $\varphi_{\beta\gamma} \colon E/\gamma \to E/\beta$ is the canonical surjective mapping such that $\varphi_\beta = \varphi_{\beta\gamma} \circ \varphi_\gamma$ for $\beta, \gamma \in \Gamma_E$, $\beta \supseteq \gamma$.

A *cylindrical measure on E* is an $\mathfrak{M}_b$-projective system $\langle \Gamma_E, E/\gamma, \varphi_{\beta\gamma}, \mathfrak{m}_\gamma \rangle$ for which there is $r \in \mathbb{R}^+$ such that $\mathfrak{m}_\gamma \in \mathfrak{M}_t(E/\gamma)$ and $\|\mathfrak{m}_\gamma\| \leq r$ for all $\gamma \in \Gamma_E$.   ∎

The name cylindrical measure comes from the representation by finitely additive measures on the algebra of "cylinders" $\varphi_\gamma^{-1}(A) \subseteq E$, where $\gamma \in \Gamma_E$, $A \in \mathsf{Bo}(E/\gamma)$. This finitely additive measure need not be a (σ-additive) measure.

Theorem 8.21. *Let E be a locally convex space, and let Γ_E, E/γ, φ_γ and $\varphi_{\beta\gamma}$ be as in Definition 8.20.*

1. *If $\mathfrak{m} \in \mathfrak{M}_u(wE)$, then $\langle \Gamma_E, E/\gamma, \varphi_{\beta\gamma}, \varphi_\gamma(\mathfrak{m}) \rangle$ is a cylindrical measure.*
2. *If $\langle \Gamma_E, E/\gamma, \varphi_{\beta\gamma}, \mathfrak{m}_\gamma \rangle$ is a cylindrical measure on E, then there is a unique $\mathfrak{m} \in \mathfrak{M}_u(wE)$ such that $\mathfrak{m}_\gamma = \varphi_\gamma(\mathfrak{m})$ for all $\gamma \in \Gamma_E$.*

Proof. The mappings $\varphi_\gamma \colon wE \to E/\gamma$ are uniformly continuous. If $\mathfrak{m} \in \mathfrak{M}_u(wE)$, then $\varphi_\gamma(\mathfrak{m}) \in \mathfrak{M}_u(E/\gamma) = \mathfrak{M}_t(E/\gamma)$ for every γ. Since $\varphi_\beta = \varphi_{\beta\gamma} \circ \varphi_\gamma$ for $\beta \supseteq \gamma$, it follows that $\langle \Gamma_E, E/\gamma, \varphi_{\beta\gamma}, \varphi_\gamma(\mathfrak{m}) \rangle$ is an $\mathfrak{M}_b$-projective system. That proves part 1.

To prove part 2, note that the uniformity of wE is induced by the mappings $\varphi_\gamma \colon wE \to E/\gamma$, $\gamma \in \Gamma_E$, and apply Theorems 6.20 and 6.29.   □

By Theorem 8.21, cylindrical measures on E may be identified with uniform measures on wE, so that again all the general results about spaces $\mathfrak{M}_u(X)$ apply to the space of cylindrical measures. Moreover, the identification shows that, although cylindrical measures are defined using the linear structure of E, they only depend on the uniform structure of wE. That suggests a natural definition of the image of a cylindrical measure under certain non-linear mappings and a definition of cylindrical measures on infinite-dimensional manifolds modelled on locally convex spaces.

8.7 Notes for Chap. 8

The space $\mathfrak{M}_{u\sigma}(X)$ in Sect. 8.1 was defined by Ferri and Neufang [55] (using a different notation) for $X = rG$, where G is a topological group. They also proved a version of Theorem 8.2: If G is an $\aleph_0$-bounded topological group, then $\mathfrak{M}_{u\sigma}(rG) = \mathfrak{M}_u(rG)$. Corollary 8.3 is due to Fedorova [53, Th.2.1].

As I note at the end of Sect. 8.1, one can find a uniform space X such that $\mathfrak{M}_u(X) \neq \mathfrak{M}_{u\sigma}(X)$ even if all cardinals are measure-free. Pelant [142] constructed a complete uniform space X for which $p_1 X$ is not complete. For any point $x \in \widehat{p_1 X} \setminus X$, we have $\partial(x) \in \mathfrak{M}_u(p_1 X) \setminus \mathfrak{M}_u(X)$ and therefore $\mathfrak{M}_u(X) \neq \mathfrak{M}_{u\sigma}(X)$.

The description of measures on a σ-algebra as uniform measures (Sect. 8.2) is pointed out by Deaibes [36, 4.5] and Frolík [72]. The statements in Theorem 8.6 and Exercise 8.7 are among basic results of abstract measure theory. They are discussed in more detail in Bogachev [11, 4.6] and Dunford and Schwartz [45, IV.9]. Sets of measures satisfying condition (iii) in Part 2 of the theorem or its equivalent form in Exercise 8.7 are called *uniformly countably additive* [11, 4.6]. See Fremlin [58, Ch.24] [59, Ch.35] for a closely related concept of *uniformly integrable* sets.

Baire measures and separable measures on completely regular spaces and cylindrical measures (Sects. 8.3, 8.4 and 8.6) were described as uniform measures by Frolík et al. [76]. Deaibes [36, 4.1.10] showed how to describe $\mathfrak{M}_\sigma(X)$ for an arbitrary uniform space X as $\mathfrak{M}_u(Y)$.

The results in Sects. 8.3 and 8.4 are part of topological measure theory. Theorems 8.11 and 8.17 show that Baire measures and separable measures are well behaved, unlike many other classes of measures on topological spaces. A proper context for these results and links to original sources are in Buchwalter [16], Bogachev [11, Ch.7], Fremlin [60], Kirk [113] and Wheeler [180].

The definition of separable measures in Sect. 8.4 is a minor variation of the original definition due to Dudley [41], who also proved several portions of Theorem 8.17. Granirer [84] investigated an equivalent property defined by continuity on bounded equicontinuous subsets of $C_b(T)$ and proved Theorem 8.16 and some of Theorem 8.17.

Wheeler [180, sec.8] includes Theorem 8.14 and other results showing that the converse is false: There exists a space T such that every separable measure on T is τ-additive although T is not paracompact. Therefore the converse to Theorem 7.22 is also false: The uniform space $X := \mathsf{F}T$ is not supercomplete (Theorem 4.20) but $\mathfrak{M}_u(X) = \mathfrak{M}_\tau(X)$.

Badrikian [5], Gel'fand and Vilenkin [79] and Schwartz [165] are the classical sources for the theory of cylindrical measures and their role in harmonic analysis and probability theory on topological vector spaces. Choquet [26] includes comparison of cylindrical measures to several other classes of functionals. Images of cylindrical measures under non-linear mappings were considered by Krée [114], [115].

Chapter 9
Direct Product and Convolution

As I noted in Sect. 6.8, historically one source of the uniform measure concept had been the study of convolution of measures on topological vector spaces and on topological groups. In this chapter I explore the connection between uniform measures and convolution in a fairly general setting that includes convolution on topological groups as a special case.

The natural way to define the convolution of two measures on a semigroup is to take the image of the direct product measure under the semigroup operation. That is the definition I adopt here for functionals on the space $U_b(X)$, after first establishing properties of direct products.

Theorem 9.11 characterizes uniform measures as exactly those functionals that satisfy a version of Fubini's theorem for direct products. Theorem 9.20 and its corollaries in Sects. 9.3 and 9.4 witness the key role of uniform measures in continuity properties of the convolution operation. In Theorems 9.29 and 9.41, topological centres in convolution semigroups are described in terms of uniform measures.

9.1 Direct Product

Let X and Y be uniform spaces, $f \in U_b(X^*Y)$ and $\mathfrak{n} \in \mathfrak{M}_b(Y)$. For every $x \in X$, the function $\backslash_y f(x,y)$ is in $U_b(Y)$ by Corollary 2.35, so that we can form $\mathfrak{n}(\backslash_y f(x,y))$. To simplify formulas, in this section I denote by $\mathfrak{n} \bowtie f$ the function on X defined by $\mathfrak{n} \bowtie f(x) := \mathfrak{n}(\backslash_y f(x,y))$, $x \in X$.

Lemma 9.1. *Let X and Y be uniform spaces, $\mathscr{F} \subseteq U_b(X^*Y)$ and $\mathfrak{A} \subseteq \mathfrak{M}_b(Y)$. If the set $\mathscr{F}$ is uniformly equicontinuous on X^*Y and $r := \sup\{\|\mathfrak{n}\| \mid \mathfrak{n} \in \mathfrak{A}\} < \infty$, then $\|\mathfrak{n} \bowtie f\|_X \leq r\|f\|_{X^*Y}$ for every $f \in \mathscr{F}$, $\mathfrak{n} \in \mathfrak{A}$, and the set $\{\mathfrak{n} \bowtie f \mid f \in \mathscr{F}, \mathfrak{n} \in \mathfrak{A}\}$ is uniformly equicontinuous on X.*

Proof. Since $\|\backslash_y f(x,y)\|_Y \leq \|f\|_{X*Y}$ for $x \in X$, we have

$$|n \times f(x)| \leq \|n\| \cdot \|\backslash_y f(x,y)\|_Y \leq r\|f\|_{X*Y}$$

for $n \in \mathfrak{A}$, and $\|n \times f\|_X \leq r\|f\|_{X*Y}$.

If $\mathscr{F} \subseteq U_b(X*Y)$ is uniformly equicontinuous, then there is $\Delta \in UP(X)$ such that $|f(x,y) - f(x',y)| \leq \Delta(x,x')$ for $f \in \mathscr{F}$, $x,x' \in X$ and $y \in Y$,

$$|n \times f(x) - n \times f(x')| = |n(\backslash_y f(x,y)) - n(\backslash_y f(x',y))|$$
$$= |n(\backslash_y f(x,y) - \backslash_y f(x',y))| \leq \|n\|\Delta(x,x') \leq r\Delta(x,x')$$

for $n \in \mathfrak{A}$; hence the set $\{n \times f \mid f \in \mathscr{F}, n \in \mathfrak{A}\}$ is uniformly equicontinuous.   □

Corollary 9.2. *Let X and Y be uniform spaces, $n \in \mathfrak{M}_b(Y)$ and $f \in U_b(X*Y)$. Then $n \times f \in U_b(X)$ and $\|n \times f\|_X \leq \|n\| \cdot \|f\|_{X*Y}$.*   □

By the corollary, $m(n \times f)$ is defined for $m \in \mathfrak{M}_b(X)$, $n \in \mathfrak{M}_b(Y)$ and $f \in U_b(X*Y)$.

Definition 9.3. *Let X and Y be uniform spaces, $m \in \mathfrak{M}_b(X)$ and $n \in \mathfrak{M}_b(Y)$. The direct product of m and n is the function $m \otimes n$ on $U_b(X*Y)$ defined by*

$$m \otimes n(f) := m(n \times f) = m(\backslash_x n(\backslash_y f(x,y)))$$

for $f \in U_b(X*Y)$.   ∎

Lemma 9.4. *Let X and Y be uniform spaces. If $m \in \mathfrak{M}_b(X)$ and $n \in \mathfrak{M}_b(Y)$, then $m \otimes n \in \mathfrak{M}_b(X*Y)$ and $\|m \otimes n\| \leq \|m\| \cdot \|n\|$.*

Proof. The linearity of $m \otimes n$ follows from the linearity of m and n, and

$$|m \otimes n(f)| = |m(n \times f)| \leq \|m\| \cdot \|n \times f\|_X \leq \|m\| \cdot \|n\| \cdot \|f\|_{X*Y}$$

for $f \in U_b(X*Y)$ by Corollary 9.2.   □

Note that if $g \in U_b(X)$, $h \in U_b(Y)$ and $f(x,y) = g(x)h(y)$ for $x \in X$, $y \in Y$, then $f \in U_b(X*Y)$ and $m \otimes n(f) = m(g) . n(h)$.

Lemma 9.5. *Let X and Y be uniform spaces, $m_0, m_1 \in \mathfrak{M}_b(X)$, $n_0, n_1 \in \mathfrak{M}_b(Y)$ and $r \in \mathbb{R}$. Then*

$$(rm_0) \otimes n_0 = m_0 \otimes (rn_0) = r(m_0 \otimes n_0)$$
$$(m_0 + m_1) \otimes n_0 = (m_0 \otimes n_0) + (m_1 \otimes n_0)$$
$$m_0 \otimes (n_0 + n_1) = (m_0 \otimes n_0) + (m_0 \otimes n_1)$$

Proof. This follows from the definition of $\otimes$ and the linearity of m_i and n_i.   □

Recall from Lemma 2.36 that, for any uniform spaces X_0, X_1 and X_2, the spaces $(X_0*X_1)*X_2$ and $X_0*(X_1*X_2)$ are naturally isomorphic.

Lemma 9.6. *Let X_i be uniform spaces and $\mathfrak{m}_i \in \mathfrak{M}_b(X_i)$ for $i = 0,1,2$, and let ψ be the uniform isomorphism from $(X_0*X_1)*X_2$ onto $X_0*(X_1*X_2)$ that maps $((x_0,x_1),x_2)$ to $(x_0,(x_1,x_2))$. Then $\psi((\mathfrak{m}_0 \otimes \mathfrak{m}_1) \otimes \mathfrak{m}_2) = \mathfrak{m}_0 \otimes (\mathfrak{m}_1 \otimes \mathfrak{m}_2)$.*

Proof. For any $f \in U_b(X_0*(X_1*X_2))$, we have

$$
\begin{aligned}
\mathfrak{m}_0 \otimes (\mathfrak{m}_1 \otimes \mathfrak{m}_2)(f) &= \mathfrak{m}_0(\backslash_{x_0}(\mathfrak{m}_1 \otimes \mathfrak{m}_2)(\backslash_{(x_1,x_2)} f(x_0,(x_1,x_2)))) \\
&= \mathfrak{m}_0(\backslash_{x_0} \mathfrak{m}_1(\backslash_{x_1} \mathfrak{m}_2(\backslash_{x_2} f(x_0,(x_1,x_2))))) \\
&= \mathfrak{m}_0(\backslash_{x_0} \mathfrak{m}_1(\backslash_{x_1} \mathfrak{m}_2(\backslash_{x_2} f \circ \psi((x_0,x_1),x_2)))) \\
&= \mathfrak{m}_0 \otimes \mathfrak{m}_1(\backslash_{(x_0,x_1)} \mathfrak{m}_2(\backslash_{x_2} f \circ \psi((x_0,x_1),x_2))) \\
&= (\mathfrak{m}_0 \otimes \mathfrak{m}_1) \otimes \mathfrak{m}_2(f \circ \psi),
\end{aligned}
$$

which proves $\psi((\mathfrak{m}_0 \otimes \mathfrak{m}_1) \otimes \mathfrak{m}_2) = \mathfrak{m}_0 \otimes (\mathfrak{m}_1 \otimes \mathfrak{m}_2)$.   □

Lemma 9.7. *For $i = 0,1$, let X_i and Y_i be uniform spaces, $\varphi_i \colon X_i \to Y_i$ uniformly continuous mappings, and $\mathfrak{m}_i \in \mathfrak{M}_b(X_i)$. Define $\varphi(x_0,x_1) := (\varphi_0(x_0),\varphi_1(x_1))$ for $x_0 \in X_0$, $x_1 \in X_1$. Then φ is a uniformly continuous mapping from X_0*X_1 to Y_0*Y_1 and $\varphi(\mathfrak{m}_0 \otimes \mathfrak{m}_1) = \varphi_0(\mathfrak{m}_0) \otimes \varphi_1(\mathfrak{m}_1)$.*

Proof. Uniform continuity of φ follows from Definition 2.33. If $f \in U_b(Y_0*Y_1)$, then

$$
\begin{aligned}
\varphi(\mathfrak{m}_0 \otimes \mathfrak{m}_1)(f) &= \mathfrak{m}_0 \otimes \mathfrak{m}_1(f \circ \varphi) = \mathfrak{m}_0(\backslash_{x_0} \mathfrak{m}_1(\backslash_{x_1} f(\varphi_0(x_0),\varphi_1(x_1)))) \\
&= \varphi_0(\mathfrak{m}_0)(\backslash_{y_0} \varphi_1(\mathfrak{m}_1)(\backslash_{y_1} f(y_0,y_1))) \\
&= \varphi_0(\mathfrak{m}_0) \otimes \varphi_1(\mathfrak{m}_1)(f),
\end{aligned}
$$

which proves $\varphi(\mathfrak{m}_0 \otimes \mathfrak{m}_1) = \varphi_0(\mathfrak{m}_0) \otimes \varphi_1(\mathfrak{m}_1)$.   □

Lemma 9.8. *Let X and Y be uniform spaces.*

1. *If $\mathfrak{m} \in \mathfrak{M}_b(X)^+$ and $\mathfrak{n} \in \mathfrak{M}_b(Y)^+$, then $\mathfrak{m} \otimes \mathfrak{n} \in \mathfrak{M}_b(X*Y)^+$.*
2. *If $\mathfrak{m} \in \mathrm{Mol}(X)$ and $\mathfrak{n} \in \mathrm{Mol}(Y)$, then $\mathfrak{m} \otimes \mathfrak{n} \in \mathrm{Mol}(X*Y)$.*
3. *If $\mathfrak{m} \in \mathfrak{M}_u(X)$ and $\mathfrak{n} \in \mathfrak{M}_u(Y)$, then $\mathfrak{m} \otimes \mathfrak{n} \in \mathfrak{M}_u(X*Y)$.*
4. *If $\mathfrak{m} \in \hat{\mathrm{p}}X$ and $\mathfrak{n} \in \hat{\mathrm{p}}Y$, then $\mathfrak{m} \otimes \mathfrak{n} \in \hat{\mathrm{p}}(X*Y)$.*
5. *If $\mathfrak{m} \in \widehat{X}$ and $\mathfrak{n} \in \widehat{Y}$, then $\mathfrak{m} \otimes \mathfrak{n} \in \widehat{X*Y}$.*

Proof. Part 1 follows directly from the definition of $\otimes$.

To prove Part 2, write $\mathfrak{m}$ and $\mathfrak{n}$ as

$$
\mathfrak{m}(g) = \sum_{x \in D} r(x) g(x) \quad \text{for } g \in U_b(X)
$$

$$
\mathfrak{n}(h) = \sum_{y \in D'} r'(y) h(y) \quad \text{for } h \in U_b(Y)
$$

with finite sets D, D' and coefficients $r(x)$, $r'(y) \in \mathbb{R}$. Then

$$\mathfrak{m} \otimes \mathfrak{n}(f) = \sum_{x \in D} \sum_{y \in D'} r(x) r'(y) f(x,y) \quad \text{for } f \in U_b(X*Y).$$

For Part 3, take any $\mathrm{UEB}(X*Y)$ set $\mathscr{F} \subseteq U_b(X*Y)$ and $\varepsilon > 0$. By Lemma 9.1, the set $\{\mathfrak{n} \rtimes f \mid f \in \mathscr{F}\}$ is $\mathrm{UEB}(X)$; thus there are $\theta > 0$ and a finite set $D_X \subseteq X$ such that

$$\forall f \in \mathscr{F} \ [\ \|\mathfrak{n} \rtimes f\|_{D_X} < \theta \Rightarrow |\mathfrak{m}(\mathfrak{n} \rtimes f)| < \varepsilon\ ].$$

By Theorem 2.34 the set $\{\backslash_y f(x,y) \mid f \in \mathscr{F}, x \in D_X\}$ is $\mathrm{UEB}(Y)$, hence there are $\theta' > 0$ and a finite set $D_Y \subseteq Y$ such that

$$\forall x \in D_X \ \forall f \in \mathscr{F} \ [\ \|\backslash_y f(x,y)\|_{D_Y} < \theta' \Rightarrow |\mathfrak{n}(\backslash_y f(x,y))| < \theta\ ].$$

Set $D := D_X \times D_Y$, and take any $f \in \mathscr{F}$ such that $\|f\|_D < \theta'$. If $x \in D_X$, then $\|\backslash_y f(x,y)\|_{D_Y} < \theta'$; therefore $|\mathfrak{n} \rtimes f(x)| = |\mathfrak{n}(\backslash_y f(x,y))| < \theta$. It follows that

$$|\mathfrak{m} \otimes \mathfrak{n}(f)| = |\mathfrak{m}(\mathfrak{n} \rtimes f)| < \varepsilon,$$

which proves that $\mathfrak{m} \otimes \mathfrak{n}$ is pointwise continuous on $\mathscr{F}$ at 0. Apply Lemma 5.25.

To prove Part 4, take any $\mathfrak{m} \in \hat{\mathfrak{p}}X$ and $\mathfrak{n} \in \hat{\mathfrak{p}}Y$. There are nets $\{x_\beta\}_\beta$ in X and $\{y_\gamma\}_\gamma$ in Y such that $\lim_\beta \partial(x_\beta) = \mathfrak{m}$ in the $U_b(X)$-weak topology and $\lim_\gamma \partial(y_\gamma) = \mathfrak{n}$ in the $U_b(Y)$-weak topology.

It is clear from the definition of $\otimes$ that in the $U_b(X*Y)$-weak topology we have $\lim_\gamma \partial(x_\beta) \otimes \partial(y_\gamma) = \partial(x_\beta) \otimes \mathfrak{n}$ for every β (see also Theorem 9.11), and $\lim_\beta \partial(x_\beta) \otimes \mathfrak{n} = \mathfrak{m} \otimes \mathfrak{n}$. Since $\partial(x_\beta) \otimes \partial(y_\gamma) = \partial((x_\beta, y_\gamma))$ and the set $\hat{\mathfrak{p}}(X*Y)$ is $U_b(X*Y)$-weakly closed in $\mathfrak{M}_b(X*Y)$, it follows that $\mathfrak{m} \otimes \mathfrak{n} \in \hat{\mathfrak{p}}(X*Y)$.

Part 5 follows from Parts 3 and 4 and from Part 3 of Theorem 6.37.   □

Exercise 9.9. Let X and Y be uniform spaces, $\mathfrak{m} \in \mathfrak{M}_t(X)$ and $\mathfrak{n} \in \mathfrak{M}_t(Y)$. Prove that $\mathfrak{m} \otimes \mathfrak{n} \in \mathfrak{M}_t(X*Y)$.   ∎

In Definition 9.3, $\mathfrak{m} \otimes \mathfrak{n}$ is obtained by applying first $\mathfrak{n}$ and then $\mathfrak{m}$. In the classical setting of countably additive measures, Fubini's theorem states that the order may be reversed:

$$\int \int f(x,y) \, d\mu(x) \, d\nu(y) = \int \int f(x,y) \, d\nu(y) \, d\mu(x)$$

For functionals on uniformly continuous functions, the analogous formula

$$\mathfrak{m}(\backslash_x \mathfrak{n}(\backslash_y f(x,y))) = \mathfrak{n}(\backslash_y \mathfrak{m}(\backslash_x f(x,y)))$$

does not necessarily hold for arbitrary $\mathfrak{m}$ and $\mathfrak{n}$, but it does hold when $\mathfrak{m} \in \mathfrak{M}_u(X)$.

Theorem 9.10 (Fubini's theorem). *Let X and Y be uniform spaces, $\mathfrak{m} \in \mathfrak{M}_u(X)$, $\mathfrak{n} \in \mathfrak{M}_b(Y)$, and $f \in U_b(X*Y)$. Then $\backslash_y \mathfrak{m}(\backslash_x f(x,y)) \in U_b(Y)$ and*

$$\mathfrak{m} \otimes \mathfrak{n}(f) = \mathfrak{n}(\backslash_y \mathfrak{m}(\backslash_x f(x,y))).$$

Proof. As the set $\{\backslash_x f(x,y) \mid y \in Y\}$ is $UEB(X)$, by Theorem 6.6 there is a net $\{\mathfrak{m}_\gamma\}_\gamma$, $\mathfrak{m}_\gamma \in Mol(X)$, such that

$$\limsup_{\gamma} {}_{y \in Y} |\mathfrak{m}(\backslash_x f(x,y)) - \mathfrak{m}_\gamma(\backslash_x f(x,y))| = 0$$

$$\lim_{\gamma} |\mathfrak{m}(\mathfrak{n} \bowtie f) - \mathfrak{m}_\gamma(\mathfrak{n} \bowtie f)| = 0$$

The theorem holds with each $\mathfrak{m}_\gamma$ in place of $\mathfrak{m}$, by the linearity of $\mathfrak{n}$. Therefore the function $\backslash_y \mathfrak{m}(\backslash_x f(x,y))$, being a $\|\cdot\|_Y$ limit of functions in $U_b(Y)$, is in $U_b(Y)$, and for all γ we have

$$\mathfrak{m}_\gamma(\mathfrak{n} \bowtie f) = \mathfrak{m}_\gamma \otimes \mathfrak{n}(f) = \mathfrak{n}(\backslash_y \mathfrak{m}_\gamma(\backslash_x f(x,y))).$$

Hence

$$\left| \mathfrak{m} \otimes \mathfrak{n}(f) - \mathfrak{n}(\backslash_y \mathfrak{m}(\backslash_x f(x,y))) \right|$$

$$\leq \lim_{\gamma} \left| \mathfrak{m}(\mathfrak{n} \bowtie f) - \mathfrak{m}_\gamma(\mathfrak{n} \bowtie f) \right| + \lim_{\gamma} \left| \mathfrak{n}(\backslash_y \mathfrak{m}_\gamma(\backslash_x f(x,y))) - \mathfrak{n}(\backslash_y \mathfrak{m}(\backslash_x f(x,y))) \right|$$

$$= 0$$

and $\mathfrak{m} \otimes \mathfrak{n}(f) = \mathfrak{n}(\backslash_y \mathfrak{m}(\backslash_x f(x,y)))$. □

Next I show that Fubini's formula holds *only* for uniform measures and is equivalent to a certain continuity property of the $\otimes$ operation.

Theorem 9.11. *For any uniform space X, the following properties of a functional $\mathfrak{m} \in \mathfrak{M}_b(X)$ are equivalent:*

(i) $\mathfrak{m} \in \mathfrak{M}_u(X)$.

(ii) *For every uniform space Y, if $\mathfrak{n} \in \mathfrak{M}_b(Y)$ and $f \in U_b(X*Y)$, then the function $\backslash_y \mathfrak{m}(\backslash_x f(x,y))$ is in $U_b(Y)$ and $\mathfrak{m} \otimes \mathfrak{n}(f) = \mathfrak{n}(\backslash_y \mathfrak{m}(\backslash_x f(x,y)))$.*

(iii) *For every uniform space Y, the mapping $\mathfrak{n} \mapsto \mathfrak{m} \otimes \mathfrak{n}$ is continuous from $\mathfrak{M}_b(Y)$ with the $U_b(Y)$-weak topology to $\mathfrak{M}_b(X*Y)$ with the $U_b(X*Y)$-weak topology.*

(iv) *For every uniform space Y, the mapping $y \mapsto \mathfrak{m} \otimes \partial_Y(y)$ is continuous from Y to $\mathfrak{M}_b(X*Y)$ with the $U_b(X*Y)$-weak topology.*

(v) *For every compact uniform space Y, the mapping $y \mapsto \mathfrak{m} \otimes \partial_Y(y)$ is continuous from Y to $\mathfrak{M}_b(X*Y)$ with the $U_b(X*Y)$-weak topology.*

Note that, in contrast, the mapping $\mathfrak{m} \mapsto \mathfrak{m} \otimes \mathfrak{n}$ from $\mathfrak{M}_b(X)$ to $\mathfrak{M}_b(X*Y)$ is continuous in the U_b-weak topologies, and also in the UEB topologies, for *every* $\mathfrak{n} \in \mathfrak{M}_b(Y)$.

Proof. (i)⇒(ii) by Theorem 9.10.

Assume (ii), and take any $f \in U_b(X^*Y)$. Then the mapping

$$\mathfrak{n} \mapsto \mathfrak{m} \otimes \mathfrak{n}(f) = \mathfrak{n}(\backslash_y \mathfrak{m}(\backslash_x f(x,y)))$$

from $\mathfrak{M}_b(Y)$ to $\mathbb{R}$ is $U_b(Y)$-weakly continuous. As this holds for all $f \in U_b(X^*Y)$, (iii) follows.

(iii)⇒(iv) because the mapping $\partial_Y : Y \to \mathfrak{M}_b(Y)$ is $U_b(Y)$-weakly continuous, and obviously (iv)⇒(v).

Next assume (v), and take any $\Delta \in \mathrm{UP}(X)$. Define Y to be $\mathrm{BLip}_b(\Delta)$ with the X-pointwise uniformity, and $f(x,h) := h(x)$ for $x \in X$ and $h \in Y$. Then $f \in U_b(X^*Y)$ and by (v) the function $h \mapsto \mathfrak{m} \otimes \partial_Y(h)(f) = \mathfrak{m}(h)$ is continuous on $\mathrm{BLip}_b(\Delta)$. That proves (v)⇒(i). □

The remaining results in this section further clarify the connection between uniform measures and the continuity of the $\otimes$ operation.

Theorem 9.12. *Let X and Y be uniform spaces. Let $\{\mathfrak{m}_\gamma\}_{\gamma \in \Gamma}$ be a net in $\mathfrak{M}_b(X)$, $\mathfrak{m} \in \mathfrak{M}_u(X)$, $\{\mathfrak{n}_\gamma\}_{\gamma \in \Gamma}$ a net in $\mathfrak{M}_b(Y)$, and $\mathfrak{n} \in \mathfrak{M}_b(Y)$. Assume that $\lim_\gamma \mathfrak{m}_\gamma = \mathfrak{m}$ in the $\mathrm{UEB}(X)$ topology on $\mathfrak{M}_b(X)$, and $\sup_\gamma \|\mathfrak{n}_\gamma\| < \infty$.*

1. *If $\lim_\gamma \mathfrak{n}_\gamma = \mathfrak{n}$ in the $\mathrm{UEB}(Y)$ topology on $\mathfrak{M}_b(Y)$, then $\lim_\gamma \mathfrak{m}_\gamma \otimes \mathfrak{n}_\gamma = \mathfrak{m} \otimes \mathfrak{n}$ in the $\mathrm{UEB}(X^*Y)$ topology on $\mathfrak{M}_b(X^*Y)$.*
2. *If $\lim_\gamma \mathfrak{n}_\gamma = \mathfrak{n}$ in the $U_b(Y)$-weak topology, then $\lim_\gamma \mathfrak{m}_\gamma \otimes \mathfrak{n}_\gamma = \mathfrak{m} \otimes \mathfrak{n}$ in the $U_b(X^*Y)$-weak topology.*

Proof. Both parts of the theorem follow from the same estimates, which I first derive for some fixed $\mathrm{UEB}(X^*Y)$ set $\mathscr{F} \subseteq U_b(X^*Y)$. The set $\{\mathfrak{n}_\gamma \ltimes f \mid f \in \mathscr{F},\ \gamma \in \Gamma\}$ is $\mathrm{UEB}(X)$ by Lemma 9.1, hence

$$\lim_\gamma \sup_{f \in \mathscr{F}} |\mathfrak{m}_\gamma(\mathfrak{n}_\gamma \ltimes f) - \mathfrak{m}(\mathfrak{n}_\gamma \ltimes f)| = 0.$$

Define $g_\gamma(x) := \sup\{|\mathfrak{n}_\gamma \ltimes f(x) - \mathfrak{n} \ltimes f(x)| \mid f \in \mathscr{F}\}$ for $\gamma \in \Gamma$ and $x \in X$. Then the set $\{g_\gamma \mid \gamma \in \Gamma\}$ is $\mathrm{UEB}(X)$.

Now suppose that the following holds for the given $\mathscr{F}$:

$$\lim_\gamma g_\gamma(x) = 0 \text{ for every } x \in X. \tag{†}$$

In that case,

$$\lim_\gamma \sup_{f \in \mathscr{F}} |\mathfrak{m}_\gamma \otimes \mathfrak{n}_\gamma(f) - \mathfrak{m} \otimes \mathfrak{n}(f)| = \lim_\gamma \sup_{f \in \mathscr{F}} |\mathfrak{m}_\gamma(\mathfrak{n}_\gamma \ltimes f) - \mathfrak{m}(\mathfrak{n} \ltimes f)|$$

$$\leq \lim_\gamma \sup_{f \in \mathscr{F}} |\mathfrak{m}_\gamma(\mathfrak{n}_\gamma \ltimes f) - \mathfrak{m}(\mathfrak{n}_\gamma \ltimes f)| + \lim_\gamma \sup_{f \in \mathscr{F}} |\mathfrak{m}(\mathfrak{n}_\gamma \ltimes f) - \mathfrak{m}(\mathfrak{n} \ltimes f)|$$

$$\leq \lim_\gamma |\mathfrak{m}|(g_\gamma) = 0$$

where the last equality holds because $|\mathfrak{m}| \in \mathfrak{M}_u(X)$.

To prove Part 1, take any $\mathsf{UEB}(X^*Y)$ set $\mathscr{F} \subseteq \mathsf{U_b}(X^*Y)$. For every $x \in X$, the set $\{\backslash_y f(x,y) \mid f \in \mathscr{F}\}$ is $\mathsf{UEB}(Y)$ (Theorem 2.34). Since $\lim_\gamma \mathfrak{n}_\gamma = \mathfrak{n}$ in the $\mathsf{UEB}(Y)$ topology, (†) follows.

To prove Part 2, take any $f \in \mathsf{U_b}(X^*Y)$. Then $\mathscr{F} := \{f\}$ is a $\mathsf{UEB}(X^*Y)$ set so that again it is enough to prove (†). For every $x \in X$, the function $\backslash_y f(x,y)$ is in $\mathsf{U_b}(Y)$, and (†) follows because $\lim_\gamma \mathfrak{n}_\gamma = \mathfrak{n}$ in the $\mathsf{U_b}(Y)$-weak topology. $\qquad\square$

Corollary 9.13. *Let X and Y be uniform spaces. Let $\{\mathfrak{m}_\gamma\}_{\gamma \in \Gamma}$ be a net in $\mathfrak{M_b}(X)^+$, $\mathfrak{m} \in \mathfrak{M_u}(X)^+$, $\{\mathfrak{n}_\gamma\}_{\gamma \in \Gamma}$ a net in $\mathfrak{M_b}(Y)^+$ and $\mathfrak{n} \in \mathfrak{M_b}(Y)^+$. If $\lim_\gamma \mathfrak{m}_\gamma(g) = \mathfrak{m}(g)$ and $\lim_\gamma \mathfrak{n}_\gamma(h) = \mathfrak{n}(h)$ for all $g \in \mathsf{U_b}(X)$, $h \in \mathsf{U_b}(Y)$, then $\lim_\gamma \mathfrak{m}_\gamma \otimes \mathfrak{n}_\gamma(f) = \mathfrak{m} \otimes \mathfrak{n}(f)$ for all $f \in \mathsf{U_b}(X^*Y)$.*

Proof. Apply Theorem 9.12 with Theorem 6.12. $\qquad\square$

Corollary 9.14. *Let X and Y be uniform spaces, $\{\mathfrak{m}_j\}_j$ a sequence in $\mathfrak{M_u}(X)$, $\mathfrak{m} \in \mathfrak{M_b}(X)$, $\{\mathfrak{n}_j\}_j$ a sequence in $\mathfrak{M_b}(Y)$ and $\mathfrak{n} \in \mathfrak{M_b}(Y)$. If $\lim_j \mathfrak{m}_j(g) = \mathfrak{m}(g)$ and $\lim_j \mathfrak{n}_j(h) = \mathfrak{n}(h)$ for all $g \in \mathsf{U_b}(X)$, $h \in \mathsf{U_b}(Y)$, then $\lim_j \mathfrak{m}_j \otimes \mathfrak{n}_j(f) = \mathfrak{m} \otimes \mathfrak{n}(f)$ for all $f \in \mathsf{U_b}(X^*Y)$.*

Proof. As $\mathfrak{m} \in \mathfrak{M_u}(X)$ by Theorem 6.19 and the set $\{\mathfrak{n}_j \mid j \in \omega\}$ is $\|\cdot\|$ bounded by Theorem P.11, the corollary follows from Theorem 9.12 and Corollary 6.18. $\qquad\square$

9.2 Convolution for Semiuniform Semigroups

When G is a topological group, the binary group operation is uniformly continuous as a mapping $\varphi \colon rG^*rG \to rG$ by Theorem 3.8. That yields a natural definition of convolution on the space $\mathfrak{M_b}(rG)$: For $\mathfrak{m}, \mathfrak{n} \in \mathfrak{M_b}(rG)$, the convolution $\mathfrak{m} \star \mathfrak{n}$ is the image $\varphi(\mathfrak{m} \otimes \mathfrak{n})$ of $\mathfrak{m} \otimes \mathfrak{n} \in \mathfrak{M_b}(rG^*rG)$ under φ (Definition 6.7).

The same approach applies in the more general setting of a semiuniform semigroup acting semiuniformly on a uniform space, as follows.

Definition 9.15. Let X be a semiuniform semigroup, and let $\varphi \colon X^*Y \to Y$ be a semiuniform action of X on a uniform space Y. For $\mathfrak{m} \in \mathfrak{M_b}(X)$ and $\mathfrak{n} \in \mathfrak{M_b}(Y)$, the *convolution of $\mathfrak{m}$ and $\mathfrak{n}$* is

$$\mathfrak{m} \star \mathfrak{n} := \varphi(\mathfrak{m} \otimes \mathfrak{n}) \in \mathfrak{M_b}(Y).$$

More precisely, $\mathfrak{m} \star \mathfrak{n}$ is the *convolution determined by the action φ*. $\qquad\blacksquare$

When the action φ is written as $\varphi(x,y) = x \boldsymbol{.} y$, the definition expands to

$$\mathfrak{m} \star \mathfrak{n}(f) = \mathfrak{m}(\backslash_x \mathfrak{n}(\backslash_y f \circ \varphi(x,y))) = \mathfrak{m}(\backslash_x \mathfrak{n}(\backslash_y f(x \boldsymbol{.} y)))$$

for $\mathfrak{m} \in \mathfrak{M_b}(X)$, $\mathfrak{n} \in \mathfrak{M_b}(Y)$ and $f \in \mathsf{U_b}(Y)$. For the semiuniform semigroup X acting on itself by $x \boldsymbol{.} y = xy$, this reduces to

$$m \star n(f) = m(\backslash_x n(\backslash_y f(xy))) = m(n \bullet f)$$

for $m, n \in \mathfrak{M}_b(X)$ and $f \in U_b(X)$. Here, $(n, f) \mapsto n \bullet f := \backslash_x n(\backslash_y f(xy))$ is the canonical action of $\mathfrak{M}_b(X)$ on $U_b(X)$.

The properties of the direct product established in the previous section now immediately produce corresponding properties of convolution.

Theorem 9.16. *Let X be a semiuniform semigroup acting semiuniformly on a uniform space Y, $m_0, m_1 \in \mathfrak{M}_b(X)$, $n_0, n_1 \in \mathfrak{M}_b(Y)$ and $r \in \mathbb{R}$. Then*

$$(rm_0) \star n_0 = m_0 \star (rn_0) = r(m_0 \star n_0)$$

$$(m_0 + m_1) \star n_0 = (m_0 \star n_0) + (m_1 \star n_0)$$

$$m_0 \star (n_0 + n_1) = (m_0 \star n_0) + (m_0 \star n_1)$$

Proof. This follows from Lemma 9.5.   □

Theorem 9.17. *Let X be a semiuniform semigroup acting semiuniformly on a uniform space Y, $m_0, m_1 \in \mathfrak{M}_b(X)$ and $n \in \mathfrak{M}_b(Y)$. Then*

$$(m_0 \star m_1) \star n = m_0 \star (m_1 \star n).$$

Proof. Apply Lemmas 9.6 and 9.7.   □

Theorem 9.18. *Let X be a semiuniform semigroup acting semiuniformly on a uniform space Y.*

1. *If $m \in \mathfrak{M}_b(X)$ and $n \in \mathfrak{M}_b(Y)$, then $\|m \star n\| \leq \|m\| \cdot \|n\|$.*
2. *If $m \in \mathfrak{M}_b(X)^+$ and $n \in \mathfrak{M}_b(Y)^+$, then $m \star n \in \mathfrak{M}_b(Y)^+$.*
3. *If $m \in \mathrm{Mol}(X)$ and $n \in \mathrm{Mol}(Y)$, then $m \star n \in \mathrm{Mol}(Y)$.*
4. *If $m \in \mathfrak{M}_u(X)$ and $n \in \mathfrak{M}_u(Y)$, then $m \star n \in \mathfrak{M}_u(Y)$.*
5. *If $m \in \hat{p}X$ and $n \in \hat{p}Y$, then $m \star n \in \hat{p}Y$.*
6. *If $m \in \hat{X}$ and $n \in \hat{Y}$, then $m \star n \in \hat{Y}$.*

Proof. Apply Lemmas 9.4 and 9.8.   □

Exercise 9.19. Let X be a semiuniform semigroup acting semiuniformly on a uniform space Y. Show that if $m \in \mathfrak{M}_t(X)$ and $n \in \mathfrak{M}_t(Y)$, then $m \star n \in \mathfrak{M}_t(Y)$.   ∎

Theorem 9.20. *Let X be a semiuniform semigroup acting semiuniformly on a uniform space Y. Let $\{m_\gamma\}_\gamma$ be a net in $\mathfrak{M}_b(X)$, $m \in \mathfrak{M}_u(X)$, $\{n_\gamma\}_\gamma$ a net in $\mathfrak{M}_b(Y)$ and $n \in \mathfrak{M}_b(Y)$.*

1. *If $\sup_\gamma \|n_\gamma\| < \infty$, $\lim_\gamma m_\gamma = m$ in the $\mathsf{UEB}(X)$ topology and $\lim_\gamma n_\gamma = n$ in the $\mathsf{UEB}(Y)$ topology, then $\lim_\gamma m_\gamma \star n_\gamma = m \star n$ in the $\mathsf{UEB}(Y)$ topology.*
2. *If $\lim_\gamma \|m_\gamma\| \leq \|m\|$, $\sup_\gamma \|n_\gamma\| < \infty$, $\lim_\gamma m_\gamma = m$ in the $U_b(X)$-weak topology and $\lim_\gamma n_\gamma = n$ in the $U_b(Y)$-weak topology, then $\lim_\gamma m_\gamma \star n_\gamma = m \star n$ in the $U_b(Y)$-weak topology.*

3. *If* $\mathfrak{m}_\gamma, \mathfrak{n}_\gamma \geq 0$ *for all* γ, $\mathfrak{m}, \mathfrak{n} \geq 0$, $\lim_\gamma \mathfrak{m}_\gamma = \mathfrak{m}$ *in the* $U_b(X)$-*weak topology and* $\lim_\gamma \mathfrak{n}_\gamma = \mathfrak{n}$ *in the* $U_b(Y)$-*weak topology, then* $\lim_\gamma \mathfrak{m}_\gamma \star \mathfrak{n}_\gamma = \mathfrak{m} \star \mathfrak{n}$ *in the* $U_b(Y)$-*weak topology.*

4. *If the nets* $\{\mathfrak{m}_\gamma\}_\gamma$ *and* $\{\mathfrak{n}_\gamma\}_\gamma$ *are sequences* $\{\mathfrak{m}_j\}_j$ *and* $\{\mathfrak{n}_j\}_j$, $\mathfrak{m}_j \in \mathfrak{M}_u(X)$ *for* $j \in \omega$, $\lim_j \mathfrak{m}_j = \mathfrak{m}$ *in the* $U_b(X)$-*weak topology and* $\lim_j \mathfrak{n}_j = \mathfrak{n}$ *in the* $U_b(Y)$-*weak topology, then* $\lim_j \mathfrak{m}_j \star \mathfrak{n}_j = \mathfrak{m} \star \mathfrak{n}$ *in the* $U_b(Y)$-*weak topology.*

5. *If* $\lim_\gamma \mathfrak{n}_\gamma = \mathfrak{n}$ *in the* $U_b(Y)$-*weak topology, then* $\lim_\gamma \mathfrak{m} \star \mathfrak{n}_\gamma = \mathfrak{m} \star \mathfrak{n}$ *in the* $U_b(Y)$-*weak topology.*

Proof. Part 1 follows from Theorem 9.12, Part 2 from Theorems 6.14 and 9.12, Part 3 from Corollary 9.13, Part 4 from Corollary 9.14 and Part 5 from Theorem 9.11. □

If S is a semitopological semigroup whose points are separated by $\mathsf{LUC}(S)$, then $U_b(rS) = \mathsf{LUC}(S)$ by Lemma 3.5, hence $\mathfrak{M}_b(rS) = \mathsf{LUC}(S)^*$. Thus the following corollary is a special case of Theorem 9.20.

Corollary 9.21. *Let S be a semitopological semigroup whose points are separated by* $\mathsf{LUC}(S)$.

1. *Let* $\mathfrak{B} \subseteq \mathsf{LUC}(S)^*$ *be a* $\|\cdot\|$ *bounded set,* $\mathfrak{m}_0 \in \mathfrak{M}_u(rS)$ *and* $\mathfrak{n}_0 \in \mathfrak{B}$. *With the* $\mathsf{UEB}(rS)$ *topology on* $\mathfrak{B}$ *and* $\mathsf{LUC}(S)^*$, *the mapping* $(\mathfrak{m}, \mathfrak{n}) \mapsto \mathfrak{m} \star \mathfrak{n}$ *from the product* $\mathsf{LUC}(S)^* \times \mathfrak{B}$ *to* $\mathsf{LUC}(S)^*$ *is jointly continuous at* $(\mathfrak{m}_0, \mathfrak{n}_0)$.

2. *Let* $\mathfrak{A}$ *be the unit sphere* $\{\mathfrak{m} \in \mathsf{LUC}(S)^* \mid \|\mathfrak{m}\| = 1\}$ *and let* $\mathfrak{B} \subseteq \mathsf{LUC}(S)^*$ *be a* $\|\cdot\|$ *bounded set,* $\mathfrak{m}_0 \in \mathfrak{A} \cap \mathfrak{M}_u(rS)$ *and* $\mathfrak{n}_0 \in \mathfrak{B}$. *With the* $\mathsf{LUC}(S)$-*weak topology on* $\mathfrak{A}$, $\mathfrak{B}$ *and* $\mathsf{LUC}(S)^*$, *the mapping* $(\mathfrak{m}, \mathfrak{n}) \mapsto \mathfrak{m} \star \mathfrak{n}$ *from* $\mathfrak{A} \times \mathfrak{B}$ *to* $\mathsf{LUC}(S)^*$ *is jointly continuous at* $(\mathfrak{m}_0, \mathfrak{n}_0)$.

3. *Let* $\mathfrak{m}_0 \in \mathfrak{M}_u(rS)^+$ *and let* $\mathfrak{n}_0 \in \mathsf{LUC}(S)^{*+}$. *With the* $\mathsf{LUC}(S)$-*weak topology on* $\mathsf{LUC}(S)^*$, *the mapping* $(\mathfrak{m}, \mathfrak{n}) \mapsto \mathfrak{m} \star \mathfrak{n}$ *from* $\mathsf{LUC}(S)^{*+} \times \mathsf{LUC}(S)^{*+}$ *to* $\mathsf{LUC}(S)^{*+}$ *is jointly continuous at* $(\mathfrak{m}_0, \mathfrak{n}_0)$.

4. *With the* $\mathsf{LUC}(S)$-*weak topology on* $\mathfrak{M}_u(rS)$, *convolution is jointly sequentially continuous as a mapping from* $\mathfrak{M}_u(rS) \times \mathfrak{M}_u(rS)$ *to* $\mathfrak{M}_u(rS)$.

5. *If* $\mathfrak{m} \in \mathfrak{M}_u(rS)$, *then the mapping* $\mathfrak{n} \mapsto \mathfrak{m} \star \mathfrak{n}$ *from* $\mathsf{LUC}(S)^*$ *to itself is continuous in the* $\mathsf{LUC}(S)$-*weak topology.* □

As in Sect. 3.1, the requirement that $\mathsf{LUC}(S)$ should separate the points of S is merely a technical assumption. It could be omitted if we allowed the uniform space rS to be non-Hausdorff. In any case, the requirement is not too restrictive in view of Exercise 3.6.

Examples 9.38 and 9.39 in Sect. 9.4 demonstrate that various assumptions cannot be omitted in Theorem 9.20 and Corollary 9.21 (and therefore also in Theorem 9.12).

For molecular measures

$$\mathfrak{m}(f) = \sum_{x \in D} r(x) f(x), \quad f \in U_b(X)$$

$$\mathfrak{n}(g) = \sum_{y \in D'} r'(y) g(y), \quad g \in U_b(Y)$$

where $D \subseteq X$, $D' \subseteq Y$ are finite sets and $r(x), r'(y) \in \mathbb{R}$, the definition of convolution yields an explicit formula:

$$\mathfrak{m} \star \mathfrak{n}(g) = \sum_{x \in D} \sum_{y \in D'} r(x) r'(y) g(x \cdot y), \quad g \in \mathsf{U}_b(Y).$$

Now let $X = Y$ and $x \cdot y = xy$ for $x, y \in X$. By Theorem 6.6, every $\mathfrak{m} \in \mathfrak{M}_u(X)$ is approximated in the UEB topology by a $\|\cdot\|$ bounded net of molecular measures, and Part 1 of Corollary 9.21 implies that finite algebraic identities satisfied by the convolution operation on $\mathrm{Mol}(X)$ are inherited by $\mathfrak{M}_u(X)$. In particular, if X is commutative, then so is $\star$ on $\mathfrak{M}_u(X)$. In contrast, $\star$ is typically not commutative on $\mathfrak{M}_b(X)$ (Exercise 9.34).

Exercise 9.22. Let X and Y be semiuniform semigroups and let $\varphi \colon X \to Y$ be a uniformly continuous homomorphism. Prove that $\varphi(\mathfrak{m} \star \mathfrak{n}) = \varphi(\mathfrak{m}) \star \varphi(\mathfrak{n})$ for $\mathfrak{m}, \mathfrak{n} \in \mathfrak{M}_b(X)$.   ∎

9.3   Topological Centres in Convolution Semigroups

For a topological group G, the space $\mathfrak{M}_b(rG) = \mathsf{LUC}(G)^*$ provides a natural setting for studying convolution. When endowed with the $\star$ operation and the $\mathsf{U}_b(rG)$-weak topology, $\mathfrak{M}_b(rG)$ is typically a large and complicated space. However, considerable amount of information about its algebraic and topological structure may be derived from a tractable description of its topological centre.

The topological centre may be defined for any semigroup that is also a topological space:

Definition 9.23. Let S be a semigroup with a topology. The *(left) topological centre of S* is
$$\Lambda(S) := \{s \in S \mid \text{the mapping } t \mapsto st \text{ is continuous on } S\}. \qquad ∎$$

Theorem 9.24. *Let X be a semiuniform semigroup. Consider $\mathfrak{M}_b(X)$ as a semigroup with the convolution operation $\star$ and the $\mathsf{U}_b(X)$-weak topology. A functional $\mathfrak{m} \in \mathfrak{M}_b(X)$ belongs to $\Lambda(\mathfrak{M}_b(X))$ if and only if $\backslash_y \mathfrak{m}(\backslash_x f(xy)) \in \mathsf{U}_b(X)$ and*

$$\mathfrak{m} \star \mathfrak{n}(f) = \mathfrak{n}(\backslash_y \mathfrak{m}(\backslash_x f(xy)))$$

for all $f \in \mathsf{U}_b(X)$ and $\mathfrak{n} \in \mathfrak{M}_b(X)$.

Proof. Assume $\mathfrak{m} \star \mathfrak{n}(f) = \mathfrak{n}(\backslash_y \mathfrak{m}(\backslash_x f(xy)))$ for all $f \in \mathsf{U}_b(X)$ and $\mathfrak{n} \in \mathfrak{M}_b(X)$. If $\lim_\gamma \mathfrak{n}_\gamma = \mathfrak{n}$ in the $\mathsf{U}_b(X)$-weak topology in $\mathfrak{M}_b(X)$, then

$$\lim_\gamma \mathfrak{m} \star \mathfrak{n}_\gamma = \lim_\gamma \mathfrak{n}_\gamma(\backslash_y \mathfrak{m}(\backslash_x f(xy))) = \mathfrak{n}(\backslash_y \mathfrak{m}(\backslash_x f(xy))) = \mathfrak{m} \star \mathfrak{n}(f),$$

which proves $\mathfrak{m} \in \Lambda(\mathfrak{M}_b(X))$.

To prove that the condition is necessary, assume $\mathfrak{m} \in \Lambda(\mathfrak{M}_b(X))$ and take any $f \in U_b(X)$. The mapping $\mathfrak{n} \mapsto \mathfrak{m} \star \mathfrak{n}$ from $\mathfrak{M}_b(X)$ to itself is linear and $U_b(X)$-weakly continuous, hence the functional $\mathfrak{n} \mapsto \mathfrak{m} \star \mathfrak{n}(f)$ from $\mathfrak{M}_b(X)$ to $\mathbb{R}$ is $U_b(X)$-weakly continuous, and there is $g \in U_b(X)$ such that $\mathfrak{m} \star \mathfrak{n}(f) = \mathfrak{n}(g)$ for all $\mathfrak{n} \in \mathfrak{M}_b(X)$. In particular, for $\mathfrak{n} = \partial(y)$, we get

$$\mathfrak{m}(\backslash_x f(xy)) = \mathfrak{m} \star \partial(y)(f) = \partial(y)(g) = g(y)$$

so that $\backslash_y \mathfrak{m}(\backslash_x f(xy)) = g \in U_b(X)$. Hence

$$\mathfrak{m} \star \mathfrak{n}(f) = \mathfrak{n}(g) = \mathfrak{n}(\backslash_y \mathfrak{m}(\backslash_x f(xy)))$$

for all $\mathfrak{n} \in \mathfrak{M}_b(X)$. □

The similarity between Theorems 9.11 and 9.24 suggests that the spaces $\mathfrak{M}_u(X)$ and $\Lambda(\mathfrak{M}_b(X))$ are related. The next lemma describes the obvious part of the relationship for $\mathfrak{M}_b(X)$ and its subsemigroups.

Lemma 9.25. *Let X be a semiuniform semigroup. Let $S \subseteq \mathfrak{M}_b(X)$ be closed under the $\star$ operation. Consider S as a semigroup with $\star$ and the $U_b(X)$-weak topology.*

1. $\Lambda(S) \supseteq S \cap \mathfrak{M}_u(X)$.
2. *If X is precompact, then $\Lambda(S) = S = S \cap \mathfrak{M}_u(X)$.*

Proof. Part 1 follows from Part 5 of Theorem 9.20.

If X is precompact, then $S = S \cap \mathfrak{M}_u(X)$ by Corollary 6.3, and

$$\Lambda(S) \subseteq S = S \cap \mathfrak{M}_u(X) \subseteq \Lambda(S)$$

by Part 1. That proves Part 2. □

In particular, for $S = \mathfrak{M}_b(X)$, we get $\mathfrak{M}_u(X) \subseteq \Lambda(\mathfrak{M}_b(X))$ for every semiuniform semigroup X and $\mathfrak{M}_u(X) = \Lambda(\mathfrak{M}_b(X))$ when X is precompact. It is easy to find instances where the equality does not hold.

Exercise 9.26. Construct a semigroup X with the discrete uniformity such that $\Lambda(\mathfrak{M}_b(X)) = \mathfrak{M}_b(X) \neq \mathfrak{M}_u(X)$. ∎

The space $\mathfrak{M}_b(X)$ contains interesting convolution semigroups other than the whole $\mathfrak{M}_b(X)$. The uniform compactification $\hat{p}X$ of X is a subsemigroup of $\mathfrak{M}_b(X)$ by Theorem 9.18. By Lemma 9.25, $\hat{X} \subseteq \Lambda(\hat{p}X)$ for every semiuniform semigroup X, and $\hat{X} = \Lambda(\hat{p}X)$ when X is precompact. By the forthcoming Corollary 9.30, the equalities $\mathfrak{M}_u(X) = \Lambda(\mathfrak{M}_b(X))$ and $\hat{X} = \Lambda(\hat{p}X)$ hold also for ambitable semigroups.

Lemma 9.27. *Let X be a semiuniform semigroup and $f \in U_b(X)$. Let $\varphi \colon \hat{p}X \to \mathbb{R}^X$ be the mapping defined by $\varphi(\mathfrak{n}) := \backslash_x \mathfrak{n}(\backslash_y f(xy)) = \mathfrak{n} \bullet f$ for $\mathfrak{n} \in \hat{p}X$.*

1. φ *is continuous from $\hat{p}X$ to $\mathbb{R}^X$ with the X-pointwise topology.*
2. $\varphi(\hat{p}X) = \overline{\mathrm{orb}}(f)$.

Proof. 1. For every $x \in X$, the mapping $\mathfrak{n} \mapsto \partial(x) \star \mathfrak{n}$ is continuous from $\hat{\mathfrak{p}}X$ to itself. As $\partial(x) \star \mathfrak{n}(f) = \mathfrak{n}(\backslash_y f(xy))$, the mapping $\mathfrak{n} \mapsto \mathfrak{n}(\backslash_y f(xy))$ from $\hat{\mathfrak{p}}X$ to $\mathbb{R}$ is continuous for each $x \in X$, and therefore the mapping $\mathfrak{n} \mapsto \backslash_x \mathfrak{n}(\backslash_y f(xy))$ is continuous from $\hat{\mathfrak{p}}X$ to $\mathbb{R}^X$.

2. Clearly $\varphi(\partial(y)) = \backslash_x f(xy)$ for every $y \in X$, hence $\varphi(\partial(X)) = \mathrm{orb}(f)$. The mapping φ is continuous by Part 1, $\hat{\mathfrak{p}}X$ is compact, and $\partial(X)$ is dense in $\hat{\mathfrak{p}}X$. It follows that $\varphi(\hat{\mathfrak{p}}X) = \overline{\mathrm{orb}}(f)$.   □

Lemma 9.28. *Let X be a semiuniform semigroup, $\mathfrak{m} \in \mathfrak{M}_b(X)$ and $f \in \mathsf{U}_b(X)$. Let the mapping $\mathfrak{n} \mapsto \mathfrak{m} \star \mathfrak{n}$ from $\hat{\mathfrak{p}}X$ to $\mathfrak{M}_b(X)$ be $\mathsf{U}_b(X)$-weakly continuous. Then $\mathfrak{m}$ is X-pointwise continuous on $\overline{\mathrm{orb}}(f)$.*

Proof. As in Lemma 9.27, let $\varphi(\mathfrak{n}) := \backslash_x \mathfrak{n}(\backslash_y f(xy))$ for $\mathfrak{n} \in \hat{\mathfrak{p}}X$. By the definition of convolution, $\mathfrak{m} \star \mathfrak{n}(f) = \mathfrak{m}(\backslash_x \mathfrak{n}(\backslash_y f(xy))) = \mathfrak{m}(\varphi(\mathfrak{n}))$. Thus $\mathfrak{m} \circ \varphi$ is continuous from $\hat{\mathfrak{p}}X$ to $\mathbb{R}$.

By Lemma 9.27, φ is continuous from $\hat{\mathfrak{p}}X$ to $\overline{\mathrm{orb}}(f)$, and $\varphi(\hat{\mathfrak{p}}X) = \overline{\mathrm{orb}}(f)$. Since $\hat{\mathfrak{p}}X$ is compact, it follows that $\mathfrak{m}$ is continuous on $\overline{\mathrm{orb}}(f)$.   □

Theorem 9.29. *Let X be a semiuniform semigroup, and let $S \subseteq \mathfrak{M}_b(X)$ be a semigroup with the $\star$ operation and the $\mathsf{U}_b(X)$-weak topology. If X is ambitable and $S \supseteq \hat{\mathfrak{p}}X$, then $\Lambda(S) = \mathfrak{M}_u(X) \cap S$.*

Proof. The inclusion $\mathfrak{M}_u(X) \cap S \subseteq \Lambda(S)$ holds by Lemma 9.25.

To prove the opposite inclusion, take any $\mathfrak{m} \in \Lambda(S)$ and any $\Delta \in \mathsf{UP}(X)$. Since $\hat{\mathfrak{p}}X \subseteq S$, the mapping $\mathfrak{n} \mapsto \mathfrak{m} \star \mathfrak{n}$ from $\hat{\mathfrak{p}}X$ to $\mathfrak{M}_b(X)$ is $\mathsf{U}_b(X)$-weakly continuous by the definition of $\Lambda(S)$. Since X is ambitable, $\mathsf{BLip}_b(\Delta) \subseteq \overline{\mathrm{orb}}(f)$ for some $f \in \mathsf{U}_b(X)$. By Lemma 9.28, $\mathfrak{m}$ is continuous on $\overline{\mathrm{orb}}(f)$ and therefore also on $\mathsf{BLip}_b(\Delta)$. Thus $\mathfrak{m} \in \mathfrak{M}_u(X)$.   □

Corollary 9.30. *Let X be an ambitable semiuniform semigroup. Then $\Lambda(\hat{\mathfrak{p}}X) = \hat{X}$ and $\Lambda(\mathfrak{M}_b(X)) = \mathfrak{M}_u(X)$.*   □

To conclude this section, I now establish a connection between topological centres and unique amenability. In the next section, this will yield results about uniquely amenable topological groups.

Definition 9.31. Let X be a uniform space. A functional $\mathfrak{m} \in \mathfrak{M}_b(X)$ is a *mean on X* iff $\mathfrak{m} \geq 0$ and $\mathfrak{m}(1) = 1$.

Let X be a semiuniform semigroup. A functional $\mathfrak{m} \in \mathfrak{M}_b(X)$ is *left-invariant* iff $\partial(x) \star \mathfrak{m} = \mathfrak{m}$ for every $x \in X$. Say that X is *uniquely (left-)amenable* iff there exists exactly one left-invariant mean in $\mathfrak{M}_b(X)$.

A semitopological semigroup, and in particular a topological group, is *uniquely amenable* iff it is uniquely amenable as a semiuniform semigroup with its right uniformity.   ∎

Lemma 9.32. *Let X be a semiuniform semigroup. If $\mathfrak{m} \in \mathfrak{M}_b(X)$ is left-invariant, then so is $\mathfrak{m}^+$.*

Proof. Take any left-invariant $m \in \mathfrak{M}_b(X)$ and $x \in X$. Since $m^+ \geq m$ and $m^+ \geq 0$, we have $\partial(x) \star m^+ \geq \partial(x) \star m = m$ and $\partial(x) \star m^+ \geq 0$; therefore $\partial(x) \star m^+ - m^+ \geq 0$. At the same time,

$$(\partial(x) \star m^+ - m^+)(1) = \partial(x) \star m^+(1) - m^+(1) = m^+(1) - m^+(1) = 0,$$

which shows that $\partial(x) \star m^+ = m^+$. □

Theorem 9.33. *Let X be a uniquely amenable semiuniform semigroup with the unique left-invariant mean m.*

1. *If $n \in \mathfrak{M}_b(X)$ is left-invariant, then $n = n(1)\,m$.*
2. *$m \in \Lambda(\mathfrak{M}_b(X))$.*

Proof. 1. If $n \in \mathfrak{M}_b(X)$ is left-invariant, then so is n^+ by Lemma 9.32. If also $n^+(1) \neq 0$, then $n^+/n^+(1)$ is a left-invariant mean; hence $n^+ = n^+(1)\,m$. Similarly $n^- = n^-(1)\,m$.

2. Take any n and n_γ in $\mathfrak{M}_b(X)$ such that $\lim_\gamma n_\gamma(f) = n(f)$ for every $f \in U_b(X)$. Since $m \star n_\gamma$ and $m \star n$ are left-invariant, from Part 1 we obtain $m \star n_\gamma = n_\gamma(1)\,m$ and $m \star n = n(1)\,m$. Since $\lim_\gamma n_\gamma(1) = n(1)$, it follows that $\lim_\gamma m \star n_\gamma(f) = m \star n(f)$ for every $f \in U_b(X)$. □

9.4 Case of Topological Groups

For the reader's convenience, I now summarize the main results of this chapter in the particular case of a topological group acting on itself by left translation.

When G is a topological group, $U_b(rG) = \mathrm{LUC}(G)$ by Lemma 3.5 and Theorem 3.9, and thus $\mathfrak{M}_b(rG) = \mathrm{LUC}(G)^*$. In this section I write G^{LUC} instead of $\widehat{\mathrm{pr}}rG$, to conform to the notation often used in abstract harmonic analysis.

By Theorem 7.20, if G is a locally compact group, then $\mathfrak{M}_u(rG) = \mathfrak{M}_t(rG)$; the latter space is also known as $M(G)$ in the literature.

By Definition 9.15, the convolution $m \star n$ of $m, n \in \mathrm{LUC}(G)^*$ satisfies

$$m \star n(f) = m(\backslash_x n(\backslash_y f(xy))) = m(n \bullet f)$$

for $f \in \mathrm{LUC}(G)$, where $n \bullet f$ is the canonical action of $\mathrm{LUC}(G)^*$ on $\mathrm{LUC}(G)$.

By Theorems 9.16, 9.17 and 9.18 and Exercise 9.19, $\mathrm{LUC}(G)^*$ with its norm $\|\cdot\|$ and the $\star$ multiplication is a Banach algebra, and each of

$$\mathrm{LUC}(G)^{*+}, \ \mathrm{Mol}(G), \ \mathfrak{M}_t(rG), \ \mathfrak{M}_u(rG), \ \widehat{rG}, \ G^{\mathrm{LUC}}$$

with the $\star$ operation is a subsemigroup of $\mathrm{LUC}(G)^*$.

As I explain in Sect. 9.2, finite algebraic identities satisfied by the convolution operation on $\mathrm{Mol}(G)$ are inherited by $\mathfrak{M}_u(rG)$. If G is commutative, then so is $\star$ on $\mathfrak{M}_u(rG)$.

Exercise 9.34. Find a commutative topological group G for which $\mathsf{LUC}(G)^*$ with the $\star$ operation is not commutative. ∎

Exercise 9.35. Let $\varphi\colon G \to G'$ be a continuous homomorphism of topological groups. Prove that $\varphi(m \star n) = \varphi(m) \star \varphi(n)$ for $m, n \in \mathsf{LUC}(G)^*$. ∎

The following corollary is simply a restatement of Corollary 9.21 for the special case of a topological group.

Corollary 9.36 (of Theorem 9.20). *Let G be a topological group.*

1. *Let $\mathfrak{B} \subseteq \mathsf{LUC}(G)^*$ be a $\|\cdot\|$ bounded set, $m_0 \in \mathfrak{M}_u(rG)$, and $n_0 \in \mathfrak{B}$. With the $\mathsf{UEB}(rG)$ topology on $\mathfrak{B}$ and $\mathsf{LUC}(G)^*$, the mapping $(m,n) \mapsto m \star n$ from $\mathsf{LUC}(G)^* \times \mathfrak{B}$ to $\mathsf{LUC}(G)^*$ is jointly continuous at (m_0, n_0).*
2. *Let $\mathfrak{A}$ be the unit sphere $\{m \in \mathsf{LUC}(G)^* \mid \|m\| = 1\}$, let $\mathfrak{B} \subseteq \mathsf{LUC}(G)^*$ be a $\|\cdot\|$ bounded set, $m_0 \in \mathfrak{A} \cap \mathfrak{M}_u(rG)$ and $n_0 \in \mathfrak{B}$. With the $\mathsf{LUC}(G)$-weak topology on $\mathfrak{A}$, $\mathfrak{B}$ and $\mathsf{LUC}(G)^*$, the mapping $(m,n) \mapsto m \star n$ from $\mathfrak{A} \times \mathfrak{B}$ to $\mathsf{LUC}(G)^*$ is jointly continuous at (m_0, n_0).*
3. *Let $m_0 \in \mathfrak{M}_u(rG)^+$ and $n_0 \in \mathsf{LUC}(G)^{*+}$. With the $\mathsf{LUC}(G)$-weak topology on $\mathsf{LUC}(G)^*$, the mapping $(m,n) \mapsto m \star n$ from $\mathsf{LUC}(G)^{*+} \times \mathsf{LUC}(G)^{*+}$ to $\mathsf{LUC}(G)^{*+}$ is jointly continuous at (m_0, n_0).*
4. *With the $\mathsf{LUC}(G)$-weak topology on $\mathfrak{M}_u(rG)$, convolution is jointly sequentially continuous as a mapping from $\mathfrak{M}_u(rG) \times \mathfrak{M}_u(rG)$ to $\mathfrak{M}_u(rG)$.*
5. *If $m \in \mathfrak{M}_u(rG)$, then the mapping $n \mapsto m \star n$ from $\mathsf{LUC}(G)^*$ to itself is continuous in the $\mathsf{LUC}(G)$-weak topology.* □

Exercise 9.37. Let G be a topological group such that the mapping $(x,y) \mapsto xy$ is uniformly continuous from the uniform product $rG \times rG$ to rG (cf. Exercise 3.11). Prove that Part 1 of Corollary 9.36 holds with any $m_0 \in \mathsf{LUC}(G)^*$. ∎

The next two examples demonstrate that various restricting assumptions cannot be omitted in Corollary 9.36 and Exercise 9.37.

Example 9.38. This example shows that convolution need not be jointly $\mathsf{LUC}(G)$-weakly continuous on $\|\cdot\|$ bounded sets of molecular measures.

Let G be the additive group of integers with the discrete uniformity. Identify the set G with $\partial(G) \subseteq G^{\mathsf{LUC}}$ (Sect. 6.5), and fix $x \in G^{\mathsf{LUC}} \setminus G$. There is a net $\{x_\gamma\}_{\gamma \in \Gamma}$ in G that converges to x in the $\mathsf{LUC}(G)$-weak topology. For every $\gamma \in \Gamma$ there is $\beta(\gamma) \geq \gamma$ such that $x_\gamma \neq x_{\beta(\gamma)}$, and clearly the net $\{x_{\beta(\gamma)}\}_{\gamma \in \Gamma}$ also converges to x in the $\mathsf{LUC}(G)$-weak topology. The mapping $\varphi\colon y \mapsto -y$ is a uniform automorphism of rG, hence it extends to a homeomorphism $\widetilde{\varphi}\colon G^{\mathsf{LUC}} \to G^{\mathsf{LUC}}$. The nets $\{-x_\gamma\}_\gamma$ and $\{-x_{\beta(\gamma)}\}_\gamma$ converge to $\widetilde{\varphi}(x)$.

Let $m_\gamma := \partial(x_\gamma) - \partial(x_{\beta(\gamma)})$ and $n_\gamma := \partial(-x_\gamma) - \partial(-x_{\beta(\gamma)})$ for $\gamma \in \Gamma$. Define $f \in \mathsf{LUC}(G)$ by $f(0) := 1$ and $f(y) := 0$ for $y \neq 0$. Then

$$m_\gamma \star n_\gamma = 2\partial(0) - \partial(x_{\beta(\gamma)} - x_\gamma) - \partial(x_\gamma - x_{\beta(\gamma)})$$

$$m_\gamma \star n_\gamma(f) = 2f(0) - f(x_{\beta(\gamma)} - x_\gamma) - f(x_\gamma - x_{\beta(\gamma)}) = 2$$

and therefore the net $\{\mathfrak{m}_\gamma \star \mathfrak{n}_\gamma\}_\gamma$ does not converge to 0 in the $\mathrm{LUC}(G)$-weak topology, while $\lim_\gamma \mathfrak{m}_\gamma = \lim_\gamma \mathfrak{n}_\gamma = 0$. ∎

Example 9.39. This example shows that, for a general functional $\mathfrak{m} \in \mathrm{LUC}(G)^{*+}$, the mapping $\mathfrak{n} \mapsto \mathfrak{m} \star \mathfrak{n}$ need not be $\mathrm{UEB}(rG)$ sequentially continuous on $\mathrm{LUC}(G)^{*+}$; in fact, not even on $\widehat{rG}$. Note that, in contrast, for every topological group G and every $\mathfrak{m} \in \mathrm{LUC}(G)^*$, the mapping $\mathfrak{n} \mapsto \mathfrak{n} \star \mathfrak{m}$ is $\mathrm{UEB}(rG)$ continuous on $\mathrm{LUC}(G)^*$. This further illustrates the asymmetry between right and left for convolution on $\mathrm{LUC}(G)^*$.

The example exhibits a topological group G, $\mathfrak{m} \in G^{\mathrm{LUC}}$ and a sequence $\{x_j\}_j$ in G such that $\lim_j x_j = \mathfrak{n} \in \widehat{rG}$ in the UEB topology and yet the sequence of $\mathfrak{m} \star x_j$ does not converge to $\mathfrak{m} \star \mathfrak{n}$ even in the $\mathrm{LUC}(G)$-weak topology.

Let G be the group of those homeomorphisms of the interval $[0,4]$ onto itself that fix the endpoint 0 (and therefore also the endpoint 4). The group operation is mapping composition, and the topology of G is that of the right-invariant metric

$$\Delta(x,y) := \sup_{t\in[0,4]} |x(t) - y(t)| \text{ for } x,y \in G.$$

Define $x_j \in G$ for $j = 1,2,\dots$ by

$$x_j(t) := \begin{cases} t/j & \text{when } 0 \le t \le 2 \\ (2 + (2j-1)(t-2))/j & \text{when } 2 < t \le 4 \end{cases}$$

The sequence $\{x_j\}_j$ is Cauchy in rG. Identify again the set G with $\partial(G) \subseteq \mathrm{LUC}(G)^*$ as in Sect. 6.5. Then $\{x_j\}_j$ is Cauchy in the UEB uniformity; let $\mathfrak{n}$ be its UEB limit in $\widehat{rG} \subseteq \mathrm{LUC}(G)^*$.

Define the function $h \in \mathrm{LUC}(G)$ by $h(z) := z(1)$ for $z \in G$. Let $\mathfrak{m}$ be a cluster point of the sequence $\{x_k^{-1}\}_k$ in G^{LUC}. For $k \ge 2j$ we have

$$h(x_k^{-1} \circ x_j) = x_k^{-1}(x_j(1)) = x_k^{-1}(1/j) = 2 + \frac{(1/j) - (2/k)}{2 - (1/k)},$$

hence $\mathfrak{m} \star x_j(h) = \lim_k h(x_k^{-1} \circ x_j) = 2 + 1/2j$. On the other hand,

$$\mathfrak{n}(\backslash_z h(y \circ z)) = \lim_j h(y \circ x_j) = \lim_j y(x_j(1)) = 0 \quad \text{for all } y \in G,$$

$$\mathfrak{m} \star \mathfrak{n}(h) = \mathfrak{m}(\backslash_y \mathfrak{n}(\backslash_z h(y \circ z))) = 0 \ne 2 = \lim_j \mathfrak{m} \star x_j(h)$$

so that the sequence $\{\mathfrak{m} \star x_j\}_j$ does not converge to $\mathfrak{m} \star \mathfrak{n}$ in the $\mathrm{LUC}(G)$-weak topology. ∎

Some properties of $\mathfrak{M}_t(rG) = \mathfrak{M}_u(rG)$ for G locally compact carry over to $\mathfrak{M}_u(rG)$ for a general topological group G. On the other hand, the following exercise is an example of an algebraic property that does not generalize this way.

Exercise 9.40. 1. Let G be a topological group, $m \in \mathfrak{M}_t(rG)^+$, $n \in LUC(G)^{*+}$ and $m \star n \in \mathfrak{M}_t(rG)$. Prove that $n \in \mathfrak{M}_t(rG)$.
2. Find a topological group G, $m \in \mathfrak{M}_u(rG)^+$ and $n \in LUC(G)^{*+} \setminus \mathfrak{M}_u(rG)$ such that $m \star n \in \mathfrak{M}_u(rG)$.   ∎

In accordance with Definition 9.23, when G is a topological group and S is a subsemigroup of $LUC(G)^*$ with $\star$ and the $LUC(G)$-weak topology, the topological centre of S is

$$\Lambda(S) := \{m \in S \mid \text{the mapping } n \mapsto m \star n \text{ is } LUC(G)\text{-weakly continuous on } S\}.$$

By Theorem 9.24, a functional $m \in LUC(G)^*$ is in $\Lambda(LUC(G)^*)$ if and only if for all $f \in LUC(G)$ and $n \in LUC(G)^*$ we have $\backslash_y m(\backslash_x f(xy)) \in LUC(G)$ and

$$m \star n(f) = n(\backslash_y m(\backslash_x f(xy))).$$

Theorem 9.41. *Let G be a topological group, and let $S \subseteq LUC(G)^*$ be a semigroup with the $\star$ operation and the $LUC(G)$-weak topology.*

1. $\Lambda(S) \supseteq S \cap \mathfrak{M}_u(rG)$.
2. *If G is precompact, then $\Lambda(S) = S = S \cap \mathfrak{M}_u(rG)$.*
3. *If G is ambitable and $S \supseteq G^{LUC}$, then $\Lambda(S) = S \cap \mathfrak{M}_u(rG)$.*

Proof. Parts 1 and 2 are a special case of Lemma 9.25. Part 3 is a special case of Theorem 9.29.   □

Corollary 9.42. *Let G be an ambitable topological group. Then $\Lambda(G^{LUC}) = \widehat{rG}$ and $\Lambda(LUC(G)^*) = \mathfrak{M}_u(rG)$.*   □

Thus we have $\Lambda(G^{LUC}) = \widehat{rG}$ and $\Lambda(LUC(G)^*) = \mathfrak{M}_u(rG)$ whenever G satisfies one of the conditions in Theorem 3.35. In particular, the two equalities hold for every $\aleph_0$-bounded group. For a locally compact group G, we get $\Lambda(G^{LUC}) = G$ because rG is complete, and $\Lambda(LUC(G)^*) = \mathfrak{M}_u(rG) = \mathfrak{M}_t(rG)$.

Corollary 9.42 may be used to characterize certain uniquely amenable groups. This is illustrated in Theorem 9.44 at the end of this section. The proof relies on the next theorem, which is of independent interest.

On every precompact topological group, there exists a non-zero left-invariant uniform measure (see the notes in Sect. 9.5). The following result is a converse.

Theorem 9.43. *Let G be a topological group for which there exists a left-invariant $m \in \mathfrak{M}_u(rG)^+$, $m \neq 0$. Then G is precompact.*

Proof. Take any uniformly discrete set Y in rG. There is $\Delta \in RP(G)$ such that $\Delta(y,y') \geq 2$ for $y,y' \in Y$, $y \neq y'$.
    For every non-empty finite set $D \subseteq G$, define $g_D \in BLip_b(\Delta)$ by

$$g_D(x) := (1 - \Delta(x,D))^+, \ x \in G.$$

Finite subsets of G are upwards directed by $\subseteq$, and $g_D \nearrow 1$. If $\mathfrak{m} \in \mathfrak{M}_u(rG)^+$, $\mathfrak{m} \neq 0$, then $\mathfrak{m}(g_D) > 0$ for some D, hence $\mathfrak{m}(g_{\{z\}}) > 0$ for at least one $z \in D$. Fix such z and for $y \in Y$ define $f_y \in \mathsf{U}_b(rG)$ by $f_y(x) := g_{\{z\}}(yx)$ for $x \in G$.

If $y, y' \in Y$ and $x \in G$ are such that $f_y(x) > 0$ and $f_{y'}(x) > 0$, then $1 - \Delta(yx, z) > 0$ and $1 - \Delta(y'x, z) > 0$,

$$\Delta(y, y') = \Delta(yx, y'x) \leq \Delta(yx, z) + \Delta(y'x, z) < 2,$$

hence $y = y'$. Thus $f_y \wedge f_{y'} = 0$ for $y, y' \in Y$, $y \neq y'$.

If $\mathfrak{m}$ is left-invariant, then

$$\mathfrak{m}(f_y) = \mathfrak{m}(\setminus_x g_{\{z\}}(yx)) = \partial(y) \star \mathfrak{m}(g_{\{z\}}) = \mathfrak{m}(g_{\{z\}}) \text{ for } y \in Y$$

so that $|Y| \mathfrak{m}(g_{\{z\}}) \leq \mathfrak{m}(1)$. It follows that Y is finite.   □

Theorem 9.44. *No ambitable topological group is uniquely amenable.*

Proof. Apply Theorems 3.24, 9.33 and 9.43 and Corollary 9.42.   □

Corollary 9.45. *Every uniquely amenable locally compact group is compact. Every uniquely amenable $\aleph_0$-bounded group is precompact.*   □

9.5 Notes for Chap. 9

The definition and properties of the direct product in Sect. 9.1 are taken from my paper [138]. Other variants of the direct product of uniform measures were defined by Khurana [111] and Zahradník [181].

Bourbaki [13, Ch.8] defined the convolution of measures on locally compact spaces as the image of the direct product measure. Hewitt and Ross [97, 19.1] and Pym [149] defined convolution for general functionals on spaces of functions on a semigroup. For a semigroup acting on itself, Definition 9.15 agrees with the definition of convolution in [97] and with that of evolution in [149]. When restricted to the uniform compactification, Definition 9.15 agrees with the operation defined by Hindman and Strauss [98, 21.43]. When cylindrical measures on a locally convex space E are identified with uniform measures on wE as in Sect. 8.6, the convolution of cylindrical measures [5, Exp.2] [165, II.II.2] is a special case of convolution of uniform measures on topological groups.

LeCam [122] noted that the space of uniform measures "arises naturally in various arguments about convolutions or Fourier transforms on linear spaces". Continuing LeCam's approach, Caby [23] proved several results about the continuity of convolution of uniform measures on commutative groups.

For general (not necessarily commutative) topological groups, Csiszár [32] defined a property equivalent to being a positive uniform measure, which he called

ρ-continuity, and used it to prove results about the continuity of convolution. He also suggested extending the definition of convolution to $m \star n$ where $m \in \mathfrak{M}_b(X)$, $n \in \mathfrak{M}_b(Y)$ and X acts on Y.

Various forms of the continuity properties in Theorem 9.20, Corollaries 9.21 and 9.36 and Exercise 9.37 are due to Caby [23], Csiszár [32, Th.1], Ferri and Neufang [55, 4.2], Neufang et al. [132], Salmi [162] and Tortrat [170]. Pym [149], [150] proved continuity results for several natural topologies on general convolution algebras.

LeCam [123] proved that convolution on $\mathfrak{M}_u(rG)$ is commutative when G is the additive group of a locally convex space; his proof works more generally for any commutative group G. It is well known [32] [97, 19.24] [149] that convolution on $\mathfrak{M}_b(rG)$ need not be commutative for commutative G (Exercise 9.34).

Theorem 9.24 is a straightforward modification of [118, L.2]. Theorem 9.29 with its proof is adapted from [140], where I proved it for topological groups.

Definition 9.31 is a simple adaptation of the concepts commonly defined for topological and semitopological semigroups; see e.g. Lau [119]. Lemma 9.32 is a special case of a general result of Namioka [130, 3.2]. Theorem 9.33 was used by Lau [118, Cor.5] in his proof that every uniquely amenable locally compact group is compact.

Example 9.38 is a special case of a general construction of Salmi [162, Th.1]. Another example of Salmi [162, p.161] may be used to show that Part 2 of Theorem 9.12 does not hold without the assumption $\sup_\gamma \|n_\gamma\| < \infty$, even when $X = Y$ is the compact group $\mathbb{R}/\mathbb{Z}$. Example 9.39 is taken from [132].

Csiszár [32] proved Part 1 in Exercise 9.40 and asked a question [32, p.36] that is answered in Part 2 of the exercise (adapted from my paper [138]).

The uniform compactification $\widehat{\mathfrak{p}rG} = G^{\mathsf{LUC}}$ of a topological group G with its right uniformity is known as the *canonical $\mathcal{L}C$-compactification* [8], *universal enveloping semigroup* [176], $\mathcal{L}\mathcal{U}\mathcal{C}$-compactification [120], or *greatest ambit* [143] of G; it features prominently in the general theory of semigroup compactifications, surveyed by Berglund et al. [8, Ch.3] and Ruppert [161, Ch.III].

Corollary 9.42 generalizes the results of Ferri and Neufang [55], Lau [118] and Lau and Pym [120], which in turn generalize a number of previous results about topological centres. For locally compact groups, the statement in the corollary was further refined by Budak et al. [19].

If G is a precompact group, then $\widehat{rG}$ is a compact group [31, 11.3.23] [159, 10.16], and by Lemma 6.9, the Haar measure on $\widehat{rG}$ defines a left-invariant $m \in \mathfrak{M}_u(rG)^+$. The converse is Theorem 9.43, which I proved in [137].

Corollary 9.45 was proved by Lau [118] for locally compact groups, and by Megrelishvili et al. [128] for $\aleph_0$-bounded groups.

Research Problem 3. This is related to Problem 1 in Sect. 3.5. Characterize, or at least find large interesting classes of, semiuniform semigroups for which $\widehat{X} = \Lambda(\widehat{\mathfrak{p}}X)$ and $\mathfrak{M}_u(X) = \Lambda(\mathfrak{M}_b(X))$. In particular, do the two equalities hold for all topological groups?

Theorem 3.35 with Corollary 9.30 provide a partial answer to the last question. The memoir by Dales et al. [33] includes an in-depth treatment of these and related problems for the case where X is a semigroup with the discrete uniformity (so that $\mathfrak{M}_b(X) = \ell_\infty(X)^*$ and $\widehat{p}X = \beta X$). ∎

If $\mathfrak{M}_u(rG) = \Lambda(\mathfrak{M}_b(rG))$ were to hold for every topological group, then, by Theorems 9.33 and 9.43, every uniquely amenable topological group would be precompact. That would answer the question asked by Megrelishvili et al. [128].

Part III
Topics from Farther Afield

This part covers several topics that complement the basic theory in Part II. The chapters that follow do not depend on each other, and may be read in any order.

Chapter 10 covers the basics of the theory of free uniform measures, in which bounded functions and bounded sets of functions used in Part II are replaced by unbounded functions and sets. In the categorical language, this yields a free functor from the category of uniform spaces to the category of complete locally convex spaces. Chapter 11 includes an application motivated by the notion of weak convergence in probability theory. Chapter 12 deals with certain automatic continuity results by which measurable functionals are necessarily continuous.

Chapter 10
Free Uniform Measures

Uniform measures are functionals on the space of bounded uniformly continuous functions. In this chapter I describe a parallel theory of functionals on the space of all (not necessarily bounded) uniformly continuous functions. The "unbounded version" of $\mathfrak{M}_u(X)$ is the space $\mathfrak{M}_F(X)$ of *free uniform measures*. The adjective *free* refers to a universal property of $\mathfrak{M}_F(X)$ that characterizes free functors.

The natural mapping $\mathfrak{M}_F(X) \to \mathfrak{M}_u(X)$ is injective, so that every free uniform measure identifies with a unique uniform measure. Theorem 10.5 characterizes those uniform measures that are free in this sense. The rest of Sect. 10.1 is devoted to analogues for $\mathfrak{M}_F(X)$ of the properties previously established for $\mathfrak{M}_u(X)$. Section 10.2 describes the universal (free) property of $\mathfrak{M}_F(X)$.

In Sect. 10.3, I characterize free uniform measures on spaces with the CDE property. The remaining sections deal with several known spaces of measures that turn out to be particular instances of $\mathfrak{M}_F(X)$ for suitably chosen uniform spaces.

10.1 Basic Properties

For any uniform space X, the space $U(X)$ of all uniformly continuous real-valued functions on X with the X-pointwise order is a Riesz space (Theorem 1.18). Linear functionals on $U(X)$ are partially ordered by the dual order $\leq$ (Definition P.18). That defines a positive cone in any space of linear functionals on $U(X)$. However, in general, such a space with the dual order need not be a vector lattice (Riesz space).

Definition 10.1. Let X be a uniform space. A linear functional $\mathfrak{m}$ on $U(X)$ is a *free uniform measure on X* iff the restriction of $\mathfrak{m}$ to every $UE(X)$ subset of $U(X)$ is continuous in the X-pointwise topology. Let $\mathfrak{M}_F(X)$ denote the space of free uniform measures on X.

The $UE(X)$ *uniformity*, or simply the UE *uniformity*, on subsets of $\mathfrak{M}_F(X)$ is the $\mathfrak{S}$-uniformity where $\mathfrak{S}$ is the set of $UE(X)$ subsets of $U(X)$. The $UE(X)$ *topology*, or simply the UE *topology*, is the topology of the UE uniformity. ∎

J. Pachl, *Uniform Spaces and Measures*, Fields Institute Monographs 30,
DOI 10.1007/978-1-4614-5058-0_11,
© Springer Science+Business Media New York 2013

By Lemma 1.20, a linear functional on $U(X)$ is a free uniform measure if and only if its restriction to every set $\mathrm{BLip}(\Delta, h)$ is X-pointwise continuous, where $\Delta \in \mathrm{UP}(X)$ and h is 1-Lipschitz for Δ, $h \geq 0$. And the $\mathrm{UE}(X)$ uniformity is the $\mathfrak{S}$-uniformity where $\mathfrak{S}$ is the set of all such sets $\mathrm{BLip}(\Delta, h)$.

Since every $\mathrm{UEB}(X)$ set is $\mathrm{UE}(X)$, the restriction to $U_b(X)$ of a free uniform measure on X is a uniform measure. The restriction mapping $\mathfrak{M}_F(X) \to \mathfrak{M}_u(X)$ is obviously linear, and it is injective: If $\mathfrak{m} \in \mathfrak{M}_F(X)$ and $\mathfrak{m}(f) = 0$ for every $f \in U_b(X)$, then $\mathfrak{m}(f) = \lim_{j \in \omega} \mathfrak{m}((-j) \vee f \wedge j) = 0$ for every $f \in U(X)$.

Thus every free uniform measure naturally identifies with a unique uniform measure on the same space; or, stated informally, every free uniform measure is a uniform measure. Evidently every molecular measure on X, considered as a linear functional on the space $U(X)$, is a free uniform measure, so that

$$\mathrm{Mol}(X) \subseteq \mathfrak{M}_F(X) \subseteq \mathfrak{M}_u(X).$$

As Example 10.6 shows, the space $\mathfrak{M}_F(X)$ need not be included in the order-bounded dual $U(X)^\sim$ of the Riesz space $U(X)$. However, by the next lemma, if $\mathfrak{M}_F(X) \subseteq U(X)^\sim$, then the natural identification of $\mathfrak{M}_F(X)$ with a subspace of $\mathfrak{M}_u(X)$ preserves lattice operations.

Lemma 10.2. *Let X be a uniform space, $\mathfrak{m}_0, \mathfrak{m}_1 \in U(X)^\sim$, and for $j = 0, 1$ let $\mathfrak{n}_j \in U_b(X)^\sim$ be the restriction of $\mathfrak{m}_j$ to $U_b(X)$.*

1. *The restriction of $\mathfrak{m}_0^+$ to $U_b(X)$ is $\mathfrak{n}_0^+$.*
2. *The restriction of $\mathfrak{m}_0 \vee \mathfrak{m}_1$ to $U_b(X)$ is $\mathfrak{n}_0 \vee \mathfrak{n}_1$.*

Proof. Apply Lemma P.19. $\square$

Exercise 10.3. Prove that these properties of a uniform space X are equivalent:

   (i) $U(X) = U_b(X)$.
  (ii) $\mathfrak{M}_F(X) = \mathfrak{M}_u(X)$.
 (iii) $\mathfrak{M}_F(X)^+ = \mathfrak{M}_u(X)^+$.
  (iv) The $\mathrm{UE}(X)$ and $\mathrm{UEB}(X)$ topologies coincide on $\mathfrak{M}_F(X)$.
   (v) The $\mathrm{UE}(X)$ and $\mathrm{UEB}(X)$ topologies coincide on $\mathfrak{M}_F(X)^+$.   ∎

The forthcoming Theorem 10.5 characterizes free uniform measures by their values on $U_b(X)$. The key step in its proof is the following lemma.

Lemma 10.4. *Let X be a uniform space, $\Delta \in \mathrm{UP}(X)$, $h: X \to \mathbb{R}^+$, and let $\mathfrak{n} \in \mathfrak{M}_u(X)$ be such that the finite $\lim_{j \in \omega} \mathfrak{n}(f \wedge j)$ exists for every $f \in \mathrm{Lip}(\Delta, h)^+$. Then $\mathfrak{n}$ is uniformly continuous on $\mathrm{BLip}(\Delta, h) \cap U_b(X)$ in the X-pointwise uniformity.*

Proof. With the aim of deriving a contradiction, suppose that $\mathfrak{n}$ is not X-pointwise uniformly continuous on the set $\mathrm{BLip}(\Delta, h) \cap U_b(X)$, and therefore not X-pointwise continuous at 0 on $2\mathrm{BLip}(\Delta, h) \cap U_b(X)$.

There are $\varepsilon > 0$ and a net $\{h_\gamma\}_{\gamma \in \Gamma}$ in $2\mathrm{BLip}(\Delta, h) \cap U_b(X)$ such that $\lim_\gamma h_\gamma = 0$ pointwise and $\mathfrak{n}(h_\gamma) > 2\varepsilon$ for all γ. Since $h_\gamma = h_\gamma^+ - h_\gamma^-$, no generality is lost by assuming $h_\gamma \geq 0$ for all γ.

First, I construct $g_k \in 2\mathsf{BLip}(\Delta, h) \cap U_b(X)$ for $k \in \omega$ such that $0 \le g_0 \le g_1 \le \cdots$ and $\mathfrak{n}(g_k) > k\varepsilon$ for all k. Choose any $\gamma(0) \in \Gamma$ and let $g_0 = h_{\gamma(0)}$. When $k \ge 1$ and $g_0, g_1, \ldots, g_{k-1}$ have been constructed, we have $\lim_\gamma g_{k-1} \wedge h_\gamma = 0$ pointwise, and the set $\{ g_{k-1} \wedge h_\gamma \mid \gamma \in \Gamma \}$ is $\mathsf{UEB}(X)$. Hence there is $\gamma(k)$ such that $\mathfrak{n}(g_{k-1} \wedge h_{\gamma(k)}) < \varepsilon$, and $g_k := g_{k-1} \vee h_{\gamma(k)}$ is as claimed because

$$\mathfrak{n}(g_k) = \mathfrak{n}(g_{k-1}) + \mathfrak{n}(h_{\gamma(k)}) - \mathfrak{n}(g_{k-1} \wedge h_{\gamma(k)}) > (k-1)\varepsilon + 2\varepsilon - \varepsilon = k\varepsilon.$$

Now $g_k \nearrow g$, where $g \in 2\mathsf{BLip}(\Delta, h)$ and $r_g := \lim_{j \in \omega} \mathfrak{n}(g \wedge j) \in \mathbb{R}$ by the assumption. Moreover, $\lim_{j \in \omega} (g \wedge j) - ((g - g_k) \wedge j) = g_k$ pointwise for every $k \in \omega$, and $0 \le (g \wedge j) - ((g - g_k) \wedge j) \le g_k$ for $j, k \in \omega$. Since $\mathfrak{n} \in \mathfrak{M}_u(X)$, it follows that $\lim_j \mathfrak{n}((g - g_k) \wedge j) = r_g - \mathfrak{n}(g_k) < r_g - k\varepsilon$ for every $k \in \omega$.

For any fixed j we have $\lim_k \mathfrak{n}((g - g_k) \wedge j) = 0$ because $(g - g_k) \searrow 0$ and the set $\{ (g - g_k) \wedge j \mid k \in \omega \}$ is $\mathsf{UEB}(X)$. Thus there are $j(i) \in \omega$ and $k(i) \in \omega$ for $i \in \omega$ such that $0 = j(0) \le j(1) \le \cdots$ and $k(0) \le k(1) \le \cdots$ and

$$\mathfrak{n}\big((g - g_{k(i)}) \wedge j(i+1) \big) - \mathfrak{n}\big((g - g_{k(i)}) \wedge j(i) \big) < -1 \qquad (\dagger)$$

for all i. Write

$$f_m := \max \big\{ (g - g_{k(i)}) \wedge j(i+1) \mid 0 \le i \le m \big\} \ \text{ for } m \in \omega,$$

and $f := \lim_m f_m$, so that $f_m, f \in 4\mathsf{BLip}(\Delta, h)^+$. Then

$$f_{m-1} \wedge \big((g - g_{k(m)}) \wedge j(m+1) \big) = (g - g_{k(m)}) \wedge j(m)$$

because $(g - g_k) \searrow 0$. That together with $(\dagger)$ yields

$$\mathfrak{n}(f_m) = \mathfrak{n}(f_{m-1}) + \mathfrak{n}\big((g - g_{k(m)}) \wedge j(m+1) \big) - \mathfrak{n}\big((g - g_{k(m)}) \wedge j(m) \big)$$
$$< \mathfrak{n}(f_{m-1}) - 1,$$

so that $\lim_m \mathfrak{n}(f_m) = -\infty$. Since $f \wedge j(m) = f_m$, this is a contradiction. $\qquad \square$

Theorem 10.5. *Let X be any uniform space. These two properties of a linear functional $\mathfrak{n}$ on $U_b(X)$ are equivalent:*

(i) $\mathfrak{n}$ is a restriction of a free uniform measure to $U_b(X)$.
(ii) $\mathfrak{n} \in \mathfrak{M}_u(X)$ and finite $\lim_{j \in \omega} \mathfrak{n}((-j) \vee f \wedge j)$ exists for every $f \in U(X)$.

Proof. Let $\mathfrak{n}$ be the restriction to $U_b(X)$ of $m \in \mathfrak{M}_F(X)$. Then obviously $\mathfrak{n} \in \mathfrak{M}_u(X)$. For any function $f \in U(X)$, the set $\{ (-j) \vee f \wedge j \mid j \in \omega \}$ is $\mathsf{UE}(X)$ and therefore $m(f) = \lim_{j \in \omega} \mathfrak{n}((-j) \vee f \wedge j)$. Thus (i) implies (ii).

For the converse, let $\mathfrak{n}$ have property (ii) and $m(f) := \lim_{j \in \omega} \mathfrak{n}((-j) \vee f \wedge j)$ for $f \in U(X)$. Take any $\Delta \in \mathsf{UP}(X)$ and any Δ-1-Lipschitz function $h \colon X \to \mathbb{R}^+$. By Lemma 10.4, $\mathfrak{n}$ is X-pointwise uniformly continuous on $\mathsf{BLip}(\Delta, h) \cap U_b(X)$.

By Theorem 2.24, $\mathfrak{n}$ agrees on $\mathsf{BLip}(\Delta,h)\cap\mathsf{U_b}(X)$ with an X-pointwise uniformly continuous mapping $\mathfrak{m}'\colon\mathsf{BLip}(\Delta,h)\to\mathbb{R}$. But then $\mathfrak{m}(f)=\mathfrak{m}'(f)$ for $f\in\mathsf{BLip}(\Delta,h)$ by the continuity of $\mathfrak{m}'$; hence $\mathfrak{m}$ is X-pointwise continuous on $\mathsf{BLip}(\Delta,h)$.

Moreover, if $f,g\in\frac{1}{2}\mathsf{BLip}(\Delta,h)$, then $f+g\in\mathsf{BLip}(\Delta,h)$ and

$$f(x)+g(x) = \lim_{j\in\omega}\left((-j)\vee f(x)\wedge j\right)+\left((-j)\vee g(x)\wedge j\right)$$

for all $x\in X$, and $\mathfrak{m}(f+g)=\mathfrak{m}'(f+g)=\mathfrak{m}(f)+\mathfrak{m}(g)$ by the continuity of $\mathfrak{m}'$. Thus $\mathfrak{m}$ is a linear functional on $\mathsf{U}(X)$ because for any $f,g\in\mathsf{U}(X)$ there are $\Delta\in\mathsf{UP}(X)$ and a Δ-1-Lipschitz function $h\colon X\to\mathbb{R}^+$ such that $f,g\in\frac{1}{2}\mathsf{BLip}(\Delta,h)$.   □

Example 10.6. This example demonstrates that free uniform measures need not be order-bounded as functionals on the Riesz space $\mathsf{U}(X)$.

Define $\mathfrak{m}\in\mathfrak{M}_\mathsf{F}(\mathbb{R})$ by

$$\mathfrak{m}(f):=\sum_{j=2}^{\infty}\frac{1}{j^2}\left(f(j)-f\left(j+\frac{1}{j}\right)\right)$$

for $f\in\mathsf{U}(\mathbb{R})$. Let $g(x):=|x|$ for $x\in\mathbb{R}$. For every $k\in\omega$, $k\geq 2$, there is $f_k\in\mathsf{U_b}(X)$ such that $0\leq f_k\leq g$, $f_k(j)=j$ and $f_k(j+1/j)=0$ for $j=2,3,\ldots,k$, and $f_k(x)=0$ for all $x\geq k+1$. Then $\mathfrak{m}(f_k)=\sum_{j=2}^{k}1/j$, and thus $\mathfrak{m}$ is not bounded on the set $\{f\in\mathsf{U}(X)\mid 0\leq f\leq g\}$.

Note that there is no $\mathfrak{m}'\in\mathfrak{M}_\mathsf{F}(\mathbb{R})^+$ such that $\mathfrak{m}'\geq\mathfrak{m}$. Therefore $\mathfrak{M}_\mathsf{F}(\mathbb{R})$ is not a Riesz space.   ∎

Exercise 10.7. As shown in the previous example, $\mathfrak{M}_\mathsf{F}(\mathbb{R})\setminus\mathsf{U}(\mathbb{R})^\sim\neq\emptyset$. Find a uniform space X for which $\mathsf{U}(X)^\sim\setminus\mathfrak{M}_\mathsf{F}(X)\neq\emptyset$.   ∎

Example 10.6 shows that the space $\mathfrak{M}_\mathsf{F}(X)$ need not be spanned by its positive cone $\mathfrak{M}_\mathsf{F}(X)^+$. Nevertheless, a number of results proved in Chap. 6 for uniform measures hold also for free uniform measures, with $\mathsf{U_b}(X)$ replaced by $\mathsf{U}(X)$ and $\mathsf{UEB}(X)$ sets replaced by $\mathsf{UE}(X)$ sets, and with similar proofs. In the rest of this section I prove several of these analogues and state others as exercises.

Exercise 10.8. For any uniform space X, prove that a linear functional on $\mathsf{U}(X)$ is a molecular measure if and only if it is continuous on $\mathsf{U}(X)$ in the X-pointwise topology.   ∎

When $x_0\in X$, all the Δ-1-Lipschitz functions f on X such that $f(x_0)=0$ are between h and $-h$, where h is the Δ-1-Lipschitz function $\backslash_x\Delta(x_0,x)$. Accordingly, the UE sets $\mathsf{BLip}(\Delta,\backslash_x\Delta(x_0,x))$ play a significant role, as illustrated by the following lemma.

Lemma 10.9. *Let X be a uniform space, $\Delta\in\mathsf{UP}(X)$ and $x_0\in X$. Then*

$$\|\partial_X(x)-\partial_X(y)\|_\mathscr{F} = \Delta(x,y)$$

for all $x,y\in X$, where $\mathscr{F}:=\mathsf{BLip}(\Delta,\backslash_x\Delta(x_0,x))$.

Proof. The argument parallels the proof of Lemma 5.12. Write $h(x) := \Delta(x_0, x)$ for $x \in X$, and $f_y(x) := \Delta(x, y) - h(y)$ for $x, y \in X$. As $-h(x) \le f_y(x) \le h(x)$ for all $x, y \in X$, the functions f_y are in $\mathscr{F} = \mathsf{BLip}(\Delta, h)$. Moreover, $f_y(x) - f_y(y) = \Delta(x, y)$ for $x, y \in X$, and

$$\|\partial_X(x) - \partial_X(y)\|_{\mathscr{F}} = \sup\{|f(x) - f(y)| \mid f \in \mathscr{F}\} \ge \Delta(x, y).$$

The opposite inequality follows from the definition of $\mathsf{BLip}(\Delta, h)$. □

Lemma 10.10. *Let X be any uniform space. Every* UE *continuous linear functional on* $\mathfrak{M}_\mathsf{F}(X)$ *is of the form* $\mathfrak{m} \mapsto \mathfrak{m}(g)$, $\mathfrak{m} \in \mathfrak{M}_\mathsf{F}(X)$, *for some* $g \in \mathsf{U}(X)$.

Thus the UE dual of $\mathfrak{M}_\mathsf{F}(X)$ naturally identifies with $\mathsf{U}(X)$.

Proof. This parallels the proof of Lemma 6.5. On every UE set, the X-pointwise topology and the $\mathfrak{M}_\mathsf{F}(X)$-weak topology coincide. By Lemma 1.20, every UE set is contained in the set $\mathsf{BLip}(\Delta, h)$ for some $\Delta \in \mathsf{UP}(X)$ and Δ-1-Lipschitz h. Apply the Mackey–Arens Theorem P.8 with $E = \mathfrak{M}_\mathsf{F}(X)$ and $F = \mathsf{U}(X)$. □

Theorem 10.11. *Let X be any uniform space.*

1. *The space $\mathfrak{M}_\mathsf{F}(X)$ is* $\mathsf{UE}(X)$ *complete.*
2. *The space $\mathsf{Mol}(X)$ is* $\mathsf{UE}(X)$ *dense in $\mathfrak{M}_\mathsf{F}(X)$.*
3. *The mapping $\partial_X \colon X \to \mathfrak{M}_\mathsf{F}(X)$ is a uniform isomorphism from X onto $\partial_X(X)$ with the* $\mathsf{UE}(X)$ *uniformity.*

Thus $\mathfrak{M}_\mathsf{F}(X)$ is a $\mathsf{UE}(X)$ completion of $\mathsf{Mol}(X)$, and if X is identified with $\partial_X(X)$, then the closure of $\partial_X(X)$ in $\mathfrak{M}_\mathsf{F}(X)$ is a completion of X.

Proof. The argument parallels the proof of Theorem 6.6.

1. Take any UE Cauchy net $\{\mathfrak{m}_\gamma\}_\gamma$ in $\mathfrak{M}_\mathsf{F}(X)$. The net $\{\mathfrak{m}_\gamma(g)\}_\gamma$ converges for every $g \in \mathsf{U}(X)$, and $\mathfrak{m}(g) := \lim_\gamma \mathfrak{m}_\gamma(g)$ defines a linear functional $\mathfrak{m}$ on $\mathsf{U}(X)$. Then $\lim_\gamma \|\mathfrak{m}_\gamma - \mathfrak{m}\|_{\mathscr{F}} = 0$ for every UE set $\mathscr{F}$. Thus $\mathfrak{m}$ is X-pointwise continuous on $\mathscr{F}$, and $\mathfrak{m} \in \mathfrak{M}_\mathsf{F}(X)$.
2. Let $\mathfrak{A}$ be the UE closure of $\mathsf{Mol}(X)$ in $\mathfrak{M}_\mathsf{F}(X)$. If there were $\mathfrak{m}_0 \in \mathfrak{M}_\mathsf{F}(X) \setminus \mathfrak{A}$, then, by Theorem P.5 and Lemma 10.10, there would be $g \in \mathsf{U}(X)$ such that $\mathfrak{m}(g) = 0$ for all $\mathfrak{m} \in \mathsf{Mol}(X)$ and $\mathfrak{m}_0(g) \ne 0$, a contradiction.

Part 3 follows from Lemma 10.9. □

Exercise 10.12. For any uniform space X, prove that the set $\mathsf{Mol}(X)^+$ is UE dense in $\mathfrak{M}_\mathsf{F}(X)^+$. ∎

Exercise 10.13. This is a version of Exercise 5.33. Let X be a metric space with metric Δ and h a Δ-1-Lipschitz function on X, $h \ge 0$. Let $\mathfrak{A}$ be the space of those linear functionals on $\mathsf{Lip}(\Delta, h)$ that are X-pointwise continuous on $\mathsf{BLip}(\Delta, h)$. Identify each $\mathfrak{m} \in \mathsf{Mol}(\mathsf{coz}(h))$ with its restriction to $\mathsf{Lip}(\Delta, h)$ so that $\mathsf{Mol}(\mathsf{coz}(h)) \subseteq \mathfrak{A}$. Prove the following.

1. The $\|\cdot\|_{\mathsf{BLip}(\Delta,h)}$ norm dual of $\mathfrak{A}$ is $\mathsf{Lip}(\Delta,h)$.
2. In the $\|\cdot\|_{\mathsf{BLip}(\Delta,h)}$ norm, the space $\mathfrak{A}$ is complete, and $\mathsf{Mol}(\mathsf{coz}(h))$ is dense in $\mathfrak{A}$. ∎

Definition 10.14. (Cf. Definition 6.7.) Let X and Y be two uniform spaces and $\varphi\colon X\to Y$ a uniformly continuous mapping. For any $\mathsf{m}\in\mathfrak{M}_{\mathsf{F}}(X)$ define $\varphi(\mathsf{m})$, the *image of* m *under* φ, by $\varphi(\mathsf{m})(g):=\mathsf{m}(g\circ\varphi)$ for $g\in\mathsf{U}(Y)$. ∎

Exercise 10.15. For any uniform spaces X and Y and any uniformly continuous mapping $\varphi\colon X\to Y$, prove the following.

1. $\varphi(\mathsf{m})\in\mathfrak{M}_{\mathsf{F}}(X)$ for every $\mathsf{m}\in\mathfrak{M}_{\mathsf{F}}(X)$.
2. $\mathsf{m}\mapsto\varphi(\mathsf{m})$ is a positive linear mapping between the partially ordered vector spaces $\mathfrak{M}_{\mathsf{F}}(X)$ and $\mathfrak{M}_{\mathsf{F}}(Y)$.
3. The mapping $\mathsf{m}\mapsto\varphi(\mathsf{m})$ is continuous when both $\mathfrak{M}_{\mathsf{F}}(X)$ and $\mathfrak{M}_{\mathsf{F}}(Y)$ are equipped with the UE topology.
4. The mapping $\mathsf{m}\mapsto\varphi(\mathsf{m})$ is continuous from $\mathfrak{M}_{\mathsf{F}}(X)$ with the $\mathsf{U}(X)$-weak topology to $\mathfrak{M}_{\mathsf{F}}(Y)$ with the $\mathsf{U}(Y)$-weak topology.
5. If $\mathsf{m}\in\mathsf{Mol}(X)$, then $\varphi(\mathsf{m})\in\mathsf{Mol}(Y)$. ∎

Theorem 10.16. *Let X be any uniform space.*

1. *Let* $\mathsf{m}\in\mathfrak{M}_{\mathsf{F}}(X)^+$, *and let* $\{\mathsf{m}_\gamma\}_\gamma$ *be a net of positive linear functionals on* $\mathsf{U}(X)$ *such that* $\lim_\gamma\mathsf{m}_\gamma(f)=\mathsf{m}(f)$ *for every* $f\in\mathsf{U}(X)$. *Then* $\lim_\gamma\mathsf{m}_\gamma=\mathsf{m}$ *in the* UE *topology.*
2. *The* $\mathsf{U}(X)$-*weak topology and the* UE *topology coincide on the positive cone* $\mathfrak{M}_{\mathsf{F}}(X)^+$.

Proof. Take any UE set $\mathscr{F}\subseteq\mathsf{U}(X)^+$, and $\varepsilon>0$. For the function $h\in\mathsf{U}(X)^+$ defined by $h(x):=\sup\{f(x)\mid f\in\mathscr{F}\}$, $x\in X$, the set $\{h\wedge j\mid j\in\omega\}$ is UE. Hence there is $j_0\in\omega$ such that $\mathsf{m}(h\wedge j_0)>\mathsf{m}(h)+\varepsilon$.

If $f\in\mathscr{F}$, then $f-f\wedge j_0\le h-h\wedge j_0$ so that $\mathsf{m}(f-f\wedge j_0)\le\mathsf{m}(h-h\wedge j_0)$ and $\mathsf{m}_\gamma(f-f\wedge j_0)\le\mathsf{m}_\gamma(h-h\wedge j_0)$ for all γ, and

$$|\mathsf{m}(f)-\mathsf{m}_\gamma(f)|\le\mathsf{m}(f-f\wedge j_0))+|\mathsf{m}(f\wedge j_0)-\mathsf{m}_\gamma(f\wedge j_0)|+\mathsf{m}_\gamma(f-f\wedge j_0))$$

$$\le\mathsf{m}(h-h\wedge j_0))+|\mathsf{m}(f\wedge j_0)-\mathsf{m}_\gamma(f\wedge j_0)|+\mathsf{m}_\gamma(h-h\wedge j_0)).$$

Let n and n_γ be the restrictions of m and m_γ to $\mathsf{U}_\mathsf{b}(X)$. Then $\mathsf{n}\in\mathfrak{M}_\mathsf{u}(X)^+$ and $\mathsf{n}_\gamma\in\mathfrak{M}_\mathsf{b}(X)^+$. The set $\{f\wedge j_0\mid f\in\mathscr{F}\}$ is $\mathsf{UEB}(X)$. Since $\lim_\gamma\mathsf{n}_\gamma=\mathsf{n}$ in the UEB topology by Theorem 6.12, for almost all γ we have

$$\sup_{f\in\mathscr{F}}|\mathsf{m}(f\wedge j_0)-\mathsf{m}_\gamma(f\wedge j_0)|<\varepsilon.$$

Now for almost all γ also

$$\mathsf{m}_\gamma(h-h\wedge j_0)\le|\mathsf{m}(h-h\wedge j_0)-\mathsf{m}_\gamma(h-h\wedge j_0)|+\mathsf{m}(h-h\wedge j_0)<2\varepsilon,$$

and therefore

$$\sup_{f\in\mathscr{F}} |m(f) - m_\gamma(f)|$$

$$\leq m(h - h \wedge j_0)) + \sup_{f\in\mathscr{F}} |m(f \wedge j_0) - m_\gamma(f \wedge j_0)| + m_\gamma(h - h \wedge j_0)) < 4\varepsilon.$$

That proves that $\lim_\gamma m_\gamma = m$ uniformly on every UE subset of $U(X)^+$ and therefore also on every UE subset of $U(X)$.

Part 2 follows from Part 1.   □

Lemma 10.17. *Let X be a uniform space, $\varepsilon > 0$, $\mathfrak{A} \subseteq \mathfrak{M}_F(X)$, $\Delta \in \mathrm{UP}(X)$, let h be a Δ-1-Lipschitz function, $h \geq 0$, and $\mathscr{F} = \mathrm{BLip}(\Delta, h)$. Assume that for every $r \in \mathbb{R}^+$ there are $m \in \mathfrak{A}$ and $f \in \mathscr{F}^+$ for which $|m((f - r)^+)| > \varepsilon$. Then there exists a linear mapping $\varphi \colon \mathfrak{M}_F(X) \to \ell_1$ such that:*

(i) $\|\varphi(m)\|_1 \leq \|m\|_{\mathscr{F}}$ for all $m \in \mathfrak{M}_F(X)$.
(ii) φ is continuous from $\mathfrak{M}_F(X)$ with the $U(X)$-weak topology to ℓ_1 with the ℓ_∞-weak topology.
(iii) The set $\varphi(\mathfrak{A}) \subseteq \ell_1$ is not precompact in the ℓ_1 norm.

Proof. The construction is similar to the one in the proofs of Lemma 5.40 and Theorem 5.41.

First, repeatedly apply the assumption about $\mathfrak{A}$ and $\mathscr{F}$ to construct $r_j \in \mathbb{R}^+$, $m_j \in \mathfrak{A}$, $f_j \in \mathscr{F}^+$ and finite sets $D_j \subseteq X$ for $j \in \omega$ such that:

(a) $r_j \geq \max \{2h(y) \mid y \in D_i\}$ and $r_j \geq r_i$ for all $i < j$.
(b) $|m_j((f_j - r_j)^+)| > \varepsilon$.
(c) If $f \in 2\mathscr{F}$ and $\|f\|_{D_j} = 0$, then $|m_j(f)| < \varepsilon/3$.

I claim that the functions $g_j \in \mathscr{F}^+$, $j \in \omega$, defined by

$$g_j(x) := \max_{y \in D_j} \left((f_j(y) - r_j)^+ - \Delta(x,y) \right)^+ \quad \text{for } x \in X$$

have the following properties:

(d) $|m_j(g_j)| > 2\varepsilon/3$.
(e) $|m_i(g_j)| < \varepsilon/3$ for all $i < j$.
(f) $g_i \wedge g_j = 0$ for all $i < j$.

Indeed, (d) follows from (b) and (c) because $\|g_j - (f_j - r_j)^+\|_{D_j} = 0$, and (e) follows from (c) because $\|g_j\|_{D_i} = 0$ for $i < j$ by (a). To verify (f), note that if $i < j$, $x \in X$ and $g_i(x) > 0$, then there is $y \in D_i$ such that

$$\Delta(x,y) < (f_i(y) - r_i)^+ < (h(y) - r_i)^+ \leq h(y)$$
$$h(x) \leq h(y) + \Delta(x,y) < 2h(y)$$

and thus $r_j > h(x)$ by (a), and $g_j(x) \leq (f_j(x) - r_j)^+ \leq (h(x) - r_j)^+ = 0$.

The expression $\varphi(\mathfrak{m})(j) := \mathfrak{m}(g_j)/2$, $\mathfrak{m} \in \mathfrak{M}_{\mathsf{F}}(X)$, $j \in \omega$, defines a linear mapping φ from $\mathfrak{M}_{\mathsf{F}}(X)$ to $\mathbb{R}^{\omega}$. If $d \in \ell_{\infty}$, $\|d\|_{\omega} \leq 1$, then the sequence of functions $\sum_{j=0}^{k} d(j)g_j/2$ indexed by $k \in \omega$ is in $\mathscr{F}$ and converges X-pointwise to the function $\sum_{j \in \omega} d(j)g_j/2$. If $\mathfrak{m} \in \mathfrak{M}_{\mathsf{F}}(X)$, then

$$\mathfrak{m}\left(\sum_{j \in \omega} d(j)g_j/2\right) = \lim_{k} \mathfrak{m}\left(\sum_{j=0}^{k} d(j)g_j/2\right) = \lim_{k} \sum_{j=0}^{k} d(j)\mathfrak{m}(g_j)/2 = \langle d, \varphi(\mathfrak{m})\rangle.$$

Thus φ maps $\mathfrak{M}_{\mathsf{F}}(X)$ into ℓ_1, and conditions (i) and (ii) hold.

The set $\varphi(\mathfrak{A})$ is not precompact in the ℓ_1 norm $\|\cdot\|_1$ because

$$\|\varphi(\mathfrak{m}_j) - \varphi(\mathfrak{m}_i)\|_1 \geq |\mathfrak{m}_j(g_j)/2 - \mathfrak{m}_i(g_j)/2| > \varepsilon/6$$

for $i < j$, by (d) and (e). $\square$

Theorem 10.18. *Let X be a uniform space. These properties of a set $\mathfrak{A} \subseteq \mathfrak{M}_{\mathsf{F}}(X)$ are equivalent:*

(i) $\mathfrak{A}$ is X-pointwise equicontinuous on every UE set $\mathscr{F} \subseteq \mathsf{U}(X)$.

(ii) $\mathfrak{A}$ is precompact in the UE uniformity.

(iii) $\mathfrak{A}$ is relatively UE compact in $\mathfrak{M}_{\mathsf{F}}(X)$.

(iv) $\mathfrak{A}$ is relatively UE countably compact in $\mathfrak{M}_{\mathsf{F}}(X)$.

(v) $\mathfrak{A}$ is relatively $\mathsf{U}(X)$-weakly compact in $\mathfrak{M}_{\mathsf{F}}(X)$.

(vi) $\mathfrak{A}$ is relatively $\mathsf{U}(X)$-weakly countably compact in $\mathfrak{M}_{\mathsf{F}}(X)$.

(vii) $\mathfrak{A}$ is UE bounded, and if $\{\mathfrak{m}_j\}_j$ is a sequence in $\mathfrak{A}$ and $\{f_i\}_i$ is a UE sequence in $\mathsf{U}_{\mathsf{b}}(X)$ and the two double limits $\lim_i \lim_j \mathfrak{m}_j(f_i)$ and $\lim_j \lim_i \mathfrak{m}_j(f_i)$ exist, then they are equal.

Proof. The proof of (i)$\Rightarrow$(ii) parallels the first part of the proof of Theorem 6.16: Take any $\mathfrak{A}$ that satisfies (i) and any UE set $\mathscr{F} \subseteq \mathsf{U}(X)$, and let $\mathfrak{A}'$ be the set of restrictions of the functionals in $\mathfrak{A}$ to $\mathscr{F}$. By Lemma 1.20, no generality is lost by assuming that $\mathscr{F}$ is of the form $\mathsf{BLip}(\Delta, h)$ and hence X-pointwise compact. The set $\mathfrak{A}' \subseteq \mathsf{C}_{\mathsf{b}}(\mathscr{F})$ is $\|\cdot\|_{\mathscr{F}}$ bounded because if $\{f_j\}_j$ is any sequence in $\mathscr{F}$, then the set $\{f_j/j \mid j \in \omega\}$ is UE and $\lim_j \sup_{\mathfrak{m} \in \mathfrak{A}'} |\mathfrak{m}(f_j/j)| = 0$ by (i). By the Ascoli Theorem P.14, the set $\mathfrak{A}'$ is $\|\cdot\|_{\mathscr{F}}$ precompact. By Theorem 2.9 with $\alpha = 0$, $\mathfrak{A}$ is UE precompact.

(ii)$\Rightarrow$(iii) because $\mathfrak{M}_{\mathsf{F}}(X)$ is complete in the UE uniformity (Theorem 10.11). Clearly, (iii)$\Rightarrow$(iv)$\Rightarrow$(vi) and (iii)$\Rightarrow$(v)$\Rightarrow$(vi). The equivalence of (iv) and (vii) follows from Theorem P.13 along with Lemma 10.10 and Part 1 of Theorem 10.11.

To prove the implication (vi)$\Rightarrow$(i), take any relatively $\mathsf{U}(X)$-weakly countably compact set $\mathfrak{A} \subseteq \mathfrak{M}_{\mathsf{F}}(X)$ and any UE set $\mathscr{F} \subseteq \mathsf{U}(X)$.

First, note that for every $\varepsilon > 0$ there is $r_{\varepsilon} \in \mathbb{R}^+$ such that $|\mathfrak{m}(f - f \wedge r_{\varepsilon})| \leq \varepsilon$ for all $\mathfrak{m} \in \mathfrak{A}$ and $f \in \mathscr{F}^+$. Indeed, if this were not so, then the image $\varphi(\mathfrak{A})$ of $\mathfrak{A}$ under the mapping $\varphi \colon \mathfrak{M}_{\mathsf{F}}(X) \to \ell_1$ in Lemma 10.17 would be relatively ℓ_{∞}-weakly countably compact but not relatively norm compact in ℓ_1, in contradiction to Theorem P.15.

Take any $\varepsilon > 0$, $f \in \mathscr{F}^+$, and any net $\{f_\gamma\}_\gamma$ in $\mathscr{F}^+$ that converges to f pointwise. As $\mathfrak{A}$ is relatively $\mathsf{U}_\mathsf{b}(X)$-weakly countably compact as a subset of $\mathfrak{M}_\mathsf{u}(X)$, it is X-pointwise equicontinuous on the UEB set $\{f \wedge r_\varepsilon \mid f \in \mathscr{F}^+\} \subseteq \mathsf{U}_\mathsf{b}(X)$ by Theorem 6.16, and

$$\lim_\gamma \sup\{|\mathfrak{m}(f_\gamma - f)| \mid \mathfrak{m} \in \mathfrak{A}\} \leq \lim_\gamma \sup\{|\mathfrak{m}(f_\gamma \wedge r_\varepsilon - f \wedge r_\varepsilon)| \mid \mathfrak{m} \in \mathfrak{A}\} + 2\varepsilon = 2\varepsilon.$$

Thus $\mathfrak{A}$ is pointwise equicontinuous on $\mathscr{F}^+$ for every $\mathsf{UE}(X)$ set $\mathscr{F}$ and therefore also on $\mathscr{F}$. □

Exercise 10.19. Show that, for every uniform space X, the space $\mathfrak{M}_\mathsf{F}(X)$ is $\mathsf{U}(X)$-weakly sequentially complete. ∎

Exercise 10.20. Let X be a uniform space and E a locally convex space. Say that a linear mapping $\overrightarrow{\mathfrak{m}} : \mathsf{U}(X) \to E$ is an *E-valued free uniform measure on X* iff the restriction of $\mathfrak{m}$ to every $\mathsf{UE}(X)$ subset of $\mathsf{U}(X)$ is continuous in the X-pointwise topology. Let $\mathfrak{M}_\mathsf{F}(X,E)$ denote the space of E-valued free uniform measures on X. Prove the following for any uniform space X and any locally convex space E:

1. A linear mapping $\overrightarrow{\mathfrak{m}} : \mathsf{U}(X) \to E$ is in $\mathfrak{M}_\mathsf{F}(X,E)$ if and only if $w \circ \overrightarrow{\mathfrak{m}} \in \mathfrak{M}_\mathsf{F}(X)$ for every $w \in E^*$.
2. A linear mapping $\overrightarrow{\mathfrak{m}} : \mathsf{U}_\mathsf{b}(X) \to E$ is a restriction to $\mathsf{U}_\mathsf{b}(X)$ of an E-valued free uniform measure if and only if finite $\lim_{j \in \omega} \overrightarrow{\mathfrak{m}}((-j) \vee f \wedge j)$ exists for every $f \in \mathsf{U}(X)$ and $\overrightarrow{\mathfrak{m}} \in \mathfrak{M}_\mathsf{u}(X,E)$. ∎

10.2 Universal Property

In this section I show that, in categorical language, $\mathfrak{M}_\mathsf{F}$ is a free functor from the category of uniform spaces to the category of complete locally convex spaces. I start with a straightforward extension of the "weak integral" notion in Definition 6.39.

Definition 10.21. Let X be a uniform space, $\mathfrak{m}$ a linear functional on the space $\mathsf{U}(X)$, and E a locally convex space. When φ is a uniformly continuous mapping from X to E with its additive uniformity, define the linear functional $\mathfrak{m}^E(\varphi)$ on E^* by $\mathfrak{m}^E(\varphi)(w) := \mathfrak{m}(w \circ \varphi)$ for $w \in E^*$. ∎

Note that if $\mathfrak{m}$ is a linear functional on $\mathsf{U}(X)$ and $\mathfrak{n}$ is its restriction to $\mathsf{U}_\mathsf{b}(X)$, and if the range of φ is bounded, then $\mathfrak{m}^E(\varphi) = \mathfrak{n}^E(\varphi)$, where $\mathfrak{n}^E(\varphi)$ is defined in Definition 6.39. Thus the two definitions agree when we identify each free uniform measure with the corresponding uniform measure, its restriction to $\mathsf{U}_\mathsf{b}(X)$. In particular, the two definitions agree for every molecular measure, whether considered as a functional on $\mathsf{U}_\mathsf{b}(X)$ or on $\mathsf{U}(X)$.

Theorem 10.22. *Let X be any uniform space. These two properties of a linear functional $\mathfrak{m}$ on $U(X)$ are equivalent:*

(i) $\mathfrak{m} \in \mathfrak{M}_F(X)$.
(ii) *For every complete locally convex space E and for every uniformly continuous mapping $\varphi: X \to E$, the functional $\mathfrak{m}^E(\varphi)$ is E-weakly continuous (and therefore identifies with an element of E).*

Proof. The proof mimics that of Theorem 6.40.

Assume $\mathfrak{m} \in \mathfrak{M}_F(X)$, and take any complete locally convex space E and any uniformly continuous mapping $\varphi: X \to E$. If a set $B \subseteq E^*$ is equicontinuous on E, then $\mathscr{F} := \{w \circ \varphi \mid w \in B\} \subseteq U(X)$ is a $UE(X)$ set; hence the restriction of $\mathfrak{m}$ to $\mathscr{F}$ is X-pointwise continuous, and the restriction of $\mathfrak{m}^E(\varphi)$ to B is E-weakly continuous. Hence $\mathfrak{m}^E(\varphi)$ is E-weakly continuous by Theorem P.12. That proves (i)$\Rightarrow$(ii).

To prove the converse, let E be the space $\mathfrak{M}_F(X)$ with the UE topology and $\varphi := \partial_X$. By Theorem 10.11, the space E is complete and φ is uniformly continuous.

Every $f \in U(X)$ defines $w_f \in E^*$ by $w_f(\mathfrak{n}) := \mathfrak{n}(f)$ for $\mathfrak{n} \in E$. If $\mathfrak{m}$ is a linear functional on $U(X)$ with property (ii), then there is $\mathfrak{n} \in E = \mathfrak{M}_F(X)$ such that $\mathfrak{m}^E(\varphi)(w) = w(\mathfrak{n})$ for every $w \in E^*$, and in particular

$$\mathfrak{m}(f) = \mathfrak{m}(w_f \circ \varphi) = \mathfrak{m}^E(\varphi)(w_f) = w_f(\mathfrak{n}) = \mathfrak{n}(f),$$

for every $f \in U(X)$, so that $\mathfrak{m} = \mathfrak{n} \in \mathfrak{M}_F(X)$. $\square$

Theorem 10.22 may be rephrased as a universal property of the space $\mathfrak{M}_F(X)$:

Theorem 10.23. *Let X be any uniform space and E a complete locally convex space with its additive uniformity. For every uniformly continuous mapping $\varphi: X \to E$ there exists a linear mapping $\widetilde{\varphi}: \mathfrak{M}_F(X) \to E$ continuous in the UE topology on $\mathfrak{M}_F(X)$ and such that $\varphi = \widetilde{\varphi} \circ \partial_X$.*

Proof. Let $\widetilde{\varphi}(\mathfrak{m}) := \mathfrak{m}^E(\varphi)$ for $\mathfrak{m} \in \mathfrak{M}_F(X)$. Then $\widetilde{\varphi}$ maps $\mathfrak{M}_F(X)$ into E by Theorem 10.22 and obviously $\varphi = \widetilde{\varphi} \circ \partial_X$.

To prove that $\widetilde{\varphi}$ is UE continuous, take any continuous seminorm α on E and let $B := \{w \in E^* \mid |w(y)| \leq \alpha(y) \text{ for all } y \in E\}$. The set $\mathscr{F} := \{w \circ \varphi \mid w \in B\}$ is $UE(X)$ and

$$\|\mathfrak{m}\|_{\mathscr{F}} = \sup\{|\mathfrak{m}(w \circ \varphi)| \mid w \in B\} = \sup\{|\widetilde{\varphi}(\mathfrak{m})(w)| \mid w \in B\} = \alpha(\widetilde{\varphi}(\mathfrak{m}))$$

for $\mathfrak{m} \in \mathfrak{M}_F(X)$, by Corollary P.10. $\square$

Exercise 10.24. Prove the following analogue of Theorem 10.23 for the space $\mathfrak{M}_u(X)$: Let X be a uniform space and E a complete locally convex space with its additive uniformity. For every uniformly continuous mapping $\varphi: X \to E$ whose range $\varphi(X)$ is bounded in E, there exists a linear mapping $\widetilde{\varphi}: \mathfrak{M}_u(X) \to E$ continuous in the UEB topology on $\mathfrak{M}_u(X)$ and such that $\varphi = \widetilde{\varphi} \circ \partial_X$.   ∎

10.3 Measures with Compact Support

In this section I characterize free uniform measures on uniform spaces with the CDE property.

Definition 10.25. For any uniform space X, let $\mathfrak{M}_C(X)$ denote the space of the linear functionals on $U(X)$ that are continuous in the compact–open topology on $U(X)$. ∎

If $\mathfrak{m} \in \mathfrak{M}_C(X)$, then the restriction of $\mathfrak{m}$ to $U_b(X)$ is in $\mathfrak{M}_t(X)$, and by Theorem 5.3 there is a tight Borel measure μ on X such that $\mathfrak{m}(f) = \int f \, d\mu$ for $f \in U_b(X)$.

Theorem 10.26. *Let X be a uniform space. These properties of a measure μ on the σ-algebra $\mathrm{Bo}(X)$ are equivalent:*

(i) μ has a compact support $K \subseteq X$.
(ii) The finite integral $\int f \, d\mu$ exists for every $f \in U(X)$, and the functional that maps each $f \in U(X)$ to $\int f \, d\mu$ is in $\mathfrak{M}_C(X)$.
(iii) The functional that maps each $f \in U_b(X)$ to $\int f \, d\mu$ is continuous on $U_b(X)$ with the compact–open topology.

Proof. Obviously, (i)⇒(ii)⇒(iii). To prove (iii)⇒(i), take any measure μ on $\mathrm{Bo}(X)$ such that the mapping $\mathfrak{m} \colon f \mapsto \int f \, d\mu$, $f \in U_b(X)$, is continuous in the compact–open topology. Then by Theorem P.30 μ is tight, therefore τ-additive, therefore it has a support.

There is a compact set $K \subseteq X$ for which $|\mathfrak{m}(f)| \leq 1$ for every $f \in U_b(X)$ such that $\|f\|_K = 0$, and therefore also $\mathfrak{m}(f) = 0$ for every $f \in U_b(X)$ such that $\|f\|_K = 0$. Thus $|\mathfrak{m}|(f) = 0$ for every $f \in U_b(X)$ such that $\|f\|_K = 0$, and $|\mu|(X \setminus K) = 0$. It follows that the support of μ is included in K; hence it is compact. □

Corollary 10.27. *Let X be a uniform space.*

1. $\mathfrak{M}_C(X) \subseteq U(X)^\sim$.
2. $\mathfrak{M}_C(X) \subseteq \mathfrak{M}_F(X)$. □

Theorem 10.28. *Let X be a complete uniform space with the CDE property. Then $\mathfrak{M}_F(X) = \mathfrak{M}_C(X)$.*

Proof. Take any $\mathfrak{m} \in \mathfrak{M}_F(X)$, and let $\mathfrak{n}$ be the restriction of $\mathfrak{m}$ to $U_b(X)$. As in Sect. 7.2, let $\rho_\mathfrak{n}$ be the unique tight Borel measure on $\hat{p}X$ such that $\mathfrak{n}(f) = \int \hat{f} \, d\rho_\mathfrak{n}$ for $f \in U_b(X)$. Let $K \subseteq \hat{p}X$ be the support of $\rho_\mathfrak{n}$ (Theorem P.27). It is enough to prove that $K \subseteq X$ because then $\mathfrak{m} \in \mathfrak{M}_C(X)$ by Theorem 10.26.

For $\Delta \in \mathrm{UP}(X)$, let $A(\Delta) := \{x \in X \mid K \cap \mathrm{int}\, \overline{\bigcirc[x, \Delta]} \neq \emptyset\}$. Here, as in Definition 7.13, interior and closure are operations in $\hat{p}X$.

I claim that for every $\Delta \in \mathrm{UP}(X)$ there exists a finite set $D \subseteq A(\Delta)$ such that $A(\Delta) \subseteq \bigcup_{y \in D} \bigcirc[y, \Delta/5]$. Suppose otherwise. Then there exist $\Delta \in \mathrm{UP}(X)$ and a sequence of points $x_j \in A(\Delta)$, $j \in \omega$, such that $\Delta(x_i, x_j) \geq 5$ for all $i, j \in \omega$, $i \neq j$. For the functions $f_j \in \mathsf{U}_b(X)^+$ defined by $f_j(x) := (2 - \Delta(x, x_j))^+$, $x \in X$, we have

$$|\mathfrak{n}|(f_j) = \int \overline{f_j} \, \mathrm{d}|\rho_\mathfrak{n}| \geq |\rho_\mathfrak{n}| \left(\mathrm{int} \, \overline{\bigodot [x_j, \Delta]} \right) > 0$$

because $\mathrm{int} \, \overline{\bigodot [x_j, \Delta]} \cap K \neq \emptyset$. Therefore there are functions $g_j \in \mathsf{U}_b(X)^+$ such that $g_j \leq f_j$ and $\mathfrak{n}(g_j) \neq 0$ for all $j \in \omega$. Let $r_j := 1/|\mathfrak{n}(g_j)|$ for $j \in \omega$, and replace $\{g_j\}_j$ by a subsequence in which all $\mathfrak{n}(g_j)$ have the same sign. The set $\{r_j g_j \mid j \in \omega\}$ is $\mathsf{UE}(X)$ by the CDE property. The function defined by $g(x) := \sup_j r_j g_j(x)$ for $x \in X$ is in $\mathsf{U}(X)$, and $\lim_j \mathfrak{m}(\max_{0 \leq i \leq j} r_i g_i) = \mathfrak{m}(g)$ because $\lim_j \max_{0 \leq i \leq j} r_i g_i = g$ pointwise. However, $|\mathfrak{m}(\max_{0 \leq i \leq j} r_i g_i)| = \sum_{i=0}^{j} |\mathfrak{n}(r_i g_i)| = j + 1$. The contradiction proves the claim.

To prove $K \subseteq X$, suppose there exists $z \in K \setminus X$. Then there is $\Delta \in \mathsf{UP}(X)$ for which $z \notin \mathscr{E}(\Delta/7)$ by Corollary 7.15, and by the preceding claim there is a finite set $D \subseteq A(\Delta)$ such that

$$A(\Delta) \subseteq \bigcup_{y \in D} \bigodot [y, \Delta/5] \subseteq \bigcup_{y \in D} \overline{\bigodot [y, \Delta/6]} \subseteq \bigcup_{y \in D} \mathrm{int} \, \overline{\bigodot [y, \Delta/7]} \subseteq \mathscr{E}(\Delta/7).$$

The set $V := \hat{\mathfrak{p}} X \setminus \bigcup_{y \in D} \overline{\bigodot [y, \Delta/6]}$ is open and $z \in V$; hence $|\rho_\mathfrak{n}|(K \cap V) > 0$.

If $x \in A(\Delta)$, then $\overline{\bigodot [x, \Delta]} \subseteq \overline{\bigodot [y, \Delta/6]}$ for some $y \in D$, and if $x \in X \setminus A(\Delta)$, then $K \cap \mathrm{int} \, \overline{\bigodot [x, \Delta]} = \emptyset$. Thus in both cases we have $K \cap V \cap \mathrm{int} \, \overline{\bigodot [x, \Delta]} = \emptyset$. Therefore $K \cap V \subseteq \hat{\mathfrak{p}} X \setminus \mathscr{E}(\Delta)$ and $|\rho_\mathfrak{n}|(\hat{\mathfrak{p}} X \setminus \mathscr{E}(\Delta)) > 0$, in contradiction to Theorem 7.14. That concludes the proof of the inclusion $K \subseteq X$. $\square$

Exercise 10.29. Let X be a complete uniform space with the CDE property. Prove that every $\mathsf{U}(X)$-weakly compact subset $\mathfrak{A}$ of $\mathfrak{M}_\mathsf{F}(X)$ is equicontinuous on $\mathsf{U}(X)$ with the compact–open topology (and thus all the functionals in $\mathfrak{A}$ are represented by tight Borel measures on a common compact subset of X).   ∎

Recall from Sect. 6.5 that if X is a uniform space, $\iota: X \hookrightarrow \hat{X}$ is the embedding of X into its completion and $f \in \mathsf{U}(X)$, then $f = \hat{f} \circ \iota$ for a unique $\hat{f} \in \mathsf{U}(\hat{X})$. The mapping $f \mapsto \hat{f}$ is a Riesz space isomorphism of $\mathsf{U}(X)$ onto $\mathsf{U}(\hat{X})$, and $\mathsf{UE}(\hat{X})$ subsets of $\mathsf{U}(\hat{X})$ are precisely the images of $\mathsf{UE}(X)$ subsets of $\mathsf{U}(X)$. Hence the mapping $\mathfrak{m} \mapsto \iota(\mathfrak{m})$ (Definition 10.14) is an isomorphism of $\mathfrak{M}_\mathsf{F}(X)$ with the $\mathsf{UE}(X)$ topology onto $\mathfrak{M}_\mathsf{F}(\hat{X})$ with the $\mathsf{UE}(\hat{X})$ topology. In the rest of this chapter, I write $\mathfrak{M}_\mathsf{F}(X) \cong \mathfrak{M}_\mathsf{F}(\hat{X})$ as a shorthand for this isomorphism (equality modulo completion).

Corollary 10.30. *Let X be a uniform space. If X has the CDE property, then* $\mathfrak{M}_\mathsf{F}(X) \cong \mathfrak{M}_\mathsf{C}(\hat{X})$.

Proof. The space $\hat{X}$ has the CDE property by Lemma 4.10. $\square$

Example 10.31. This modification of Example 7.19 exhibits a complete uniform space X and a positive free uniform measure on X whose restriction to $U_b(X)$ is not in $\mathfrak{M}_\sigma(X)$. Hence $\mathfrak{M}_F(X)^+ \neq \mathfrak{M}_C(X)^+$.

Let T be the closed interval $[0,1]$ with its usual compact topology and Q the set of irrational numbers in T. The construction proceeds as in Example 7.19, with one change: Now the mappings $\psi_q \colon T \setminus \{q\} \to \mathbb{R}$, $q \in Q$, are

$$\psi_q(x) := \begin{cases} 1/\sqrt{q-x} & \text{if } 0 \leq x < q \\ -1/\sqrt{x-q} & \text{if } q < x \leq 1 \end{cases}$$

which makes them integrable with respect to the Lebesgue measure μ on $\mathrm{Bo}(T)$.

Define X to be the uniform space on the set $T \setminus Q$ whose uniformity is induced by the set $\{\psi_q \mid q \in Q\}$ of mappings from $T \setminus Q$ to $\mathbb{R}$. The same argument as in Example 7.19 demonstrates that X is complete.

Fix a function $f \in U(X)$. I claim that there are a finite set $D \subseteq Q$ and $r, r' \in \mathbb{R}^+$ such that $|f(x)| \leq r + r' \max_{q \in D} |\psi_q(x)|$ for all $x \in X$. By Theorem 2.6, there are a finite set $D \subseteq Q$ and $\theta > 0$ such that if $x, y \in X$ and $\max\{|\psi_q(x) - \psi_q(y)| \mid q \in D\} < \theta$, then $|f(x) - f(y)| < 1$. Write $\Delta(x,y) := \max\{|\psi_q(x) - \psi_q(y)| \mid q \in D\}$ for x and y in $T \setminus D$. The set $T \setminus D$ is a finite union of intervals I of two types:

1. There is $q_I \in D$ such that $\Delta(x,y) = \psi_{q_I}(y) - \psi_{q_I}(x)$ for all $x, y \in I$, $x < y$.
2. There is $t \in \mathbb{R}^+$ such that $\Delta(x,y) \leq t(y-x)$ for all $x, y \in I$, $x < y$.

Since each such I is an interval disjoint from D, every ψ_q, $q \in D$, is increasing on I. It is enough to prove the claim for f restricted to $I \cap X$ for each interval I of one of the two types. Take any $x, y \in I \cap X$, $x < y$. There are $k \geq 1$ and $x_0, x_1, \ldots, x_k \in I \cap X$ such that $x = x_0 < x_1 < \cdots < x_k = y$ and

$$\Delta(x_0, x_1) < \theta, \qquad \theta/2 < \Delta(x_{i-1}, x_i) < \theta \text{ for } 2 \leq i \leq k.$$

Then $|f(x_i) - f(x_{i-1})| < 1$ for $i = 1, \ldots, k$; hence $|f(x) - f(y)| < k$.

If I is an interval of type (1), then $(k-1)\theta/2 < \psi_{q_I}(y) - \psi_{q_I}(x)$ because ψ_{q_I} is increasing on I and

$$|f(x) - f(y)| < k < 1 + 2\left(\psi_{q_I}(y) - \psi_{q_I}(x)\right)/\theta.$$

That proves the claim with $r := |f(y_0)| + 1 + 2|\psi_{q_I}(y_0)|/\theta$, where y_0 is an arbitrarily chosen point in $I \cap X$ and $r' := 2/\theta$.

If I is of type (2), then $(k-1)\theta/2 < t(y-x)$, and

$$|f(x) - f(y)| < k < 1 + 2t(y-x)/\theta \leq 1 + 2t/\theta.$$

That proves the claim with $r := |f(y_0)| + 1 + 2t/\theta$, where y_0 is an arbitrarily chosen point in $I \cap X$, and $r' := 0$.

The projective system construction in Example 7.19 yields $\mathfrak{m} \in \mathfrak{M}_u(X)^+ \setminus$ $\mathfrak{M}_\sigma(X)$ such that $\mathfrak{m}(|\psi_q|X| \wedge j) = \int (|\psi_q| \wedge j) \, d\mu$ for all $q \in Q$, $j \in \omega$. By the claim above and Theorem 10.5, $\mathfrak{m}$ is the restriction of a free uniform measure on X.   ∎

10.4   Instances of Free Uniform Measures

The rest of this chapter parallels Chap. 8, where various spaces of measures are obtained as instances of $\mathfrak{M}_u(X)$. Here I follow the same approach to obtain several noteworthy instances of $\mathfrak{M}_F(X)$.

All the results proved on preceding pages for general spaces $\mathfrak{M}_F(X)$ apply to each specific instance. Although I do not explicitly list such consequences in what follows, the reader is encouraged to do so. A general result is often better understood when seen through special cases.

I start with a brief example. The next three sections then deal with other spaces of measures in more detail.

Example 10.32. Let X be a Banach space, and consider X as a metric space with the metric of its norm $\|\cdot\|$. Say that a measure $\mu \geq 0$ on the σ-algebra $\mathsf{Bo}(X)$ is *of order* 1 iff the function $x \mapsto \|x\|$, $x \in X$, is μ-integrable.

For every $f \in \mathsf{U}(X)$ there are $r, r' \in \mathbb{R}^+$ such that $|f(x)| \leq r + r'\|x\|$ for all $x \in X$. By Theorems 5.28 and 10.5, the cone of positive tight Borel measures (i.e. Radon measures) on X of order 1 identifies with $\mathfrak{M}_F(X)^+$.   ∎

10.5   Measures on Abstract σ-Algebras

Let S be a non-empty set and Σ a σ-algebra on S. This section answers the following question: For which measures on Σ are all Σ-measurable real-valued functions integrable?

As in Example 2.3 and Sect. 8.2, assume that Σ separates the points of S and let $\mathscr{U}(\Sigma)$ denote the uniformity on S induced by the mappings $\varphi \colon S \to \mathbb{N}$ such that $\varphi^{-1}(j) \in \Sigma$ for every $j \in \mathbb{N}$.

Lemma 10.33. *Let Σ be a point-separating σ-algebra of subsets of a non-empty set S. Let X be the uniform space on the point set S with the uniformity $\mathscr{U}(\Sigma)$. Then every compact subset of $\widehat{X}$ is finite.*

Proof. In view of Theorem P.35, it is enough to prove that $\widehat{X}$ is a P-space. Take any $z \in \widehat{X}$, and let V_j, $j \in \omega$, be open neighbourhoods of z. By Theorem 1.21, there are $f_j \in \mathsf{U}_b(\widehat{X})$, $j \in \omega$, such that $f_j(z) = 0$ and $f_j(x) = 1$ for $x \in \widehat{X} \setminus V_j$. Without loss of generality, assume that $0 \leq f_j \leq 1$ and $f_0 \leq f_1 \leq f_2 \leq \cdots$ (if necessary replace f_j by $1 \wedge \max\{f_i^+ \mid 0 \leq i \leq j\}$). Write $f(x) := \lim_j f_j(x)$ for $x \in \widehat{X}$, and let g and g_j be the restrictions of f_j and f to S, so that $f_j = \widehat{g}_j$.

Clearly g is Σ-measurable, and therefore $g \in U_b(X)$. The set $\{g_j \mid j \in \omega\}$ is $UEB(X)$ because X is inversion-closed (Theorem 4.7); hence $m(g) = \lim_j m(g_j)$ for every $m \in \mathfrak{M}_u(X)$. In particular, $\widehat{g}(x) = \lim_j \widehat{g_j}(x) = \lim_j f_j(x) = f(x)$ for every $x \in \widehat{X}$, which means that $f = \widehat{g} \in U_b(\widehat{X})$. The open set $\{x \in \widehat{X} \mid f(x) < 1\}$ is included in all V_j, $j \in \omega$. □

Exercise 10.34. For a non-empty set S and a point-separating σ-algebra Σ on S, let again X be the uniform space on the point set S with the uniformity $\mathscr{U}(\Sigma)$.

1. For $S = \mathbb{R}$ and $\Sigma = Bo(\mathbb{R})$, show that X is complete.
2. Find S and Σ such that X is not complete.   ∎

Theorem 10.35. *Let Σ be a point-separating σ-algebra of subsets of a non-empty set S. Let X be the uniform space on the point set S with the uniformity $\mathscr{U}(\Sigma)$. The following properties of a measure μ on Σ are equivalent:*

(i) Every Σ-measurable real-valued function on S is μ-integrable.
(ii) There is $m \in \mathrm{Mol}(\widehat{X})$ such that $\int f \, d\mu = m(\widehat{f})$ for every $f \in \ell_\infty(S, \Sigma)$.

Proof. As in Sect. 8.2, the integral $n(f) := \int f \, d\mu$ for $f \in U_b(X) = \ell_\infty(S, \Sigma)$ defines $n \in \mathfrak{M}_u(X)$. By Theorem 10.5, if μ has property (i), then n is a restriction to $U_b(X)$ of some $m \in \mathfrak{M}_F(X)$. It follows that $\int f \, d\mu = m(f)$ for every $f \in U(X)$. The space X has the CDE property by Theorem 4.9, and $\mathfrak{M}_F(X) \cong \mathrm{Mol}(\widehat{X})$ by Corollary 10.30 and Lemma 10.33. That proves (i)$\Rightarrow$(ii).

To prove the converse, take any Σ-measurable function $f \colon S \to \mathbb{R}$. If μ has property (ii), then the finite limits

$$\lim_{j \in \omega} \int f^+ \wedge j \, d|\mu| = \lim_{j \in \omega} |m|((\widehat{f})^+ \wedge j)$$

$$\lim_{j \in \omega} \int f^- \wedge j \, d|\mu| = \lim_{j \in \omega} |m|((\widehat{f})^- \wedge j)$$

exist; hence f is μ-integrable.   □

10.6   Riesz Measures on Completely Regular Spaces

Definition 10.36. Let T be a completely regular topological space. A *Riesz measure on T* is a measure μ on the σ-algebra $\sigma(C(T))$ such that every function in $C(T)$ is μ-integrable.   ∎

Note that a measure μ on the σ-algebra $\sigma(C(T))$ is a Riesz measure if and only if $\lim_{j \in \omega} \int (-j) \vee f \wedge j \, d|\mu|$ exists and is finite for every $f \in C(T)$. Indeed, the condition is obviously necessary. Conversely, if $f \in C(T)$ and the finite limits $\lim_{j \in \omega} \int f^+ \wedge j \, d|\mu|$ and $\lim_{j \in \omega} \int f^- \wedge j \, d|\mu|$ exist, then f is μ-integrable.

Recall from Sect. 8.3 that $C(T) = U(cFT)$ and that measures on the σ-algebra $\sigma(C(T))$ naturally identify with uniform measures on the space cFT. By the next theorem, Riesz measures on T naturally identify with free uniform measures on the same space cFT. In the identification, the $U(cFT)$-weak topology on $\mathfrak{M}_F(cFT)$ becomes the $C(T)$-weak topology on the space of Riesz measures on T that arises from the duality $\langle \mu, f \rangle = \int f \, d\mu = \lim_{j \in \omega} \int (-j) \vee f \wedge j \, d\mu$ for a Riesz measure μ and $f \in C(T)$.

Theorem 10.37. *Let T be a completely regular topological space. The following properties of a linear functional $\mathfrak{m}$ on the space $C(T)$ are equivalent:*

(i) There is a Riesz measure μ on T such that $\mathfrak{m}(f) = \int f \, d\mu$ for all $f \in C(T)$.
(ii) $\mathfrak{m} \in \mathfrak{M}_F(cFT)$.
(iii) The functional on $U(\widehat{cFT})$ defined by $\hat{f} \mapsto \mathfrak{m}(f)$, $f \in C(T)$, is in $\mathfrak{M}_C(\widehat{cFT})$.
(iv) $\mathfrak{m} \in C(T)^\sim$.
(v) If $\{f_j\}_j$ is a sequence in $C(T)$ such that $f_j \searrow 0$, then $\lim_j \mathfrak{m}(f_j) = 0$.

The implication (iv)$\Rightarrow$(v) is a special case of the following lemma.

Lemma 10.38. *Let X be an inversion-closed uniform space, $\mathfrak{m} \in U(X)^\sim$, and let $\{f_j\}_j$ be a sequence in $U(X)$ such that $f_j \searrow 0$. Then $\lim_j \mathfrak{m}(f_j) = 0$.*

Proof. Assume $\mathfrak{m} \geq 0$, and take any $\varepsilon > 0$. The sum $f(x) := \sum_{j \in \omega} (f_j(x) - \varepsilon)^+$ is finite for every $x \in X$. For any $r \in \mathbb{R}$, we have

$$\{x \in X \mid f(x) > r\} = \bigcup_{j \in \omega} \left\{ x \in X \mid \sum_{i=0}^{j} (f_i(x) - \varepsilon)^+ > r \right\}$$

$$\{x \in X \mid f(x) < r\} = \bigcup_{j \in \omega} \left\{ x \in X \mid f_j(x) < \varepsilon \text{ and } \sum_{i=0}^{j} (f_i(x) - \varepsilon)^+ < r \right\}$$

which means that f is cozero-continuous, and $f \in U(X)$ by Theorem 4.7. Therefore $\lim_j \mathfrak{m}((f_j - \varepsilon)^+) = 0$. Since $\mathfrak{m}(f_j) \leq \varepsilon \mathfrak{m}(1) + \mathfrak{m}((f_j - \varepsilon)^+)$, it follows that $\lim_j \mathfrak{m}(f_j) = 0$. Since every element of $U(X)^\sim$ is a difference of two positive functionals, the assumption $\mathfrak{m} \geq 0$ causes no loss of generality. □

Proof of Theorem 10.37. (i)$\Rightarrow$(ii) follows from the equality $\mathfrak{M}_\sigma(cFT) = \mathfrak{M}_u(cFT)$ established in Sect. 8.3 and from Theorem 10.5. We have (ii)$\Rightarrow$(iii) by Corollary 10.30, (iii)$\Rightarrow$(iv) by Corollary 10.27 and (iv)$\Rightarrow$(v) by Lemma 10.38.

Now assume that (v) holds. The restriction of $\mathfrak{m}$ to $C_b(T)$ is in $\mathfrak{M}_\sigma(cFT)$, hence in $\mathfrak{M}_u(cFT)$. For every $f \in C(T)^+$ we have $(f - f \wedge j) \searrow 0$; hence $\mathfrak{m} \in \mathfrak{M}_F(cFT)$ by Theorem 10.5. That proves (v)$\Rightarrow$(ii).

Next, assume that (ii) holds, and let $\mathfrak{n}$ be the restriction of $\mathfrak{m}$ to $C_b(T)$. Then $\mathfrak{n} \in \mathfrak{M}_u(cFT) = \mathfrak{M}_\sigma(cFT)$; hence there is a measure μ such that $\mathfrak{n}(f) = \int f \, d\mu$ for all $f \in C_b(T)$. By the already established implication (ii)$\Rightarrow$(iii) the finite limits

$$\lim_{j\in\omega} \mathfrak{n}((-j)\vee f\wedge j) = \lim_{j\in\omega}\int(-j)\vee f\wedge j\,d\mu$$

$$\lim_{j\in\omega} |\mathfrak{n}|((-j)\vee f\wedge j) = \lim_{j\in\omega}\int(-j)\vee f\wedge j\,d|\mu|$$

exist for every $f\in C(T)$. That together with Theorem 10.5 proves (ii)$\Rightarrow$(i). □

Let again T be a completely regular space, and this time consider the uniform space $\mathsf{F}T$. Recall from Sect. 8.4 that separable measures on $\sigma(C(T))$ naturally identify with uniform measures on the space $\mathsf{F}T$. By the next theorem, separable Riesz measures on T naturally identify with free uniform measures on $\mathsf{F}T$.

Theorem 10.39. *Let T be a completely regular topological space. The following properties of a linear functional $\mathfrak{m}$ on the space $C(T)$ are equivalent:*

(i) There is a separable Riesz measure μ on T such that $\mathfrak{m}(f) = \int f\,d\mu$ for all $f\in C(T)$.

(ii) $\mathfrak{m}\in\mathfrak{M}_{\mathsf{F}}(\mathsf{F}T)$.

(iii) The functional on $\mathsf{U}(\widehat{\mathsf{F}T})$ defined by $\widehat{f}\mapsto\mathfrak{m}(f)$, $f\in C(T)$, is in $\mathfrak{M}_{\mathsf{C}}(\widehat{\mathsf{F}T})$.

Proof. If (i) holds, then the restriction of $\mathfrak{m}$ to $C_{\mathsf{b}}(T)$ is in $\mathfrak{M}_{\mathsf{u}}(\mathsf{F}T)$ by Theorem 8.15 and $\mathfrak{m}\in\mathfrak{M}_{\mathsf{F}}(\mathsf{F}T)$ by Theorem 10.5. That proves (i)$\Rightarrow$(ii).

To prove (ii)$\Rightarrow$(i), assume $\mathfrak{m}\in\mathfrak{M}_{\mathsf{F}}(\mathsf{F}T)$. Then also $\mathfrak{m}\in\mathfrak{M}_{\mathsf{F}}(\mathsf{c}\mathsf{F}T)$ because the space $\mathsf{F}T$ is finer than $\mathsf{c}\mathsf{F}T$. By Theorem 10.37, there is a Riesz measure μ on T such that $\mathfrak{m}(f) = \int f\,d\mu$ for all $f\in C(T)$ and μ is separable by Theorem 8.15.

Finally, (ii)$\Leftrightarrow$(iii) by Corollary 10.30. □

If T is a completely regular space such that every measure on the σ-algebra $\sigma(C(T))$ is separable, then of course every Riesz measure on T is separable. The next two exercises show that the converse depends on set theory assumptions.

Exercise 10.40. Assuming the cardinal $2^{\aleph_0}$ is measure-free, prove that if T is a completely regular space such that every Riesz measure on T is separable, then every measure on the σ-algebra $\sigma(C(T))$ is separable. ∎

Exercise 10.41. Assuming the cardinal $2^{\aleph_0}$ is not measure-free, find a completely regular space T such that every Riesz measure on T is separable, but there is a measure on the σ-algebra $\sigma(C(T))$ that is not separable. ∎

10.7   Cylindrical Measures of Type 1

Recall from Sect. 8.6 that cylindrical measures on a locally convex space E naturally identify with uniform measures on the uniform space $\mathsf{w}E$.

Definition 10.42. Let E be a locally convex space and $\mathfrak{m}\in\mathfrak{M}_{\mathsf{u}}(\mathsf{w}E)^+$. Say that $\mathfrak{m}$ is *of type 1* iff the finite $\lim_{j\in\omega}\mathfrak{m}(|h|\wedge j)$ exists for every $h\in E^*$. ∎

For $\mathfrak{m}\in\mathfrak{M}_u(wE)^+$, let $\langle\Gamma_E, E/\gamma, \varphi_{\beta\gamma}, \mathfrak{m}_\gamma\rangle$ be the cylindrical measure on E defined by $\mathfrak{m}$; that is, $\mathfrak{m}_\gamma = \varphi_\gamma(\mathfrak{m})$ for all $\gamma\in\Gamma_E$, where $\varphi_\gamma\colon E\to E/\gamma$ are the quotient mappings. Say that this cylindrical measure is of type 1 iff $\mathfrak{m}$ is.

For every $\gamma\in\Gamma_E$ the functional $\mathfrak{m}_\gamma\in\mathfrak{M}_t(E/\gamma)^+$ is represented by a tight Borel measure $\mu_\gamma\geq 0$ on the finite-dimensional space E/γ (Theorem 5.3). Clearly $\mathfrak{m}$ is of type 1 if and only if for every $\gamma\in\Gamma_E$ and for every linear functional h on E/γ the function $|h|$ is μ_γ-integrable.

To get another equivalent formulation, for every $\gamma\in\Gamma_E$ of codimension 1 fix an arbitrary linear isomorphism between the one-dimensional space E/γ and $\mathbb{R}$. This isomorphism maps the measure μ_γ to a tight Borel measure $\mu_\gamma'\geq 0$ on $\mathbb{R}$. Clearly $\mathfrak{m}$ is of type 1 if and only if the function $x\mapsto|x|$ is μ_γ'-integrable for every γ of codimension 1.

Theorem 10.43. *Let E be a locally convex space and $\mathfrak{m}\in\mathfrak{M}_u(wE)^+$. Then $\mathfrak{m}$ is of type 1 if and only if $\mathfrak{m}$ is (the restriction to $\mathsf{U}_b(wE)$ of) a free uniform measure on wE.*

Proof. If $\mathfrak{m}$ is the restriction of a free uniform measure and $h\in E^*$, then the finite $\lim_j\mathfrak{m}(|h|\wedge j)$ exists because $|h|\in\mathsf{U}(wE)$.

To prove the converse, take any $f\in\mathsf{U}(wE)$. By Theorem 2.6, there are $\theta > 0$ and a finite set $D\subseteq E^*$ such that for all $x,y\in E$ with $\max_{h\in D}|h(x)-h(y)| < \theta$ we have $|f(x)-f(y)| < 1$. Thus

$$|f(x)| < f(0) + 1 + \max_{h\in D}|h(x)|/\theta$$

for all $x\in E$. If $\mathfrak{m}$ is of type 1, then $\lim_j\mathfrak{m}(|h|\wedge j)$ is finite for every $h\in D$; therefore $\lim_j\mathfrak{m}(|f|\wedge j)$ is finite, and $\mathfrak{m}$ is the restriction of a free uniform measure by Theorem 10.5.   □

For a locally convex space E and $\mathfrak{m}\in\mathfrak{M}_F(wE)$, Theorem 10.22 applied to the embedding $\varphi\colon wE\hookrightarrow\widehat{wE}$ yields $x_\mathfrak{m}\in\widehat{wE}$ such that $\hat{h}(x_\mathfrak{m}) = \mathfrak{m}(h)$ for every $h\in E^*$. Here, in accordance with the notation in Sect. 6.5, $\hat{h}$ is the unique continuous extension of $h\in E^*$ to the completion $\widehat{wE}$. Clearly $x_\mathfrak{m}$ is unique for each $\mathfrak{m}$; it is called the *resultant* of $\mathfrak{m}$. Thus every cylindrical measure of type 1 on E has a resultant in $\widehat{wE}$.

10.8 Notes for Chap. 10

Raĭkov [153] introduced the free locally convex space of a uniform space X and constructed it as $\mathrm{Mol}(X)$ with the $\mathrm{UE}(X)$ topology. He also pointed out the similar construction of Arens and Eells [3].

The space $\mathfrak{M}_F(X)$ is a completion of $\mathrm{Mol}(X)$ with the $\mathrm{UE}(X)$ topology; hence it is a free *complete* locally convex space of X. In the language of categories, the universal property in Theorem 10.23 states that $\mathfrak{M}_F$ is a free functor from the category of uniform spaces to the category of complete locally convex spaces. The Arens–Eells construction yields an analogous universal property for preduals of Lipschitz spaces [178, 2.2.4].

Motivated by Raĭkov's work, Berezanskiĭ [7] and Fedorova [50] established the basic properties of $\mathfrak{M}_F(X)$ (and also of the related space $\mathfrak{M}_u(X)$, as noted in Sect. 6.8). The particular case $\mathfrak{M}_F(FT)$, for a completely regular space T, has been extensively studied; Buchwalter [16, s.3] includes a detailed discussion and references (with $M(T)$ denoting $\mathfrak{M}_F(FT)$).

Theorem 10.5 is in my paper [134]; it generalizes special cases proved earlier by Berezanskiĭ [7, s.8] and Berruyer and Ivol [9]. The proof here incorporates a simplification suggested by Buchwalter (personal communication). Example 10.6 is also in [134]. The proof of Theorem 10.18 here is a simplified version of my original proof [136]. Vector-valued free uniform measures were investigated by Khurana and Cosalante [112].

The duality $\langle \mathfrak{A}, \mathrm{Lip}(\Delta, h) \rangle$ in Exercise 10.13 is a more general version of that in Exercise 5.33. The case $h := {}_{\backslash x}\Delta(x_0, x)$, where x_0 is a distinguished point in X, has a prominent role in the theory of Lipschitz spaces [178].

The property $\mathfrak{M}_F(X) \cong \mathfrak{M}_C(\widehat{X})$ was treated in detail by Deaibes [36, 4.5], who also proved Theorem 10.28 and its corollary in [38, 2.6] along with several results about other spaces of functionals on $U_b(X)$ and $U(X)$ for uniform spaces with the CDE property (called *spaces of type* (σ_1^∞) in [38]). Theorem 10.35 is due to Deaibes [37]. The results in Exercise 10.34 follow from Hewitt's characterization of realcompact spaces [96]. Related results about Borel-complete spaces were proved by Hager et al. [93].

For a completely regular space T, the completion $\widehat{cFT}$ is (one of the homeomorphic versions of) the *Hewitt realcompactification* of T. The equivalence of (i), (iii), (iv) and (v) in Theorem 10.37 is due to Hewitt [96]. The equivalence of (ii) and (iii) in Theorem 10.39 is due to Haydon [95]. Parts of Theorems 10.37 and 10.39 were also proved by Berruyer and Ivol [9] and Kirk [113]. In [135], I described Riesz and separable Riesz measures as instances of free uniform measures.

Radon measures of order 1 (Example 10.32) and cylindrical measures of type 1 (Sect. 10.7) are covered by Badrikian [5, Exp.12] along with other orders and types. Choquet [26] used the resultant of a cylindrical measure.

Chapter 11
Approximation of Probability Distributions

In this chapter I apply the theory developed in Part II to questions motivated by probability theory. Section 11.1 contains a dual characterization of seminorms $\|\cdot\|_\Delta$ on a large subspace of $\mathfrak{M}_b(X)$. In Sect. 11.2, I discuss properties of asymptotic approximation for nets of probability distributions and apply Corollary 5.43 to show that certain notions of approximation are all equivalent on sequences of distributions.

11.1 The Kantorovich–Rubinshteĭn Theorem

If X is a uniform space and $\Delta \in \mathsf{UP}_b(X)$, then Δ is a uniformly continuous function on the uniform product $X \times X$ by Lemma 2.27; thus $\mathfrak{n}(\Delta)$ is defined for every $\mathfrak{n} \in \mathfrak{M}_b(X \times X)$.

Lemma 11.1. *Let X be a uniform space, $\Delta \in \mathsf{UP}(X)$, $\mathfrak{n} \in \mathfrak{M}_b(X \times X)^+$, and let π_0 and π_1 be the two canonical projections from $X \times X$ to X. Then*

$$\|\pi_0(\mathfrak{n}) - \pi_1(\mathfrak{n})\|_\Delta \leq \mathfrak{n}(2 \wedge \Delta).$$

Proof. If $f \in \mathsf{BLip}_b(\Delta)$ and $x, y \in X$, then

$$|f \circ \pi_0((x,y)) - f \circ \pi_1((x,y))| = |f(x) - f(y)| \leq 2 \wedge \Delta(x,y)$$
$$|\mathfrak{n}(f \circ \pi_0) - \mathfrak{n}(f \circ \pi_1)| \leq \mathfrak{n}(|f \circ \pi_0 - f \circ \pi_1|) \leq \mathfrak{n}(2 \wedge \Delta),$$

and thus $\|\pi_0(\mathfrak{n}) - \pi_1(\mathfrak{n})\|_\Delta \leq \mathfrak{n}(2 \wedge \Delta)$. □

The lemma has a probabilistic interpretation, aided by the following definition.

J. Pachl, *Uniform Spaces and Measures*, Fields Institute Monographs 30,
DOI 10.1007/978-1-4614-5058-0_12,
© Springer Science+Business Media New York 2013

Definition 11.2. For a uniform space X, let

$$\mathfrak{M}_t^{+1}(X) := \{\mathfrak{m} \in \mathfrak{M}_t(X) \mid \mathfrak{m} \geq 0 \text{ and } \mathfrak{m}(1) = 1\}.$$

By Theorem 5.3, the functionals in $\mathfrak{M}_t^{+1}(X)$ are those represented by tight probability measures on $\mathsf{Bo}(X)$. When the measure representing $\mathfrak{m}$ is the distribution of an X-valued random variable ζ, the functional $\mathfrak{m} \in \mathfrak{M}_t^{+1}(X)$ itself is also said to be the *(probability) distribution of* ζ. ∎

Clearly $\mathfrak{m} \in \mathfrak{M}_t^{+1}(X)$ is the distribution of ζ if and only if $\mathsf{E}(f(\zeta)) = \mathfrak{m}(f)$ for every $f \in \mathsf{U_b}(X)$.

A special case of Lemma 11.1 states that if ζ_0 and ζ_1 are X-valued random variables whose distributions are $\mathfrak{m}_0, \mathfrak{m}_1 \in \mathfrak{M}_t^{+1}(X)$ and if $2 \wedge \Delta(\zeta_0, \zeta_1)$ is also a random variable (i.e. a measurable function on the underlying probability space), then

$$\|\mathfrak{m}_0 - \mathfrak{m}_1\|_\Delta \leq \mathsf{E}(2 \wedge \Delta(\zeta_0, \zeta_1)).$$

The main result of this section is the following theorem. It supplies a partial converse to the estimate in Lemma 11.1.

Theorem 11.3. *Let X be a uniform space, $\Delta \in \mathsf{UP}(X)$, $\mathfrak{m}_0, \mathfrak{m}_1 \in \mathfrak{M}_b(X)^+$ and let $\mathfrak{m}_0(1) = \mathfrak{m}_1(1)$. Let π_0 and π_1 be the two canonical projections from $X \times X$ to X. Then there exists $\mathfrak{n} \in \mathfrak{M}_b(X \times X)^+$ such that $\mathfrak{m}_0 = \pi_0(\mathfrak{n})$, $\mathfrak{m}_1 = \pi_1(\mathfrak{n})$ and*

$$\|\mathfrak{m}_0 - \mathfrak{m}_1\|_\Delta = \mathfrak{n}(2 \wedge \Delta).$$

As in Lemma 5.12, the constant 2 in the formula for $\|\mathfrak{m}_0 - \mathfrak{m}_1\|_\Delta$ comes from the condition $\|f\|_X \leq 1$ imposed on functions $f \in \mathsf{BLip_b}(\Delta)$.

The proof uses the following alternative expression for $\|\mathfrak{m}_0 - \mathfrak{m}_1\|_\Delta$.

Lemma 11.4. *Let X be a uniform space, $\Delta \in \mathsf{UP}(X)$, and*

$$\mathscr{F} := \{(f_0, f_1) \mid f_0, f_1 \in \mathsf{U_b}(X) \text{ and } f_0(x) + f_1(y) \leq 2 \wedge \Delta(x, y) \text{ for all } x, y \in X\}.$$

Let $\mathfrak{m}_0, \mathfrak{m}_1 \in \mathfrak{M}_b(X)^+$ and $\mathfrak{m}_0(1) = \mathfrak{m}_1(1)$. Then

$$\|\mathfrak{m}_0 - \mathfrak{m}_1\|_\Delta = \sup \{\mathfrak{m}_0(f_0) + \mathfrak{m}_1(f_1) \mid (f_0, f_1) \in \mathscr{F}\}.$$

Proof. Take any $f \in \mathsf{BLip_b}(\Delta)$. Then $\mathfrak{m}_0(f) - \mathfrak{m}_1(f) = \mathfrak{m}_0(f_0) + \mathfrak{m}_1(f_1)$ where $f_0 := f$, $f_1 := -f$, and $(f_0, f_1) \in \mathscr{F}$. Hence

$$\|\mathfrak{m}_0 - \mathfrak{m}_1\|_\Delta \leq \sup \{\mathfrak{m}_0(f_0) + \mathfrak{m}_1(f_1) \mid (f_0, f_1) \in \mathscr{F}\}.$$

To prove the opposite inequality, take any $(f_0, f_1) \in \mathscr{F}$ and define

$$f(x) := \inf_{y \in X} ((2 \wedge \Delta(x, y)) - f_1(y)).$$

The function f, being the infimum of Δ-1-Lipschitz functions, is Δ-1-Lipschitz. Moreover, $\sup f - \inf f \le 2$, which means that there is a constant $r \in \mathbb{R}$ for which $f + r \in \mathsf{BLip}_b(\Delta)$. Next, observe that $f_0 \le f \le -f_1$, and therefore

$$\mathfrak{m}_0(f+r) - \mathfrak{m}_1(f+r) = \mathfrak{m}_0(f) - \mathfrak{m}_1(f) \ge \mathfrak{m}_0(f_0) + \mathfrak{m}_1(f_1).$$

It follows that $\|\mathfrak{m}_0 - \mathfrak{m}_1\|_\Delta \ge \mathfrak{m}_0(f_0) + \mathfrak{m}_1(f_1)$.   □

Proof of Theorem 11.3. Let $\mathscr{H} \subseteq \mathsf{U}_b(X \times X)$ be the set of the functions on $X \times X$ of the form $f_0 \circ \pi_0 + f_1 \circ \pi_1$, where $f_0, f_1 \in \mathsf{U}_b(X)$. Such a representation of $h \in \mathscr{H}$ is not unique; however, if $h = f_0 \circ \pi_0 + f_1 \circ \pi_1 = f_0' \circ \pi_0 + f_1' \circ \pi_1$ with $f_0, f_1, f_0', f_1' \in \mathsf{U}_b(X)$, then the function $f_0 \circ \pi_0 - f_0' \circ \pi_0 = f_1' \circ \pi_1 - f_1 \circ \pi_1$ is constant on $X \times X$, and thus the definition

$$\mathfrak{n}'(h) := \mathfrak{m}_0(f_0) + \mathfrak{m}_1(f_1)$$

does not depend on the choice of f_0 and f_1. Evidently $\mathscr{H}$ is a vector subspace of $\mathsf{U}_b(X \times X)$, and $\mathfrak{n}'$ is a linear functional on $\mathscr{H}$.

For $f \in \mathsf{U}_b(X \times X)$, define

$$\alpha(f) := \inf\{r\mathfrak{m}_0(1) + r'\|\mathfrak{m}_0 - \mathfrak{m}_1\|_\Delta \mid r, r' \in \mathbb{R}^+ \text{ and } |f| \le r + r'(2 \wedge \Delta)\}.$$

Then α is a seminorm on $\mathsf{U}_b(X \times X)$. Note that $\alpha(f) \le \mathfrak{m}_0(1)\|f\|_{X \times X}$ for every $f \in \mathsf{U}_b(X \times X)$ and $\alpha(2 \wedge \Delta) \le \|\mathfrak{m}_0 - \mathfrak{m}_1\|_\Delta$.

I claim that $|\mathfrak{n}'(h)| \le \alpha(h)$ for every $h \in \mathscr{H}$. Take any $h \in \mathscr{H}$ and $r, r' \in \mathbb{R}^+$ such that $|h| \le r + r'(2 \wedge \Delta)$. Then $h - r \le r'(2 \wedge \Delta)$, and $\mathfrak{n}'(h - r) \le r'\|\mathfrak{m}_0 - \mathfrak{m}_1\|_\Delta$ by Lemma 11.4. Thus $\mathfrak{n}'(h) \le r\mathfrak{m}_0(1) + r'\|\mathfrak{m}_0 - \mathfrak{m}_1\|_\Delta$. The same argument with $-h$ in place of h establishes $\mathfrak{n}'(-h) \le r\mathfrak{m}_0(1) + r'\|\mathfrak{m}_0 - \mathfrak{m}_1\|_\Delta$, which proves the claim.

By the Hahn–Banach Theorem P.9, there is a linear functional $\mathfrak{n}$ on $\mathsf{U}_b(X \times X)$ that agrees with $\mathfrak{n}'$ on $\mathscr{H}$ and such that $|\mathfrak{n}(f)| \le \alpha(f)$ for every $f \in \mathsf{U}_b(X \times X)$. Then $\mathfrak{n} \ge 0$ because

$$\mathfrak{m}_0(1) = \mathfrak{n}(1) \le |\mathfrak{n}|(1) = \sup\{\mathfrak{n}(f) \mid f \in \mathsf{U}_b(X \times X), \|f\|_{X \times X} \le 1\}$$

$$\le \sup\{\alpha(f) \mid f \in \mathsf{U}_b(X \times X), \|f\|_{X \times X} \le 1\} \le \mathfrak{m}_0(1),$$

and evidently $\mathfrak{m}_0 = \pi_0(\mathfrak{n})$ and $\mathfrak{m}_1 = \pi_1(\mathfrak{n})$. Finally, $\|\mathfrak{m}_0 - \mathfrak{m}_1\|_\Delta \le \mathfrak{n}(2 \wedge \Delta)$ by Lemma 11.1, and $\mathfrak{n}(2 \wedge \Delta) \le \alpha(2 \wedge \Delta) \le \|\mathfrak{m}_0 - \mathfrak{m}_1\|_\Delta$.   □

Corollary 11.5. *Let X be a uniform space, $\Delta \in \mathsf{UP}(X)$, $\mathfrak{m}_0, \mathfrak{m}_1 \in \mathfrak{M}_u(X)^+$ and $\mathfrak{m}_0(1) = \mathfrak{m}_1(1)$. Let π_0 and π_1 be the two canonical projections from $X \times X$ to X. Then there exists $\mathfrak{n} \in \mathfrak{M}_u(X \times X)^+$ such that $\mathfrak{m}_0 = \pi_0(\mathfrak{n})$, $\mathfrak{m}_1 = \pi_1(\mathfrak{n})$ and*

$$\|\mathfrak{m}_0 - \mathfrak{m}_1\|_\Delta = \mathfrak{n}(2 \wedge \Delta).$$

If $\mathfrak{m}_0, \mathfrak{m}_1 \in \mathfrak{M}_t(X)^+$, then there exists such $\mathfrak{n}$ in $\mathfrak{M}_t(X \times X)^+$.

Proof. Apply Theorem 11.3 and Lemmas 6.26 and 6.27.   □

Again Corollary 11.5 yields a probabilistic interpretation: If $m_0, m_1 \in \mathfrak{M}_t^{+1}(X)$, then there exist X-valued random variables ζ_0 and ζ_1 with distributions m_0 and m_1 such that $2 \wedge \Delta(\zeta_0, \zeta_1)$ is also a random variable and

$$\|m_0 - m_1\|_\Delta = \mathsf{E}(2 \wedge \Delta(\zeta_0, \zeta_1)).$$

In fact, let ν be the tight Borel probability measure representing the functional $n \in \mathfrak{M}_t(X \times X)^+$ in the corollary, and let ζ_0 and ζ_1 be the two canonical projections π_0 and π_1 on the probability space $(X \times X, \mathrm{Bo}(X \times X), \nu)$.

Corollary 11.6. *Let X be a uniform space, $\Delta \in \mathrm{UP}(X)$, and let $m \in \mathfrak{M}_b(X)$ be such that $m(1) = 0$. Let π_0 and π_1 be the two canonical projections from $X \times X$ to X. Then there exists $n \in \mathfrak{M}_b(X \times X)^+$ such that $m = \pi_0(n) - \pi_1(n)$ and $\|m\|_\Delta = n(2 \wedge \Delta)$.*

If $m \in \mathfrak{M}_u(X)^+$, then there exists such n in $\mathfrak{M}_u(X \times X)^+$. If $m \in \mathfrak{M}_t(X)^+$, then there exists such n in $\mathfrak{M}_t(X \times X)^+$.

Proof. The first statement follows from Theorem 11.3 with m^+ and m^- in place of m_0 and m_1. The other statements follow the same way from Corollary 11.5.   □

Corollary 11.6 is an alternative description of seminorms $\|\cdot\|_\Delta$ on the subspace $\{m \in \mathfrak{M}_b(X) \mid m(1) = 0\}$ of codimension 1 in $\mathfrak{M}_b(X)$.

11.2   Asymptotic Approximation of Probability Distributions

The $\mathsf{C}_b(X)$-weak topology on $\mathfrak{M}_t^{+1}(X)$ features in many limit theorems in probability theory, where it is called the *weak topology*. When a net of probability distributions converges in the weak topology, its limit serves as a *weak approximation* of the distributions in the net.

A topology on the set $\mathfrak{M}_t^{+1}(X)$ is not enough if we wish to reason about the more general notion of asymptotic approximation for *divergent* nets of probability distributions. Thus we are led to consider uniform structures on $\mathfrak{M}_t^{+1}(X)$ rather than mere topologies.

Definition 11.7. *Let $\mathscr{U}$ be a uniform structure on a set S. A net $\{m_\gamma\}_\gamma$ in S is an asymptotic $\mathscr{U}$-approximation for another net $\{n_\gamma\}_\gamma$ in S iff $\lim_\gamma \Delta(m_\gamma, n_\gamma) = 0$ for every pseudometric $\Delta \in \mathscr{U}$.*   ∎

In the rest of this section, I deal with the case of a metric space X. I shall describe a class of uniform structures on $\mathfrak{M}_t^{+1}(X)$ that appear to be reasonable candidates for extending to divergent nets the traditional notion of weak approximation for convergent nets. Although there is more than one uniform structure in this class, I show that they all yield the same notion of asymptotic approximation for sequences.

Of course, it is highly subjective and context-dependent what uniformities should be "reasonable" for this purpose. In the sequel, I replace that subjective notion by the two properties in Definition 11.9.

Lemma 11.8. *Let I_k denote the closed interval $[-k,k] \subseteq \mathbb{R}$ and $\|\cdot\|_\Delta$ the usual metric on I_k (the restriction of $\Delta_\mathbb{R}$ to I_k).*

1. *The $C_b(I_k)$-weak and the $U_b(I_k)$-weak uniformities and the $\|\cdot\|_\Delta$ uniformity coincide on $\mathfrak{M}_t^{+1}(I_k)$.*
2. *The uniformity from Part 1 is the only one compatible with the $C_b(I_k)$-weak topology on $\mathfrak{M}_t^{+1}(I_k)$.*

Proof. Since I_k is compact, so is $\mathfrak{M}_t^{+1}(I_k)$ with the $C_b(I_k)$-weak topology. Both statements follow from Corollary 1.7 and Corollaries 5.17 (or 5.19) and 5.37 (or 5.39). $\qquad\square$

Definition 11.9. Let X be a metric space with metric Δ. Consider the following properties of a uniform structure $\mathcal{U}$ on $\mathfrak{M}_t^{+1}(X)$:

(A1) If $\{\zeta_j\}_j$ and $\{\zeta'_j\}_j$ are two sequences of X-valued random variables such that $\lim_j \Delta(\zeta_j, \zeta'_j) = 0$ almost surely, then the sequence of distributions of ζ_j is an asymptotic $\mathcal{U}$-approximation for the sequence of distributions of ζ'_j.

(A2) For every interval $I_k := [-k,k] \subseteq \mathbb{R}$, $k = 1, 2, \ldots$, if a mapping $\varphi \colon X \to I_k$ is uniformly continuous, then so is the mapping $\mathfrak{M}_b(\varphi)$ from $\mathfrak{M}_t^{+1}(X)$ with $\mathcal{U}$ to $\mathfrak{M}_t^{+1}(I_k)$ with the unique uniformity in Lemma 11.8. $\qquad\blacksquare$

Property (A1) states that asymptotic approximation (understood almost surely) for sequences of X-valued random variables implies asymptotic approximation for the corresponding distributions.

(A2) is a functorial property of the assignment $X \mapsto (\mathfrak{M}_t^{+1}(X), \mathcal{U})$ for uniformly continuous mappings $\varphi \colon X \to I_k$, assuming that on convergent nets in $\mathfrak{M}_t^{+1}(I_k)$ the asymptotic approximation agrees with the weak approximation.

As a special case of (A1), if $\{x_j\}_j$ and $\{x'_j\}_j$ are two sequences of points in X such that $\lim_j \Delta(x_j, x'_j) = 0$, then $\{\partial(x_j)\}_j$ is an asymptotic $\mathcal{U}$-approximation for $\{\partial(x'_j)\}_j$.

Exercise 11.10. Let X be $\mathbb{R}$ with the metric $\Delta_\mathbb{R}$. Show that the $C_b(X)$-weak uniformity on $\mathfrak{M}_t^{+1}(X)$ does not satisfy (A1). $\qquad\blacksquare$

In contrast to Exercise 11.10, two other uniformities from Sects. 5.3 and 5.4 do satisfy (A1) and (A2):

Theorem 11.11. *For every metric space X with metric Δ, the $U_b(X)$-weak uniformity and the $\|\cdot\|_\Delta$ uniformity on $\mathfrak{M}_t^{+1}(X)$ have properties (A1) and (A2).*

Proof. Let $\mathfrak{m}_j, \mathfrak{m}'_j \in \mathfrak{M}_t^{+1}(X)$, $j \in \omega$, be the distributions of X-valued random variables ζ_j, ζ'_j such that $\lim_j \Delta(\zeta_j, \zeta'_j) = 0$ almost surely. Then

$$\lim_j \|\mathfrak{m}_j - \mathfrak{m}'_j\|_\Delta \leq \lim_j \mathsf{E}(2 \wedge \Delta(\zeta_j, \zeta'_j)) = 0$$

by part 2 of Theorem P.36 and by Lemma 11.1. Hence the $\|\cdot\|_\Delta$ uniformity has property (A1) and therefore so does the $U_b(X)$-weak uniformity.

(A2) follows from Lemma 11.8 and the definition of $\mathfrak{M}_b(\varphi)$.   □

Next, I prove that every uniformity on $\mathfrak{M}_t^{+1}(X)$ satisfying (A1) and (A2) is between the two uniformities in Theorem 11.11.

Theorem 11.12. *Let X be a metric space with metric Δ, and $\mathcal{U}$ a uniform structure on $\mathfrak{M}_t^{+1}(X)$.*

1. *If $\mathcal{U}$ has property (A1), then it is coarser than the $\|\cdot\|_\Delta$ uniformity on $\mathfrak{M}_t^{+1}(X)$.*
2. *If $\mathcal{U}$ has property (A2), then it is finer than the $U_b(X)$-weak uniformity on $\mathfrak{M}_t^{+1}(X)$.*

Proof. To prove Part 1, assume that $\mathcal{U}$ is not coarser than the $\|\cdot\|_\Delta$ uniformity on $\mathfrak{M}_t^{+1}(X)$. That means that there are $\Delta' \in \mathcal{U}$, $\varepsilon > 0$ and $\mathfrak{m}_j, \mathfrak{m}'_j \in \mathfrak{M}_t^{+1}(X)$ for $j = 1, 2, \ldots$ such that $\|\mathfrak{m}_j - \mathfrak{m}'_j\|_\Delta < 1/j$ and $\Delta'(\mathfrak{m}_j, \mathfrak{m}'_j) \geq \varepsilon$ for all j. By Corollary 11.5, there are X-valued random variables ζ_j, ζ'_j whose distributions are $\mathfrak{m}_j, \mathfrak{m}'_j$, respectively, and such that $\lim_j E(2 \wedge \Delta(\zeta_j, \zeta'_j)) = 0$. By Part 1 of Theorem P.36, there are subsequences $\{\zeta_{j(i)}\}_i$ and $\{\zeta'_{j(i)}\}_i$ such that $\lim_i \Delta(\zeta_{j(i)}, \zeta'_{j(i)}) = 0$ almost surely. Thus $\mathcal{U}$ does not have property (A1).

2. Assume that $\mathcal{U}$ has property (A2) and take any $f \in U_b(X)$. Fix $k \in \{1, 2, \ldots\}$ for which $\|f\|_X \leq k$, so that f maps X into $I_k := [-k, k]$.

Let $g \in U_b(I_k)$ be the function $g : x \mapsto x$. By (A2) and Lemma 11.8, the mapping $\mathfrak{M}_b(f)$ is uniformly continuous from $\mathfrak{M}_t^{+1}(X)$ with the uniformity $\mathcal{U}$ to $\mathfrak{M}_t^{+1}(I_k)$ with the $U_b(I_k)$-weak uniformity. Thus the mapping

$$\mathfrak{m} \mapsto \mathfrak{M}_b(f)(\mathfrak{m})(g) = \mathfrak{m}(g \circ f) = \mathfrak{m}(f)$$

from $\mathfrak{M}_t^{+1}(X)$ with $\mathcal{U}$ to $\mathbb{R}$ is uniformly continuous. That proves that $\mathcal{U}$ is finer than the $U_b(X)$-weak uniformity.   □

Corollary 11.13. *Let X be a metric space with metric Δ, and let $\mathcal{U}$ be a uniform structure on $\mathfrak{M}_t^{+1}(X)$. If $\mathcal{U}$ has properties (A1) and (A2), then it is compatible with the $C_b(X)$-weak topology on $\mathfrak{M}_t^{+1}(X)$.*

Proof. Apply Theorem 11.12 and Lemma 11.8.   □

Corollary 11.14. *Let X be a metric space with metric Δ, and $\mathcal{U}$ a uniform structure on $\mathfrak{M}_t^{+1}(X)$. If $\mathcal{U}$ has properties (A1) and (A2), then the following statements are equivalent for any two sequences $\{\mathfrak{m}_j\}_j$ and $\{\mathfrak{n}_j\}_j$ in $\mathfrak{M}_t^{+1}(X)$:*

(i) $\lim_j \mathfrak{m}_j(f) - \mathfrak{n}_j(f) = 0$ *for every* $f \in U_b(X)$.
(ii) *The sequence $\{\mathfrak{m}_j\}_j$ is an asymptotic $\mathcal{U}$-approximation of the sequence $\{\mathfrak{n}_j\}_j$.*
(iii) $\lim_j \|\mathfrak{m}_j - \mathfrak{n}_j\|_\Delta = 0$.

Thus all uniformities on $\mathfrak{M}_t^{+1}(X)$ that satisfy (A1) and (A2) yield the same notion of asymptotic approximation for sequences (but not necessarily for nets) of probability distributions.

Proof. By Corollary 5.43, (i) implies (iii). By Theorem 11.12, (iii) implies (ii) and (ii) implies (i).   □

11.3 Notes for Chap. 11

Theorem 11.3 is similar to Dudley's generalization [43, 20.1][44, 11.8] of the Kantorovich–Rubinshteĭn theorem [105]. Rachev and Rüschendorf [152] and Villani [174], [175] describe many related results in the theory of optimal transport.

Uniform structures on probability distributions were investigated by Dudley [42] [44, 11.7] and D'Aristotile, Diaconis and Freedman [34]. Davydov and Rotar [35] showed that the previously neglected $U_b(X)$-weak uniformity is a reasonable extension of the traditional notion of weak topology on probability distributions and proved the equivalence of (i) and (iii) in Corollary 11.14. Property (A1) in Definition 11.9 and its use in Theorem 11.12 are due to van Handel (private communication).

By a result of Dudley [44, 11.7.1], condition (iii) in Corollary 11.14 is equivalent to the following: For $j \in \omega$, there are X-valued random variables ζ_j and ζ_j' whose distributions are m_j and n_j, respectively, and such that $\lim_j \Delta(\zeta_j, \zeta_j') = 0$ almost surely.

Further development of the theory of uniformities on probability distributions should benefit from research on probability metrics, surveyed by Dudley [43], Gibbs and Su [80], Rachev [151], Villani [174], [175] and Zolotarev [184]; however, so far the focus in that area has been on quantitative results for the rate of convergence of convergent sequences, not on general uniform structures or even the uniform structures defined by probability metrics.

Chapter 12
Measurable Functionals

In this chapter I describe an approach to automatic continuity of functionals on $U_b(X)$: To prove that a linear functional $\mathfrak{m}$ on $U_b(X)$ is a uniform measure, it is sometimes enough to prove that $\mathfrak{m}$ is measurable with respect to a suitable σ-algebra on $U_b(X)$; and measurability is often easier to establish than continuity. Section 12.3 includes an application to generalized (i.e. measurable) centres in convolution semigroups.

12.1 Saturated Spaces

On certain spaces, measurable linear functionals are continuous. This useful property is captured in the following definition.

Definition 12.1. Let E and F be two vector spaces in duality, and Σ a σ-algebra on F. The space E is *saturated with respect to* Σ, or Σ-*saturated* for short, iff every Σ-measurable linear functional on F is E-weakly continuous (and thus identifies with an element of E). ∎

Of particular interest are the σ-algebra $\sigma(E)$ on F obtained when the elements of E are identified with linear functionals on F, and the σ-algebra $\mathrm{Bo}(\mathrm{w}_E F)$ of the Borel sets in the E-weak topology on F. If E is $\mathrm{Bo}(\mathrm{w}_E F)$-saturated, then it is also $\sigma(E)$-saturated, and that in turn implies that it is F-weakly sequentially complete. In this sense, saturation is a strengthening of weak sequential completeness.

Exercise 12.2. Show that if X is an Alexandroff uniform space, then the space $\mathfrak{M}_\sigma(X)$ is saturated with respect to the σ-algebra of $\mathfrak{M}_\sigma(X)$-weakly CBP sets on the space $U_b(X)$. Therefore $\mathfrak{M}_\sigma(X)$ is also saturated with respect to $\sigma(\mathfrak{M}_\sigma(X))$ and the σ-algebra of $\mathfrak{M}_\sigma(X)$-weakly Borel sets on $U_b(X)$. ∎

By Theorem 6.19, the space $\mathfrak{M}_u(X)$ is $U_b(X)$-weakly sequentially complete for every uniform space X. That raises the question whether, or under what

J. Pachl, *Uniform Spaces and Measures*, Fields Institute Monographs 30,
DOI 10.1007/978-1-4614-5058-0_13,
© Springer Science+Business Media New York 2013

conditions, $\mathfrak{M}_u(X)$ is saturated with respect to $\sigma(\mathfrak{M}_u(X))$, or even with respect to the σ-algebras of $\mathfrak{M}_u(X)$-weakly Borel or $\mathfrak{M}_u(X)$-weakly CBP sets on $U_b(X)$. Partial answers are provided by Corollary 12.5 and Theorem 12.8.

In the next lemma and theorem, X-pointwise CBP measurability of a function from a set $\mathscr{F} \subseteq \mathbb{R}^X$ to $\mathbb{R}$ means the CBP measurability on $\mathscr{F}$ equipped with the X-pointwise topology.

Lemma 12.3. *Let X be a uniform space, S a non-empty set and $\mathfrak{m}$ a linear functional on $U_b(X)$. Let a mapping $\varphi\colon X \to \ell_1(S)$ be uniformly continuous from X to $\ell_1(S)$ with the $\|\cdot\|_1$ norm and such that $\|\varphi(x)\|_1 \leq 1$ for every $x\in X$. Assume that the restriction of $\mathfrak{m}$ to every $\mathsf{UEB}(X)$ set is X-pointwise CBP measurable. Then*

$$\sum_{s\in S} |\mathfrak{m}(\varphi_s)| < \infty \quad and \quad \sum_{s\in A} \mathfrak{m}(\varphi_s) = \mathfrak{m}\left(\sum_{s\in A}\varphi_s\right) \quad for \ \ A\subseteq S$$

where $\varphi_s(x) := \varphi(x)(s)$ for $s\in S$, $x\in X$.

Proof. Since φ is uniformly continuous in the $\|\cdot\|_1$ norm, for every $A \subseteq S$ the function $\sum_{s\in A}\varphi_s$ is in $U_b(X)$ and the set $\{\sum_{s\in A}\varphi_s \mid A \subseteq S\}$ is $\mathsf{UEB}(X)$. The expression $\psi(I_A) := \mathfrak{m}(\sum_{s\in A}\varphi_s)$, $A \subseteq S$, defines a finitely additive CBP measurable mapping $\psi\colon \{0,1\}^S \to \mathbb{R}$, and the conclusion follows from Theorem P.41. $\square$

Theorem 12.4. *Let X be a uniform space with the (ℓ_1) property, and let $\mathfrak{m}$ be a linear functional on $U_b(X)$ such that the restriction of $\mathfrak{m}$ to every $\mathsf{UEB}(X)$ set is X-pointwise CBP measurable. Then $\mathfrak{m}\in\mathfrak{M}_u(X)$.*

Proof. First I prove that $\mathfrak{m}\in\mathfrak{M}_b(X)$. Indeed, if $\mathfrak{m}$ were not bounded on the $\|\cdot\|_X$ unit ball in $U_b(X)$, then there would exist functions $g_j\in U_b(X)$, $j\in\omega$, such that $\|g_j\|_X \leq 1/2^{j+1}$ and $|\mathfrak{m}(g_j)| \geq 1$ for all j. By Lemma 12.3 with $\varphi\colon X \to \ell_1$ defined by $\varphi(x)(j) := g_j(x)$ for $x\in X$, $j\in\omega$, we would get $\sum_j |\mathfrak{m}(g_j)| < \infty$, in contradiction to $|\mathfrak{m}(g_j)| \geq 1$.

Take any $\Delta\in\mathsf{UP}(X)$ and any net $\{f_\gamma\}_\gamma$ of functions $f_\gamma\in\mathsf{BLip}_b(\Delta)$ such that $\lim_\gamma f_\gamma(x) = 0$ for all $x\in X$. Fix an arbitrary $\varepsilon > 0$. There is a partition of unity φ on X that is uniformly continuous from X to $\ell_1(S)$ with the $\|\cdot\|_1$ norm and such that $\Delta\text{-diam}(\text{coz}(\varphi_s)) < \varepsilon$ for every $s\in S$.

For each $s\in S$, choose a point $x_s\in X$ such that $\Delta(x,x_s) < \varepsilon$ whenever $x\in X$ and $\varphi_s(x) \neq 0$. Then

$$\left|f_\gamma(x) - \sum_{s\in S} f_\gamma(x_s)\cdot\varphi_s(x)\right| \leq \sum_{s\in S}\varphi_s(x)\cdot|f_\gamma(x) - f_\gamma(x_s)| \leq \sum_{s\in S}\varphi_s(x)\cdot\Delta(x,x_s) < \varepsilon$$

for all γ and all $x\in X$.

For a fixed γ, define $\varphi'\colon X \to \ell_1(S)$ by $\varphi'(x)(s) := f_\gamma(x_s)\cdot\varphi_s(x)$ for $x\in X$, $s\in S$, and apply Lemma 12.3 with φ' in place of φ to get

$$\sum_{s\in S} f_\gamma(x_s)\mathfrak{m}(\varphi_s) = \mathfrak{m}\left(\sum_{s\in S} f_\gamma(x_s)\varphi_s\right).$$

By Lemma 12.3, there is a finite set $D \subseteq S$ such that $\sum_{s \in S \setminus D} |\mathrm{m}(\varphi_s)| < \varepsilon$. For almost all γ we have $|f_\gamma(x_s)| < \varepsilon$ when $s \in D$, and

$$
\begin{aligned}
|\mathrm{m}(f_\gamma)| &\leq \left| \mathrm{m}\left(f_\gamma - \sum_{s \in S} f_\gamma(x_s) \varphi_s \right) \right| + \left| \sum_{s \in S} f_\gamma(x_s) \mathrm{m}(\varphi_s) \right| \\
&\leq \|\mathrm{m}\| \varepsilon + \sum_{s \in D} |f_\gamma(x_s)| \cdot |\mathrm{m}(\varphi_s)| + \sum_{s \in S \setminus D} |f_\gamma(x_s)| \cdot |\mathrm{m}(\varphi_s)| \\
&\leq \|\mathrm{m}\| \varepsilon + \|\mathrm{m}\| \varepsilon + \varepsilon = (2\|\mathrm{m}\| + 1)\varepsilon.
\end{aligned}
$$

As this holds for every $\varepsilon > 0$, we conclude that $\mathrm{m} \in \mathfrak{M}_u(X)$. □

Corollary 12.5. *If X is a uniform space with the (ℓ_1) property, then the space $\mathfrak{M}_u(X)$ is saturated with respect to the σ-algebra of $\mathfrak{M}_u(X)$-weakly CBP subsets of $\mathsf{U}_b(X)$.*

Thus, for those uniform spaces, $\mathfrak{M}_u(X)$ is saturated also with respect to the $\mathfrak{M}_u(X)$-weakly Borel σ-algebra, as well as $\sigma(\mathfrak{M}_u(X))$, on $\mathsf{U}_b(X)$.

Proof. On every $\mathsf{UEB}(X)$ set $\mathscr{F}$, the X-pointwise topology and the $\mathfrak{M}_u(X)$-weak topology coincide. Therefore the restriction of any $\mathfrak{M}_u(X)$-weakly CBP measurable function on $\mathsf{U}_b(X)$ to $\mathscr{F}$ is X-pointwise CBP measurable. Apply Theorem 12.4. □

12.2 Functionals of Baire Class 1

Let E and F be two vector spaces in duality. In some sense, the simplest measurable functionals on F, after E-weakly continuous ones, are those of Baire class 1. The main result of this section states that for every uniform space X, the functionals of Baire class 1 on $\mathsf{U}_b(X)$ with the $\mathfrak{M}_u(X)$-weak topology are in $\mathfrak{M}_u(X)$.

The following lemma is the key step in the proof.

Lemma 12.6. *Let X be a complete metric space with metric Δ, and $\mathrm{m} \in \mathfrak{M}_b(X)$ such that $\mathrm{m} \notin \mathfrak{M}_t(X)$. Then there is $k \in \omega$ for which the restriction of m to $\mathsf{BLip}_b(k\Delta)$ is not X-pointwise continuous at any $h_0 \in \mathsf{BLip}_b(k\Delta)$.*

Proof. As in Sect. 7.2, m is represented by the tight Borel measure ρ_m on $\hat{\mathrm{p}}X$. By Part 2 of Corollary 7.15, the set X is Borel in $\hat{\mathrm{p}}X$; hence $\rho_\mathrm{m} = \mu + \nu$ for unique tight Borel measures μ and ν on $\hat{\mathrm{p}}X$ such that $|\mu|(\hat{\mathrm{p}}X \setminus X) = 0$ and $|\nu|(X) = 0$. Let $\mathrm{n} \in \mathfrak{M}_b(X)$ be the functional represented by ν. Then

$$
\|\mathrm{n}\| = \|\nu\| = |\nu|(\hat{\mathrm{p}}X) = |\nu|(\hat{\mathrm{p}}X \setminus X) > 0
$$

by Theorem 7.9 because $\mathrm{m} \notin \mathfrak{M}_t(X)$. I shall prove that there is $k \in \omega$ for which the restriction of n to $\mathsf{BLip}_b(k\Delta)$ is nowhere X-pointwise continuous.

By Part 2 of Corollary 7.15, there is $k_0 \in \omega$ such that $|v|(\text{Œ}(k_0\Delta)) < \|\mathfrak{n}\|/4$. By Lemma 5.20, there are $k \in \omega$ and a function $g \in \mathsf{BLip_b}(k\Delta)$ such that $k > 3k_0$ and $\mathfrak{n}(g) > 3\|\mathfrak{n}\|/4$.

Now take any $h_0 \in \mathsf{BLip_b}(k\Delta)$ and any X-pointwise neighbourhood of h_0 in $\mathsf{BLip_b}(k\Delta)$ of the form

$$V = \{h \in \mathsf{BLip_b}(k\Delta) \mid \|h - h_0\|_D < \varepsilon\}$$

for $\varepsilon > 0$ and a finite set $D \subseteq X$. Without loss of generality, assume that $\mathfrak{n}(h_0) \leq 0$ (if not then proceed with $-h_0$ in place of h_0).

Let $A := \{x \in X \mid \Delta(x, D) \geq 2/k\}$ and define the function $h' : D \cup A \to [-1, 1]$ by $h'(x) := h_0(x)$ for $x \in D$ and $h'(x) := g(x)$ for $x \in A$. Since $|h'(x) - h'(y)| \leq k\Delta(x,y)$ for $x, y \in D \cup A$, by Lemma 2.21 there is a function $h \in \mathsf{BLip_b}(k\Delta)$ that extends h'. Moreover, $h \in V$ because $h(x) = h_0(x)$ for $x \in D$.

I claim that the closure $\overline{A}$ of A in $\hat{\mathsf{p}}X$ contains $\hat{\mathsf{p}}X \setminus \text{Œ}(k_0\Delta)$. To prove that, define $f_D \in \mathsf{BLip_b}(k_0\Delta)$ by $f_D(y) := 1 \wedge k_0\Delta(y, D)$, $y \in X$. If $z \in \hat{\mathsf{p}}X \setminus \text{Œ}(k_0\Delta)$ then $z \notin \bigcup_{x \in D} \mathrm{int}\,\odot[x, k_0\Delta]$; hence $\overline{f_D}(z) = 1$. There is a net $\{z_\gamma\}_\gamma$ in X that converges to z in $\hat{\mathsf{p}}X$; then $\lim_\gamma 1 \wedge k_0\Delta(z_\gamma, D) = \lim_\gamma f_D(z_\gamma) = 1$. Since $k > 3k_0$, for almost all γ we have $k\Delta(z_\gamma, D) > 2$. Thus $z \in \overline{A}$, which proves the claim.

Clearly $\overline{h}(z) = \overline{g}(z)$ for all $z \in \overline{A}$ so that $\overline{h}(z) = \overline{g}(z)$ for all $z \in \hat{\mathsf{p}}X \setminus \text{Œ}(k_0\Delta)$ by the claim in the previous paragraph. Hence

$$|\mathfrak{n}(h - g)| \leq \int_{\hat{\mathsf{p}}X} |\overline{h} - \overline{g}| \, \mathrm{d}|v| = \int_{\text{Œ}(k_0\Delta)} |\overline{h} - \overline{g}| \, \mathrm{d}|v| < \|\mathfrak{n}\|/2$$

$$\mathfrak{n}(h) \geq \mathfrak{n}(g) - |\mathfrak{n}(h - g)| > 3\|\mathfrak{n}\|/4 - \|\mathfrak{n}\|/2 = \|\mathfrak{n}\|/4$$

This demonstrates that every neighbourhood of h_0 in $\mathsf{BLip_b}(k\Delta)$ contains a function h such that $\mathfrak{n}(h) - \mathfrak{n}(h_0) > \|\mathfrak{n}\|/4$. Thus the restriction of $\mathfrak{n}$ to $\mathsf{BLip_b}(k\Delta)$ is not continuous at h_0.

Now, $\mathfrak{m} - \mathfrak{n} \in \mathfrak{M}_t(X)$; hence the restriction of $\mathfrak{m} - \mathfrak{n}$ to $\mathsf{BLip_b}(k\Delta)$ is continuous, and it follows that the restriction of $\mathfrak{m}$ to $\mathsf{BLip_b}(\Delta)$ is nowhere continuous. □

Theorem 12.7. *Let X be any uniform space and $\mathfrak{m} \in \mathfrak{M}_b(X)$. Assume that for every $\Delta \in \mathsf{UP}(X)$ there is $h \in \mathsf{BLip_b}(\Delta)$ for which the restriction of $\mathfrak{m}$ to $\mathsf{BLip_b}(\Delta)$ is continuous at h. Then $\mathfrak{m} \in \mathfrak{M}_u(X)$.*

Proof. Let $\varphi : X \to Z$ be a uniformly continuous mapping to a complete metric space Z with metric Δ'. For $k \in \omega$ we have

$$\mathsf{BLip_b}(\overset{\leftarrow}{\varphi}(k\Delta')) = \{h \circ \varphi \mid h \in \mathsf{BLip_b}(k\Delta')\}.$$

If the restriction of $\mathfrak{m}$ to $\mathsf{BLip_b}(\overset{\leftarrow}{\varphi}(k\Delta'))$ is continuous at $h \circ \varphi$, then the restriction of $\varphi(\mathfrak{m})$ to $\mathsf{BLip_b}(k\Delta')$ is continuous at h. Thus $\varphi(\mathfrak{m}) \in \mathfrak{M}_t(Z)$ by Lemma 12.6 and $\mathfrak{m} \in \mathfrak{M}_u(X)$ by Part 2 in Theorem 6.10. □

Theorem 12.8. *Let X be any uniform space, and let $\mathfrak{m}$ be a linear functional on $U_b(X)$ such that the restriction of $\mathfrak{m}$ to every $UEB(X)$ set is of Baire class 1 in the X-pointwise topology. Then $\mathfrak{m} \in \mathfrak{M}_u(X)$.*

This yields another proof of Theorem 6.19: The space $\mathfrak{M}_u(X)$ is $U_b(X)$-weakly sequentially complete.

Proof. First, note that $\mathfrak{m} \in \mathfrak{M}_b(X)$ for the same reason as in the proof of Theorem 12.4: If $\mathfrak{m}$ were not bounded on the $\|\cdot\|_X$ unit ball in $U_b(X)$, then there would exist functions $g_j \in U_b(X)$, $j \in \omega$, such that $\|g_j\|_X \leq 1/2^{j+1}$ and $|\mathfrak{m}(g_j)| \geq 1$ for all j. By Lemma 12.3 with $\varphi \colon X \to \ell_1$ defined by $\varphi(x)(j) := g_j(x)$ for $x \in X$, $j \in \omega$, we would get $\sum_j |\mathfrak{m}(g_j)| < \infty$, in contradiction to $|\mathfrak{m}(g_j)| \geq 1$.

By Theorem P.37, for every $\Delta \in UP(X)$ the restriction of $\mathfrak{m}$ to $\mathrm{BLip}_b(\Delta)$ is continuous at each element of a dense subset of $\mathrm{BLip}_b(\Delta)$. Apply Theorem 12.7. $\qquad\square$

12.3   Generalized Centres in Convolution Semigroups

When we expand our attention from continuous to measurable mappings, it is reasonable to introduce a modification of Definition 9.23 for the case of subsemigroups of $\mathfrak{M}_b(X)$ with the convolution operation and the $U_b(X)$-weak topology:

Definition 12.9. When X is a semiuniform semigroup and $S \subseteq \mathfrak{M}_b(X)$ is a semigroup for the $\star$ operation, let

$$\Lambda^{CBP}(S) := \{\mathfrak{m} \in S \mid \forall f \in U_b(X) \text{ the function } \mathfrak{n} \mapsto \mathfrak{m} \star \mathfrak{n}(f)$$

$$\text{is } U_b(X)\text{-weakly CBP-measurable on } S\}. \qquad\blacksquare$$

Obviously $\Lambda(S) \subseteq \Lambda^{CBP}(S)$. From Lemma 9.25 we get $S \cap \mathfrak{M}_u(X) \subseteq \Lambda^{CBP}(S)$ for every X and $\Lambda^{CBP}(S) = S = S \cap \mathfrak{M}_u(X)$ for every precompact X. In particular, $\Lambda^{CBP}(\mathfrak{M}_b(X)) = \mathfrak{M}_u(X)$ and $\Lambda^{CBP}(\hat{\mathfrak{p}}X) = \hat{X}$ for every precompact X.

Next I show, using the results in Sect. 12.1, that $\Lambda^{CBP}(\mathfrak{M}_b(X)) = \mathfrak{M}_u(X)$ and $\Lambda^{CBP}(\hat{\mathfrak{p}}X) = \hat{X}$ also for a class of semiuniform semigroups X that includes all locally compact groups.

Lemma 12.10. *Let X be a semiuniform semigroup, $\mathfrak{m} \in \mathfrak{M}_b(X)$ and $f \in U_b(X)$. If the restriction of the function $\mathfrak{n} \mapsto \mathfrak{m} \star \mathfrak{n}(f)$ to the compact space $\hat{\mathfrak{p}}X$ is CBP measurable, then the restriction of $\mathfrak{m}$ to $\overline{\mathrm{orb}}(f)$ is X-pointwise CBP measurable.*

Proof. This parallels the proof of Lemma 9.28. Let $\varphi(\mathfrak{n}) := \setminus_x\mathfrak{n}(\setminus_y f(xy))$ for $\mathfrak{n} \in \hat{\mathfrak{p}}X$. By Lemma 9.27, φ is continuous from $\hat{\mathfrak{p}}X$ to $\overline{\mathrm{orb}}(f)$ and $\varphi(\hat{\mathfrak{p}}X) = \overline{\mathrm{orb}}(f)$. Since $\mathfrak{m} \star \mathfrak{n}(f) = \mathfrak{m}(\setminus_x\mathfrak{n}(\setminus_y f(xy))) = \mathfrak{m}(\varphi(\mathfrak{n}))$, the function $\mathfrak{m} \circ \varphi$ is CBP measurable on $\hat{\mathfrak{p}}X$, and $\mathfrak{m}$ is CBP measurable on $\overline{\mathrm{orb}}(f)$ by Corollary P.39. $\qquad\square$

Theorem 12.11. *Let X be an ambitable semiuniform semigroup with the (ℓ_1) property and $S \subseteq \mathfrak{M}_b(X)$ a semigroup for the $\star$ operation such that $S \supseteq \widehat{p}X$. Then $\Lambda^{\mathsf{CBP}}(S) = \mathfrak{M}_u(X) \cap S$.*

Proof. This parallels the proof of Theorem 9.29. Clearly $\mathfrak{M}_u(X) \cap S \subseteq \Lambda^{\mathsf{CBP}}(S)$ by Lemma 9.25. To prove the opposite inclusion, take any $\mathfrak{m} \in \Lambda^{\mathsf{CBP}}(S)$ and $\Delta \in \mathsf{UP}(X)$. As X is ambitable, there is $f \in \mathsf{U}_b(X)$ such that $\mathsf{BLip}_b(\Delta) \subseteq \overline{\mathsf{orb}}(f)$. Since $\widehat{p}X \subseteq S$, the restriction of the function $\mathfrak{n} \mapsto \mathfrak{m} \star \mathfrak{n}(f)$ to $\widehat{p}X$ is CBP measurable. By Lemma 12.10, $\mathfrak{m}$ is X-pointwise CBP measurable when restricted to $\overline{\mathsf{orb}}(f)$, hence also when restricted to $\mathsf{BLip}_b(\Delta)$. Therefore $\mathfrak{m} \in \mathfrak{M}_u(X)$ by Theorem 12.4. $\qquad\square$

Corollary 12.12. *Let X be an ambitable semiuniform semigroup with the (ℓ_1) property. Then $\Lambda^{\mathsf{CBP}}(\widehat{p}X) = \widehat{X}$ and $\Lambda^{\mathsf{CBP}}(\mathfrak{M}_b(X)) = \mathfrak{M}_u(X)$.* $\qquad\square$

Corollary 12.13. *Let G be a locally compact group. Then $\Lambda^{\mathsf{CBP}}(\widehat{p}rG) = G$ and $\Lambda^{\mathsf{CBP}}(\mathfrak{M}_b(rG)) = \mathfrak{M}_t(rG)$.* $\qquad\square$

By the following exercise, the "Baire class 1 centre" of $\mathfrak{M}_b(X)$ is $\mathfrak{M}_u(X)$ whenever X is an ambitable semiuniform semigroup.

Exercise 12.14. Let X be an ambitable semiuniform semigroup. Let $\mathfrak{m} \in \mathfrak{M}_b(X)$ be such that for every $f \in \mathsf{U}_b(X)$ the function $\mathfrak{n} \mapsto \mathfrak{m} \star \mathfrak{n}(f)$ on $\mathfrak{M}_b(X)$ with the $\mathsf{U}_b(X)$-weak topology is of Baire class 1. Prove that $\mathfrak{m} \in \mathfrak{M}_u(X)$. $\qquad\blacksquare$

12.4 Notes for Chap. 12

Building on the prior work of Christensen [27], in a joint work [28] we used saturated spaces as a tool for proving continuity of invariant functionals. Corollary 12.5 is a version of [28, Th.2]. Its proof here is similar to applications of partitions of unity in topological measure theory [17], [167].

Continuity results about functionals of Baire class 1 in Sect. 12.2 are due to Schachermayer [163]. In [141], I proved the results of Sect. 12.3 for locally compact groups.

By a result of Glasner [82], if G is any countable discrete group and $\mathfrak{m}$ is an element of $\beta G = \widehat{p}rG$ for which the mapping $\mathfrak{n} \mapsto \mathfrak{m} \star \mathfrak{n}$ is Borel measurable on βG, then $\mathfrak{m} \in G$. Thus the "Borel-measurable centre" of βG is G. This follows from Corollary 12.13. However, Glasner's proof also yields a stronger statement (for countable discrete groups G):

$$G = \{\mathfrak{m} \in \beta G \mid \forall f \in \ell_\infty(G) \text{ the function } \mathfrak{n} \mapsto \mathfrak{m} \star \mathfrak{n}(f) \text{ from } \beta G \text{ to } \mathbb{R}$$

$$\text{is universally Radon measurable}\}.$$

Here, a function is said to be universally Radon measurable iff the preimage of every open set is measurable with respect to every tight Borel (i.e. Radon) measure.

Thus the "universally Radon measurable centre" of βG is G. In contrast, a parallel statement about the generalized centre of the space $\ell_\infty(G)^* = \mathfrak{M}_b(rG)$,

$$\ell_1(G) = \{\mathfrak{m} \in \ell_\infty(G)^* \mid \forall f \in \ell_\infty(G) \text{ the function } \mathfrak{n} \mapsto \mathfrak{m} \star \mathfrak{n}(f) \text{ from } \ell_\infty(G)^* \text{ to } \mathbb{R}$$

$$\text{is universally Radon measurable}\}$$

for all countable discrete groups G (or even for the additive group of integers), is neither provable nor disprovable in the ZFC set theory; this is an application of a circle of ideas recently completed by Larson [117]. The underlying theory of *medial limits* is covered by Fremlin [61, 538Q].

Research Problem 4. Do Corollaries 12.5 and 12.12 hold more generally, with the (ℓ_1) property omitted or replaced by a weaker property?

As I already mentioned in Sect. 4.4, infinite-dimensional normed spaces do not have the (ℓ_1) property by a theorem of Zahradník [183].

The problem might be easier with a smaller σ-algebra on $U_b(X)$ (therefore a stronger notion of measurability) and in particular with $\sigma(\mathfrak{M}_u(X))$ instead of the σ-algebra of $\mathfrak{M}_u(X)$-weakly CBP sets. By a general theorem of Edgar [46, 2.3], $\sigma(\mathfrak{M}_u(X))$ is the same as the σ-algebra generated by $\mathfrak{M}_u(X)$-weakly continuous functions on $U_b(X)$. In [28], we conjectured $\mathfrak{M}_t(X)$ to be saturated with respect to the σ-algebra $\sigma(\mathfrak{M}_t(X))$ on $U_b(X)$ for every complete separable metric space X. In view of Theorem 6.10, if the conjecture were true, then $\mathfrak{M}_u(X)$ would be saturated with respect to the σ-algebra $\sigma(\mathfrak{M}_u(X))$ for every uniform space X; then Corollaries 12.5 and 12.12 would hold with the (ℓ_1) property omitted and with $\sigma(\mathfrak{M}_u(X))$ measurability in place of CBP measurability.

Schachermayer's example [163] marks a direction in which Corollary 12.5 cannot be strengthened: For the space c_0 with the metric Δ defined by the sup norm, there is a linear functional $\mathfrak{n} \in \mathfrak{M}_b(c_0)$ whose restriction to $\mathrm{BLip}_b(\Delta)$ is Baire-property measurable and yet $\mathfrak{n}$ is not in $\mathfrak{M}_t(c_0) = \mathfrak{M}_u(c_0)$.   ∎

Hints to Exercises

Chapter 1

Exercise 1.27. Apply Theorem 1.26.

Chapter 2

Exercise 2.5. The topology of a locally convex space is generated by continuous seminorms.

Exercise 2.12. If X is a separable metric space, $\mathsf{U}(X) = \mathsf{U}_b(X)$, and X is not precompact, then $\mathsf{p}_1 X \neq \mathsf{c}X$.

Exercise 2.13. —

Exercise 2.15. —

Exercise 2.18. —

Exercise 2.23. $f(x) := x^2$ and $\Delta(x,y) := |f(x) - f(y)|$ for $x, y \in \mathbb{N}$.

Exercise 2.28. Use Theorem 2.6.

Exercise 2.29. Use Theorem 2.6.

Exercise 2.30. Use Theorem 2.6.

Exercise 2.31. Every uniformly continuous image of a Cauchy net is Cauchy. Use Exercise 2.30.

Exercise 2.37. —

Chapter 3

Exercise 3.2. —

Exercise 3.4. $\Delta(x,y) := |x^2 - y^2|$ for integers x, y.

J. Pachl, *Uniform Spaces and Measures*, Fields Institute Monographs 30,
DOI 10.1007/978-1-4614-5058-0,
© Springer Science+Business Media New York 2013

Exercise 3.6. To prove that $xy \sim xy'$ when $y \sim y'$, show that if $\Delta \in RP(S)$, then the pseudometric defined by $\Delta_x(y,z) := \Delta(xy,xz)$ for $y,z \in S$ belongs to $RP(S)$.

Exercise 3.11. Let G be the group G of bijective mappings from ω to itself with the ω-pointwise topology. The mapping $x \mapsto x^{-1}$ is not uniformly continuous from rG to rG.

Exercise 3.13. —

Exercise 3.15. Take any uniform space X such that $C_b(X) \neq U_b(X)$ with the semigroup operation $(x,y) \mapsto x$.

Exercise 3.16. —

Exercise 3.17. $\Delta'(x,y) := \Delta(x,y) \vee \sup\{\Delta(xz,yz) \mid z \in X\}$.

Exercise 3.23. To prove (ii)$\Leftrightarrow$(iii), use the homeomorphism $x \mapsto x^{-1}$.

Exercise 3.29. Apply Lemma 3.28, using the commutativity of addition and the homogeneity of the norm.

Exercise 3.36. Modify the construction in Sect. 3.4 to prove an analogue of Lemma 3.28. More details are in [139].

Chapter 4

Exercise 4.2. For $f,g \in U(X)^+$, we have $1/(1+f)(1+g) \in U_b(X)$ by Theorem 1.18; hence $(1+f)(1+g) \in U(X)$ if X is inversion-closed.

Exercise 4.3. Use the uniformity in Example 2.3.

Exercise 4.12. If a closed set $L \subseteq X$ is a limit of a net of singleton sets in HX, then Δ-diam$(L) = 0$ for every $\Delta \in UP(X)$.

Exercise 4.18. If a bounded metric Δ metrizes X, then Δ^H metrizes HX. For the second statement, use Theorem 4.13.

Exercise 4.19. If $\varepsilon > 0$, $D \subseteq X$ and $X = \bigcup_{x \in D} \odot[x, \Delta/\varepsilon]$, then for every non-empty $A \subseteq X$ there is $D_A \subseteq D$ such that $\Delta^H(A,D_A) \leq \varepsilon$. Thus the hyperspace of every precompact space is precompact. Apply Theorem 4.16.

Exercise 4.22. Index all the points of X as x_α where $\alpha < |X|$. Then define $f_\alpha(x) := \sup\{(1 - 2\Delta(x,x_\beta))^+ \mid \beta < \alpha\}$ for $x \in X$ and $\varphi_\alpha := f_{\alpha+1} - f_\alpha$.

Chapter 5

Exercise 5.6. Use Part 3 of Theorem P.21.

Exercise 5.15. —

Exercise 5.29. For a Cauchy sequence $\{x_j\}_j$ in X that does not converge, define $m(f) = \lim_j f(x_j)$, $f \in U_b(X)$.

Exercise 5.31. Take the point masses for a Cauchy sequence that does not converge.

Exercise 5.32. Take $\mathfrak{m}_j := \sum_{i=1}^{j}(\partial(0) - \partial(1/i^2))$, $j = 1, 2, \ldots$.

Exercise 5.33. Mimic the proofs of Lemma 5.23 and Corollary 5.24 for parts 1 and 2. For Part 3, use Theorem 5.28.

Exercise 5.35. Apply Theorem 5.28 with $\widehat{X}$ in place of X.

Exercise 5.44. Take the point masses for a Cauchy sequence that does not converge.

Chapter 6

Exercise 6.17. Take any compact space X that is not sequentially compact, and $\mathfrak{A} := \partial_X(X) \subseteq \mathfrak{M}_u(X)$.

Exercise 6.23. For $X_0 = X_1 = \{0, 1\}$, there are $\mathfrak{m}, \mathfrak{n} \in \mathrm{Mol}(X_0 \times X_1)^+$, $\mathfrak{m} \neq \mathfrak{n}$, such that $\pi_0(\mathfrak{m}) = \pi_0(\mathfrak{n})$ and $\pi_1(\mathfrak{m}) = \pi_1(\mathfrak{n})$. For example, $\mathfrak{m} := \partial((0,0)) + \partial((1,1))$ and $\mathfrak{n} := \partial((0,1)) + \partial((1,0))$.

Exercise 6.24. Find $\mathfrak{m} \in \mathfrak{M}_b(X_0 \times X_1) \setminus \mathfrak{M}_u(X_0 \times X_1)$ so that $\pi_i(\mathfrak{m}) = 0$ for $i = 0, 1$.

Exercise 6.30. Construct an $\mathfrak{M}_b$-projective system $\langle \omega, X_i, \varphi_{ij}, \mathfrak{m}_i \rangle$ with finite sets X_i and $\|\mathfrak{m}_i\| \geq i$ for $i \in \omega$.

Exercise 6.35. $\mathsf{U}_b(\mathsf{p}X) = \mathsf{U}_b(X)$ by Theorem 2.10; hence $\mathfrak{M}_b(\mathsf{p}X) = \mathfrak{M}_b(X)$.

Exercise 6.41. Prove property (ii) for molecular measures, and use approximation by molecular measures in Theorem 6.6.

Exercise 6.42. For any $\Delta \in \mathsf{UP}(X)$, let E be the Banach space $\mathsf{C}(T)$ with the norm $\|\cdot\|_T$, where T is $\mathsf{BLip}_b(\Delta)$ with the X-pointwise topology.

Chapter 7

Exercise 7.2. Use Lemma 6.9 and Theorem 6.37.

Exercise 7.16. $X \in \mathsf{Bo}(\widehat{\mathsf{p}}X)$ by Part 2 of Corollary 7.15. Apply Theorems 7.9 and 7.14.

Exercise 7.17. Define $\mu_0(A) := \inf\{\rho_m(A \cap \mathcal{E}(\Delta)) \mid \Delta \in \mathsf{UP}(X)\}$ for $A \in \mathsf{Bo}(\widehat{\mathsf{p}}X)$.

Exercise 7.23. $\widehat{X} \cap \mathfrak{M}_\tau(X) = X$.

Exercise 7.24. Combine Exercises 7.2 and 7.23.

Exercise 7.26. $\mathsf{U}(\mathsf{c}X) = \mathsf{U}(X)$; hence $\mathfrak{M}_b(\mathsf{c}X) = \mathfrak{M}_b(X)$, $\mathfrak{M}_\sigma(\mathsf{c}X) = \mathfrak{M}_\sigma(X)$ and $\widehat{\mathsf{p}}\mathsf{c}X = \widehat{\mathsf{p}}X$. From Theorems 7.21 and 7.25, we get $\mathfrak{M}_u(\mathsf{c}X) = \mathfrak{M}_u(X)$, and therefore $X = \mathfrak{M}_u(X) \cap \widehat{\mathsf{p}}X = \mathfrak{M}_u(\mathsf{c}X) \cap \widehat{\mathsf{p}}\mathsf{c}X$ (set equality).

Exercise 7.30. Use Theorem 7.9.

Exercise 7.32. $z \in \mathcal{E}^\infty(\overleftarrow{\varphi}\Delta)$ if and only if the limit $\lim_{x \to z} \varphi(x)$ exists in Z.

Exercise 7.34. For the positive cone, use Theorem 7.33 and Corollary 6.13. For spheres, use Lemma P.22 and Corollary 6.15.

Chapter 8

Exercise 8.7. The space of finite linear combinations of characteristic functions I_E, $E \in \Sigma$, is $\|\cdot\|_X$ dense in $\ell_\infty(S, \Sigma)$. For the case of a sequence of measures in Part 1 and for Part 3, use Theorem P.23.

Exercise 8.8. —

Exercise 8.10. Consider the image of the Lebesgue measure on $[0, 1]$ under the mapping $x \mapsto (x, -x)$ into the Sorgenfrey plane.

Exercise 8.12. Use the construction in Sect. 8.2 with the σ-algebra of countable and co-countable subsets in a set of cardinality $\aleph_1$.

Exercise 8.13. For a sequence in X converging to an irrational number, consider the corresponding sequence of point masses.

Chapter 9

Exercise 9.9. If compact sets $K \subseteq X$ and $K' \subseteq Y$ approximate $\mathfrak{m}$ and $\mathfrak{n}$, respectively, then the set $K \times K' \subseteq X^*Y$ approximates $\mathfrak{m} \otimes \mathfrak{n}$.

Exercise 9.19. Use Exercise 9.9.

Exercise 9.22. Apply Definitions 6.7 and 9.15 and Lemma 9.7.

Exercise 9.26. On an infinite set, define $xy = y$.

Exercise 9.34. For the additive group G of integers with the discrete uniformity, consider two elements in $\widehat{\mathfrak{pr}}G$: A cluster point of the sequence $\{\partial(j)\}_{j \in \omega}$ and a cluster point of the sequence $\{\partial(-j)\}_{j \in \omega}$.

Exercise 9.35. Use Exercise 9.22 with Theorems 3.7 and 3.9.

Exercise 9.37. Modify the proof of Part 1 of Theorem 9.12, using the uniform product $rG \times rG$ instead of the semiuniform product rG^*rG. Show that the modified functions g_γ converge to 0 uniformly.

Exercise 9.40. 1. If compact sets K_0 and K_1 approximate $\mathfrak{m}$ and $\mathfrak{m} \star \mathfrak{n}$, respectively, then the compact set $K_0^{-1}K_1$ approximates $\mathfrak{n}$.

2. Let G be a topological group for which there is a Cauchy sequence $\{x_j\}_j$ such that the set $\{x_j^{-1} \mid j \in \omega\}$ is uniformly discrete in rG. Let $\mathfrak{m}$ be the limit of $\partial(x_j)$ in $\widehat{rG}$, and let $\mathfrak{n}$ be a cluster point of $\partial(x_j^{-1})$ in $\widehat{\mathfrak{pr}}G$. For more details, see [138, 4.5].

Chapter 10

Exercise 10.3. For (iii)$\Rightarrow$(i) and (v)$\Rightarrow$(i), assume there are $f \in U(X)$ and $x_j \in X$, $j = 1, 2, \ldots$, such that $f(x_j) > j^2$. Define $\mathfrak{m} := \sum_j \partial(x_j)/j^2$ and $\mathfrak{m}_j := \partial(x_j)/j$.

Exercise 10.7. On the unit ball in an infinite-dimensional normed space, every uniformly continuous function is bounded.

Exercise 10.8. Mimic the proof of Lemma 5.11.

Exercise 10.12. Adapt the proof of part 2 of Theorem 10.11.

Exercise 10.13. Mimic the proofs of Lemma 10.10 and Theorem 10.11.

Exercise 10.15. Mimic the proof of Lemma 6.8.

Exercise 10.19. Mimic the proof of Theorem 6.19.

Exercise 10.20. For part 1, repeat the proof of Theorem 6.44 with the help of Theorem 10.18. For part 2, use Part 1 and Theorem 10.5.

Exercise 10.24. Mimic the proof of Theorem 10.23.

Exercise 10.29. Let $Y \subseteq X$ be the union of the supports of the measures on $\mathrm{Bo}(X)$ representing the functionals in $\mathfrak{A}$. Using Theorem 10.18, prove that $\overline{Y}$ is compact.

Exercise 10.34. For Part 1, note that $\mathfrak{M}_u(\mathbb{R}) = \mathfrak{M}_\sigma(\mathbb{R})$ and $\mathfrak{M}_u(X) = \mathfrak{M}_\sigma(X)$; cf. Exercise 7.26. For part 2, use the σ-algebra of countable and co-countable subsets in a set of cardinality $\aleph_1$.

Exercise 10.40. Apply Theorems 8.16 and 10.37 and the following observation: If $2^{\aleph_0}$ is measure-free and a cardinal κ is not, then κ is measurable by a $\{0, 1\}$-valued measure.

Exercise 10.41. A discrete space of cardinality $2^{\aleph_0}$.

Chapter 11

Exercise 11.10. For the points $x_j := j$ and $x'_j = j + (1/j)$, $j = 1, 2, \ldots$, there is a function $f \in C_b(\mathbb{R})$ such that $f(x_j) - f(x'_j) = 1$ for all j.

Chapter 12

Exercise 12.2. Let $\mathfrak{m}$ be a CBP measurable linear functional on $U_b(X)$ and $\{f_j\}_j$ a sequence in $U_b(X)$, $f_j \searrow 0$. As X is Alexandroff, the function $\sum_{j \in A}(f_j - f_{j+1})$ is in $U_b(X)$ for every $A \subseteq \omega$. Apply Theorem P.41 with $\psi(I_A) := \mathfrak{m}(\sum_{j \in A}(f_j - f_{j+1}))$.

Exercise 12.14. Proceed as in the proof of Lemma 12.10 and Theorem 12.11, using Lemma P.40 and Theorem 12.8 instead of Corollary P.39 and Theorem 12.4.

References

1. Aguayo-Garrido, J.: Weakly compact operators and the Dunford-Pettis property on uniform spaces. Ann. Math. Blaise Pascal **5**(2), 1–6 (1998)
2. Alexandroff, A.D.: Additive set-functions in abstract spaces. Rec. Math. [Mat. Sbornik] N. S. **8**(50), 307–348 (1940); **9**(51), 563–628 (1941); **13**(55), 169–238 (1943)
3. Arens, R.F., Eells Jr., J.: On embedding uniform and topological spaces. Pacific J. Math. **6**, 397–403 (1956)
4. Arkhangel′skiĭ, A.V.: Classes of topological groups. Uspekhi Mat. Nauk **36**(3 [219]), 127–146, 255 (1981). In Russian. English translation: Russian Math. Surveys 36, 3 (1981), 151–174
5. Badrikian, A.: Séminaire sur les fonctions aléatoires linéaires et les mesures cylindriques. Lecture Notes in Mathematics, vol. 139. Springer, Berlin (1970)
6. Bentley, H.L., Herrlich, H., Hušek, M.: The historical development of uniform, proximal, and nearness concepts in topology. In: Handbook of the History of General Topology, vol. 2, pp. 577–629. Kluwer Academic, Dordrecht/Boston/London (1998)
7. Berezanskiĭ, I.A.: Measures on uniform spaces and molecular measures. Trudy Moskov. Mat. Obšč. **19**, 3–40 (1968). In Russian. English translation: Trans. Moscow Math. Soc. 19 (1968), 1–40
8. Berglund, J.F., Junghenn, H.D., Milnes, P.: Analysis on Semigroups. Wiley, New York (1989)
9. Berruyer, J., Ivol, B.: Espaces de mesures et compactologies. Publ. Dép. Math. (Lyon) **9**(1), 1–35 (1972)
10. Billingsley, P.: The invariance principle for dependent random variables. Trans. Am. Math. Soc. **83**, 250–268 (1956)
11. Bogachev, V.I.: Measure Theory, vol. I, II. Springer, Berlin (2007)
12. Bourbaki, N.: Éléments de Mathématique. Topologie générale. Ch. I et II. Hermann & Cie., Paris (1940)
13. Bourbaki, N.: Éléments de Mathématique. Intégration. Ch. 7 et 8. Hermann, Paris (1963)
14. Bouziad, A., Troallic, J.P.: A precompactness test for topological groups in the manner of Grothendieck. Topol. Proc. **31**(1), 19–30 (2007)
15. Buchwalter, H.: Topologies et compactologies. Publ. Dép. Math. (Lyon) **6**(2), 1–74 (1969)
16. Buchwalter, H.: Fonctions continues et mesures sur un espace complètement régulier. In: Summer School on Topological Vector Spaces (Univ. Libre Bruxelles, Brussels, 1972). Lecture Notes in Mathematics, vol. 331, pp. 183–202. Springer, Berlin (1973)
17. Buchwalter, H.: Le rôle des partitions continues de l'unité dans la théorie des mesures scalaires ou vectorielles. In: Vector space measures and applications (Proc. Conf., Univ. Dublin, Dublin, 1977), vol. I. Lecture Notes in Mathematics, vol. 644, pp. 83–95. Springer, Berlin (1978)

18. Buchwalter, H., Pupier, R.: Complétion d'un espace uniforme et formes linéaires. C. R. Acad. Sci. Paris Sér. A–B **273**, A96–A98 (1971)
19. Budak, T., Işık, N., Pym, J.S.: Minimal determinants of topological centres for some algebras associated with locally compact groups. Bull. Lond. Math. Soc. **43**, 495–506 (2011)
20. Caby, E.: Convergence of measures on uniform spaces. Thesis, University of California, Berkeley (1976)
21. Caby, E.: The weak convergence of uniform measures. J. Multivariate Anal. **9**(1), 130–137 (1979)
22. Caby, E.: Extension of uniform measures. Proc. Am. Math. Soc. **89**(3), 433–439 (1983)
23. Caby, E.: A note on the convolution of probability measures. Proc. Am. Math. Soc. **99**(3), 549–554 (1987)
24. Čech, E.: Topological Spaces. Czechoslovak Academy of Sciences, Prague (1966)
25. Choquet, G.: Étude des spaces uniformes à partir de la notion d'écart. Enseignement Mathematique (2) **11**, 170–174 (1965)
26. Choquet, G.: Mesures coniques, affines et cylindriques. In: Symposia Mathematica, vol. II (INDAM, Rome, 1968), pp. 145–182. Academic, London (1969)
27. Christensen, J.P.R.: Topology and Borel structure. North-Holland Mathematics Studies, vol. 10. North-Holland, Amsterdam (1974)
28. Christensen, J.P.R., Pachl, J.: Measurable functionals on function spaces. Ann. Inst. Fourier (Grenoble) **31**(2), 137–152 (1981)
29. Cooper, J.B.: Saks spaces and applications to functional analysis. North-Holland Mathematics Studies, vol. 139, 2nd edn. North-Holland, Amsterdam (1987)
30. Cooper, J.B., Schachermayer, W.: Uniform measures and co-Saks spaces. In: Functional Analysis, Holomorphy, and Approximation Theory (Rio de Janeiro, 1978). Lecture Notes in Mathematics, vol. 843, pp. 217–246. Springer, Berlin (1981)
31. Császár, Á.: General Topology. Adam Hilger Ltd., Bristol (1978)
32. Csiszár, I.: On the weak* continuity of convolution in a convolution algebra over an arbitrary topological group. Studia Sci. Math. Hungar. **6**, 27–40 (1971)
33. Dales, H.G., Lau, A.T.M., Strauss, D.: Banach algebras on semigroups and on their compactifications. Mem. Am. Math. Soc. **205**(966) (2010)
34. D'Aristotile, A., Diaconis, P., Freedman, D.: On merging of probabilities. Sankhyā Ser. A **50**(3), 363–380 (1988)
35. Davydov, Y., Rotar, V.: On asymptotic proximity of distributions. J. Theor. Probab. **22**(1), 82–99 (2009)
36. Deaibes, A.: Espaces uniformes et espaces de mesures. Publ. Dép. Math. (Lyon) **12**(4), 1–166 (1975)
37. Deaibes, A.: Caractérisation des mesures qui intègrent toutes les fonctions réelles mesurables. Publ. Dép. Math. (Lyon) **15**(3), 75–80 (1978)
38. Deaibes, A.: Mesures sur les espaces uniformes de type (σ_1^∞). Publ. Dép. Math. (Lyon) **15**(3), 63–73 (1978)
39. Deaibes, A.: Mesures uniformes maximales. Comment. Math. Univ. Carolin. **21**(3), 551–562 (1980)
40. Deaibes, A., Pupier, R.: Sur la sommabilité de familles de fonctions uniformément continues. Comment. Math. Univ. Carolin. **18**(4), 741–753 (1977)
41. Dudley, R.M.: Convergence of Baire measures. Studia Math. **27**, 251–268 (1966); Correction: Studia Math. **51**, 275 (1974)
42. Dudley, R.M.: Distances of probability measures and random variables. Ann. Math. Statist. **39**, 1563–1572 (1968)
43. Dudley, R.M.: Probabilities and metrics. Matematisk Institut, Aarhus Universitet, Aarhus (1976). Lecture Notes Series No. 45
44. Dudley, R.M.: Real analysis and probability. Cambridge Studies in Advanced Mathematics, vol. 74. Cambridge University Press, Cambridge (2002); Revised reprint of the 1989 original
45. Dunford, N., Schwartz, J.T.: Linear Operators. I. General Theory. Interscience Publishers, New York (1958)

46. Edgar, G.A.: Measurability in a Banach space. Indiana Univ. Math. J. **26**(4), 663–677 (1977)
47. Engelking, R.: General Topology, 2nd edn. Heldermann Verlag, Berlin (1989)
48. Fabian, M., Habala, P., Hájek, P., Montesinos Santalucía, V., Pelant, J., Zizler, V.: Functional Analysis and Infinite-Dimensional Geometry. Springer, New York (2001)
49. Fedorova, V.P.: A dual characterization of the completion and completeness of a uniform space. Mat. Sb. (N.S.) **64**(106), 631–639 (1964); In Russian
50. Fedorova, V.P.: Linear functionals and Daniell integral on spaces of uniformly continuous functions. Mat. Sb. (N.S.) **74**(116), 191–201 (1967); In Russian. English translation: Math. USSR – Sbornik **3**, 177–185 (1967)
51. Fedorova, V.P.: Dual equivalents of ultracompleteness and paracompactness. Mat. Sb. (N.S.) **76** (118), 566–572 (1968); In Russian
52. Fedorova, V.P.: Daniell integrals on an ultracomplete uniform space. Mat. Zametki **16**, 601–610 (1974); In Russian. English translation: Math. Notes **16**, 950–955 (1974)
53. Fedorova, V.P.: Integral representations of functionals on spaces of uniformly continuous functions. Sibirsk. Mat. Zh. **23**(5), 205–218, 225 (1982); In Russian. English translation: Siber. Math. J. **23**, 753–762 (1983)
54. Fedorova, V.P.: Linear functionals on spaces of uniformly continuous functions, and abstract measures. Mat. Zametki **36**(5), 743–754, 799 (1984); In Russian. English translation: Math. Notes **36**(5), 872–877 (1984)
55. Ferri, S., Neufang, M.: On the topological centre of the algebra $LUC(G)^*$ for general topological groups. J. Funct. Anal. **244**(1), 154–171 (2007)
56. Fortet, R., Mourier, E.: Convergence de la répartition empirique vers la répartition théorique. Ann. Sci. Ecole Norm. Sup. (3) **70**, 267–285 (1953)
57. Fremlin, D.H.: Measure theory, vol. 1. The Irreducible Minimum, 2nd edn. Torres Fremlin, Colchester (2011)
58. Fremlin, D.H.: Measure theory, vol. 2. Broad Foundations, 2nd edn. Torres Fremlin, Colchester (2010)
59. Fremlin, D.H.: Measure theory, vol. 3. Measure Algebras. Torres Fremlin, Colchester (2004). Corrected second printing
60. Fremlin, D.H.: Measure theory, vol. 4. Topological Measure Spaces, Parts I, II. Torres Fremlin, Colchester (2006). Corrected second printing
61. Fremlin, D.H.: Measure theory, vol. 5. Set-Theoretic Measure Theory, Parts I, II. Torres Fremlin, Colchester (2008)
62. Fremlin, D.H.: Topological Spaces After Forcing (2011). (Version of 16.6.11). http://www.essex.ac.uk/maths/people/fremlin/n05622.pdf
63. Frolík, Z.: Mesures uniformes. C. R. Acad. Sci. Paris Sér. A–B **277**, A105–A108 (1973)
64. Frolík, Z.: Représentation de Riesz des mesures uniformes. C. R. Acad. Sci. Paris Sér. A–B **277**, A163–A166 (1973)
65. Frolík, Z.: Measurable uniform spaces. Pacific J. Math. **55**, 93–105 (1974)
66. Frolík, Z.: Uniform maps into normed spaces. Ann. Inst. Fourier (Grenoble) **24**(3), 43–55 (1974)
67. Frolík, Z. (ed.): Seminar Uniform Spaces 1973–1974. Mathematical Institute, Czechoslovak Academy of Sciences, Prague (1975)
68. Frolík, Z.: Three technical tools in uniform spaces. In: Seminar Uniform Spaces 1973–1974, pp. 3–26. Mathematical Institute, Czechoslovak Academy of Sciences, Prague (1975)
69. Frolík, Z.: Measure-fine uniform spaces I. In: Measure Theory (Oberwolfach, 1975). Lecture Notes in Mathematics, vol. 541, pp. 403–413. Springer, Berlin (1976)
70. Frolík, Z. (ed.): Seminar Uniform Spaces 1975–1976. Mathematical Institute, Czechoslovak Academy of Sciences, Prague (1976)
71. Frolík, Z.: Three uniformities associated with uniformly continuous functions. In: Symposia Mathematica, vol. XVII (INDAM, Rome, 1973), pp. 69–80. Academic, London (1976)
72. Frolík, Z.: Recent development of theory of uniform spaces. In: General Topology and Its Relations to Modern Analysis and Algebra, IV (Proc. Fourth Prague Topological Sympos., Part A, Prague, 1976). Lecture Notes in Mathematics, vol. 609, pp. 98–108. Springer, Berlin (1977)

73. Frolík, Z. (ed.): Seminar Uniform Spaces 1976–1977. Mathematical Institute, Czechoslovak Academy of Sciences, Prague (1978)
74. Frolík, Z.: Measure-fine uniform spaces II. In: Measure Theory (Oberwolfach, 1981). Lecture Notes in Mathematics, vol. 945, pp. 34–41. Springer, Berlin (1982)
75. Frolík, Z.: Existence of l_∞-partitions of unity. Rend. Sem. Mat. Univ. Politec. Torino **42**(1), 9–14 (1984)
76. Frolík, Z., Pachl, J., Zahradník, M.: Examples of uniform measures. In: Proceedings of the Conference on Topology and Measure I (Zinnowitz, 1974), pp. 139–152. Ernst-Moritz-Arndt University, Greifswald (1978)
77. Frolík, Z., Pelant, J., Vilímovský, J.: On hedgehog–topologically fine uniform spaces. In: Seminar Uniform Spaces 1975–1976, pp. 75–86. Mathematical Institute, Czechoslovak Academy of Sciences, Prague (1976)
78. Frolík, Z., Pelant, J., Vilímovský, J.: Extensions of uniformly continuous functions. Bull. Acad. Polon. Sci. Sér. Sci. Math. Astronom. Phys. **26**(2), 143–148 (1978)
79. Gel'fand, I.M., Vilenkin, N.Y.: Generalized Functions, vol. 4: Applications of Harmonic Analysis. Academic, New York (1964)
80. Gibbs, A.L., Su, E.S.: On choosing and bounding probability metrics. Internat. Statist. Rev. **70**(3), 419–435 (2002)
81. Gillman, L., Jerison, M.: Rings of continuous functions. Graduate Texts in Mathematics, vol. 43. Springer, New York (1976); Reprint of the 1960 edition [Van Nostrand]
82. Glasner, E.: On two problems concerning topological centers. Topol. Proc. **33**, 29–39 (2009)
83. Gnedenko, B.V., Kolmogorov, A.N.: Limit distributions for sums of independent random variables. Gosudarstv. Izdat. Tehn.-Teor. Lit., Moscow-Leningrad (1949); In Russian
84. Granirer, E.: On Baire measures on D-topological spaces. Fund. Math. **60**, 1–22 (1967)
85. Granirer, E.E., Leinert, M.: On some topologies which coincide on the unit sphere of the Fourier–Stieltjes algebra $B(G)$ and of the measure algebra $M(G)$. Rocky Mountain J. Math. **11**(3), 459–472 (1981)
86. Grothendieck, A.: Critères de compacité dans les espaces fonctionnels généraux. Am. J. Math. **74**, 168–186 (1952)
87. Guran, I.I.: On topological groups close to being Lindelöf. Dokl. Akad. Nauk SSSR **256**(6), 1305–1307 (1981); In Russian. English translation: Soviet Math. Dokl. **23**(1), 173–175 (1981)
88. Hager, A.W.: On inverse-closed subalgebras of $C(X)$. Proc. Lond. Math. Soc. (3) **19**, 233–257 (1969)
89. Hager, A.W.: Measurable uniform spaces. Fund. Math. **77**(1), 51–73 (1972)
90. Hager, A.W.: Some nearly fine uniform spaces. Proc. Lond. Math. Soc. (3) **28**, 517–546 (1974)
91. Hager, A.W.: Real-valued functions on Alexandroff (zero-set) spaces. Comment. Math. Univ. Carolinae **16**(4), 755–769 (1975)
92. Hager, A.W.: Cozero fields. Confer. Sem. Mat. Univ. Bari **175**, 1–23 (1980)
93. Hager, A.W., Reynolds, G.D., Rice, M.D.: Borel-complete topological spaces. Fund. Math. **75**(2), 135–143 (1972)
94. van Handel, R.: Uniform observability of hidden Markov models and filter stability for unstable signals. Ann. Appl. Probab. **19**(3), 1172–1199 (2009)
95. Haydon, R.: Sur les espaces $M(T)$ et $M^\infty(T)$. C. R. Acad. Sci. Paris Sér. A-B **275**, A989–A991 (1972)
96. Hewitt, E.: Linear functionals on spaces of continuous functions. Fund. Math. **37**, 161–189 (1950)
97. Hewitt, E., Ross, K.A.: Abstract Harmonic Analysis, vol. I, 2nd edn. Springer, Berlin (1979)
98. Hindman, N., Strauss, D.: Algebra in the Stone-Čech compactification. de Gruyter Expositions in Mathematics, vol. 27. Walter de Gruyter & Co., Berlin (1998)
99. Isbell, J.R.: On finite-dimensional uniform spaces. Pacific J. Math. **9**, 107–121 (1959)
100. Isbell, J.R.: Uniform spaces. Mathematical Surveys, No. 12. American Mathematical Society, Providence, RI (1964)

101. Isbell, J.R.: Insufficiency of the hyperspace. Proc. Cambridge Philos. Soc. **62**, 685–686 (1966)
102. Jech, T.: Set Theory, 3rd edn. Springer, Berlin (2003)
103. Kalton, N.J.: The Orlicz–Pettis theorem. In: Proceedings of the Conference on Integration, Topology, and Geometry in Linear Spaces (Univ. North Carolina, Chapel Hill, N.C., 1979). Contemprary Mathematics, vol. 2, pp. 91–100. American Mathematical Society, Providence, RI (1980)
104. Kalton, N.J.: Spaces of Lipschitz and Hölder functions and their applications. Collect. Math. **55**(2), 171–217 (2004)
105. Kantorovich, L.V., Rubinshteĭn, G.S.: On a space of completely additive functions. Vestnik Leningrad. Univ. **13**(7), 52–59 (1958); In Russian
106. Katětov, M.: On real-valued functions in topological spaces. Fund. Math. **38**, 85–91 (1951); Correction: Fund. Math. **40**, 203–205 (1953)
107. Katětov, M.: On a category of spaces. In: General Topology and its Relations to Modern Analysis and Algebra (Proc. Sympos., Prague, 1961), pp. 226–229. Academic, New York (1962)
108. Katětov, M.: On certain projectively generated continuity structures. In: Sierpiński, W., Kuratowski, K. (eds.) Simposio di Topologia (Università di Messina, 27–30 Aprile, 1964), vol. I, pp. 47–50. Edizioni Oderisi, Gubbio (1965); Correction: Comment. Math. Univ. Carolin. **6**, 251–255 (1965)
109. Kelley, J.L.: General topology. Graduate Texts in Mathematics, No. 27. Springer, New York (1975); Reprint of the 1955 edition [Van Nostrand]
110. Khurana, S.S.: Uniform measures on vector-valued functions. Publ. Math. Debrecen **55**(1–2), 73–82 (1999)
111. Khurana, S.S.: Product of uniform measures. Ric. Mat. **57**(2), 203–208 (2008)
112. Khurana, S.S., Colasante, M.L.: Vector-valued free uniform measures. Atti Sem. Mat. Fis. Univ. Modena **47**(2), 429–439 (1999)
113. Kirk, R.B.: Complete topologies on spaces of Baire measure. Trans. Am. Math. Soc. **184**, 1–29 (1973)
114. Krée, P.: Équations linéaires à coefficients aléatoires. In: Symposia Mathematica, vol. VII (INDAM, Rome, 1970), pp. 515–546. Academic, London (1971)
115. Krée, P.: Images de probabilités cylindriques par certaines applications non linéaires. Accouplement de processus linéaires. C. R. Acad. Sci. Paris Sér. A-B **274**, A342–A345 (1972)
116. Kuratowski, K.: Topology, vol. I, New edition. Academic, New York (1966)
117. Larson, P.B.: The filter dichotomy and medial limits. J. Math. Log. **9**(2), 159–165 (2009)
118. Lau, A.T.M.: Continuity of Arens multiplication on the dual space of bounded uniformly continuous functions on locally compact groups and topological semigroups. Math. Proc. Cambridge Philos. Soc. **99**(2), 273–283 (1986)
119. Lau, A.T.M.: Amenability of semigroups. In: The Analytical and Topological Theory of Semigroups, pp. 313–334. Walter de Gruyter, Berlin (1990)
120. Lau, A.T.M., Pym, J.: The topological centre of a compactification of a locally compact group. Math. Z. **219**(4), 567–579 (1995)
121. LeCam, L.: Convergence in distribution of stochastic processes. Univ. Calif. Publ. Statist. **2**, 207–236 (1957)
122. LeCam, L.: Note on a Certain Class of Measures (1970). Unpublished manuscript. http://www.stat.berkeley.edu/users/rice/LeCam/papers/classmeasures.pdf
123. LeCam, L.: Remarques Sur le Théorème Limite Central dans les Espaces Localement Convexes, pp. 233–249. Éditions Centre Nat. Recherche Sci., Paris (1970)
124. LeCam, L.: Some special results of measure theory. Technical Report 265, Department of Statistics, University of California, Berkeley (1990)
125. Lévy, P.: Calcul des Probabilités. Gauthier-Villars, Paris (1925)
126. Marxen, D.: Uniform semigroups. Math. Ann. **202**, 27–36 (1973)
127. McKennon, K.: Multipliers, positive functionals, positive-definite functions, and Fourier–Stieltjes transforms. Mem. Amer. Math. Soc. 111. American Mathematical Society, Providence, RI (1971)

128. Megrelishvili, M.G., Pestov, V.G., Uspenskij, V.V.: A note on the precompactness of weakly almost periodic groups. In: Nuclear Groups and Lie groups (Madrid, 1999). Res. Exp. Math., vol. 24, pp. 209–216. Heldermann, Lemgo (2001)

129. Megrelishvili, M.G.: Compactifications of semigroups and semigroup actions. Topology Proc. **31**, 611–650 (2007)

130. Namioka, I.: On certain actions of semi-groups on L-spaces. Studia Math. **29**, 63–77 (1967)

131. Neufang, M.: A unified approach to the topological centre problem for certain Banach algebras arising in abstract harmonic analysis. Arch. Math. (Basel) **82**(2), 164–171 (2004)

132. Neufang, M., Pachl, J., Salmi, P.: Uniformly equicontinuous sets, right multiplier topology and continuity of convolution. arXiv:1202.4350v1 (2012)

133. Neumann, B.H.: Groups covered by permutable subsets. J. Lond. Math. Soc. **29**, 236–248 (1954)

134. Pachl, J.: Free uniform measures. Comment. Math. Univ. Carolin. **15**, 541–553 (1974)

135. Pachl, J.: Free uniform measures on sub-inversion-closed spaces. Comment. Math. Univ. Carolin. **17**(2), 291–306 (1976)

136. Pachl, J.: Measures as functionals on uniformly continuous functions. Pacific J. Math. **82**(2), 515–521 (1979)

137. Pachl, J.: Uniform measures on topological groups. Compositio Math. **45**(3), 385–392 (1982)

138. Pachl, J.: Uniform measures and convolution on topological groups. arXiv:math/0608139v4 (2006)

139. Pachl, J.: Semiuniform semigroups and convolution. arXiv:0811.3576v2 (2008)

140. Pachl, J.: Ambitable topological groups. Topol. Appl. **156**(13), 2200–2208 (2009)

141. Pachl, J.: Measurable centres in convolution semigroups. Topol. Appl. **159**(10–11), 2649–2653 (2012)

142. Pelant, J.: Reflections not preserving completeness. In: Seminar Uniform Spaces 1973–1974, pp. 235–240. Math. Institute, Czechoslovak Academy of Sciences, Prague (1975)

143. Pestov, V.: Dynamics of infinite-dimensional groups. University Lecture Series, vol. 40. American Mathematical Society, Providence, RI (2006)

144. Pol, R.: Remark on the restricted Baire property in compact spaces. Bull. Acad. Polon. Sci. Sér. Sci. Math. Astronom. Phys. **24**(8), 599–603 (1976)

145. Prokhorov, Yu.V.: Convergence of random processes and limit theorems in probability theory. Teor. Veroyatnost. i Primenen. **1**, 177–238 (1956); In Russian. English translation: Theor. Probab. Appl. **1**, 3 (1956), 157–214

146. Protasov, I.: Combinatorics of numbers. Mathematical Studies Monograph Series, vol. 2. VNTL, L'viv (1997)

147. Pták, V.: An extension theorem for separately continuous functions and its application to functional analysis. Czechoslovak Math. J. **14**(89), 562–581 (1964)

148. Pták, V.: Algebraic extensions of topological spaces. In: Contributions to Extension Theory of Topological Structures (Proc. Sympos., Berlin, 1967), pp. 179–188. Deutsch. Verlag Wissensch., Berlin (1969)

149. Pym, J.S.: The convolution of linear functionals. Proc. Lond. Math. Soc. (3) **14**, 431–444 (1964)

150. Pym, J.S.: The convolution of functionals on spaces of bounded functions. Proc. Lond. Math. Soc. (3) **15**, 84–104 (1965)

151. Rachev, S.T.: Probability Metrics and the Stability of Stochastic Models. Wiley, Chichester (1991)

152. Rachev, S.T., Rüschendorf, L.: Mass Transportation Problems, vol. I. Springer, New York (1998)

153. Raĭkov, D.A.: Free locally convex spaces for uniform spaces. Mat. Sb. (N.S.) **63**(105), 582–590 (1964); In Russian

154. Raĭkov, D.A.: Duality method in the theory of uniform spaces. In: Proceedings of the Fourth All-Union Topology Conf. (Tashkent, 1963), pp. 155–162. FAN, Tashkent (1967); In Russian

155. Rice, M.D.: Subcategories of uniform spaces. Trans. Am. Math. Soc. **201**, 305–314 (1975)

156. Rice, M.D.: Composition properties in uniform spaces. Acta Math. Acad. Sci. Hungar. **30**(3–4), 189–195 (1977)

157. Rice, M.D.: Uniform ideas in analysis. Real Anal. Exchange **6**(2), 139–185 (1980/81)

158. Riesz, F.: Sur les opérations fonctionnelles linéaires. C. R. Math. Acad. Sci. Paris **149**, 974–977 (1909)

159. Roelcke, W., Dierolf, S.: Uniform Structures on Topological Groups and their Quotients. McGraw-Hill, New York (1981)

160. Rogers, C.A., Jayne, J.E.: K-analytic sets. In: Analytic Sets (London Math. Soc. Conf., University College London, July 1978), pp. 1–181. Academic, New York (1980)

161. Ruppert, W.: Compact semitopological semigroups: an intrinsic theory. Lecture Notes in Mathematics, vol. 1079. Springer, Berlin (1984)

162. Salmi, P.: Joint continuity of multiplication on the dual of the left uniformly continuous functions. Semigroup Forum **80**(1), 155–163 (2010)

163. Schachermayer, W.: Measurable and continuous linear functionals on spaces of uniformly continuous functions. In: Measure Theory (Oberwolfach, 1981). Lecture Notes in Mathematics, vol. 945, pp. 155–166. Springer, Berlin (1982)

164. Schaefer, H.H.: Topological Vector Spaces. Springer, New York (1971); Third printing corrected

165. Schwartz, L.: Radon measures on arbitrary topological spaces and cylindrical measures. Oxford University Press, London (1973). Tata Institute of Fundamental Research Studies in Mathematics, No. 6

166. Semadeni, Z.: Banach spaces of continuous functions, vol. I. PWN—Polish Scientific Publishers, Warsaw (1971); Monografie Matematyczne, Tom 55

167. Sentilles, D., Wheeler, R.F.: Linear functionals and partitions of unity in $C_b(X)$. Duke Math. J. **41**, 483–496 (1974)

168. Tomášek, S.: On a certain class of Λ-structures. I, II. Czechoslovak Math. J. **20**(95), 1–18, 19–33 (1970)

169. Topsøe, F.: Topology and measure. Lecture Notes in Mathematics, vol. 133. Springer, Berlin (1970)

170. Tortrat, A.: Sur la continuité de l'opération convolution dans un demi-groupe topologique X. C. R. Acad. Sci. Paris Sér. A–B **272**, A588–A591 (1971)

171. Uspenskij, V.V.: Compactifications of topological groups. In: Proceedings of the Ninth Prague Topological Symposium (2001), pp. 331–346 (electronic). Topol. Atlas, North Bay, Ontario (2002)

172. Uspenskij, V.V.: On subgroups of minimal topological groups. Topol. Appl. **155**(14), 1580–1606 (2008)

173. Varadarajan, V.S.: Measures on topological spaces. Mat. Sb. (N.S.) **55**(97), 35–100 (1961); In Russian. English translation: Amer. Math. Soc. Transl. (2) **48**, 161–228 (1965)

174. Villani, C.: Topics in optimal transportation. Graduate Studies in Mathematics, vol. 58. American Mathematical Society, Providence, RI (2003)

175. Villani, C.: Optimal transport. Grundlehren der Mathematischen Wissenschaften [Fundamental Principles of Mathematical Sciences], vol. 338. Springer, Berlin (2009)

176. de Vries, J.: Elements of topological dynamics. Mathematics and Its Applications, vol. 257. Kluwer Academic, Dordrecht (1993)

177. Waelbroeck, L.: Some theorems about bounded structures. J. Funct. Anal. **1**, 392–408 (1967)

178. Weaver, N.: Lipschitz Algebras. World Scientific, River Edge, NJ (1999)

179. Weil, A.: Sur les espaces à structure uniforme et sur la topologie générale. Hermann & Cie, Paris (1937)

180. Wheeler, R.F.: A survey of Baire measures and strict topologies. Expo. Math. **1**(2), 97–190 (1983)

181. Zahradník, M.: Projective limits of uniform measures. Thesis, Charles University, Prague (1974); In Czech

182. Zahradník, M.: Inversion closed uniform spaces have the Daniell property. In: Seminar Uniform Spaces 1973–1974, pp. 233–234. Math. Institute, Czechoslovak Academy of Sciences, Prague (1975)
183. Zahradník, M.: l_1-continuous partitions of unity on normed spaces. Czechoslovak Math. J. **26(101)**(2), 319–329 (1976)
184. Zolotarev, V.M.: Modern Theory of Summation of Random Variables. VSP, Utrecht (1997)

Notation Index

J. Pachl, *Uniform Spaces and Measures*, Fields Institute Monographs 30,
DOI 10.1007/978-1-4614-5058-0,
© Springer Science+Business Media New York 2013

Author Index

A

Aguayo–Garrido, J., 98
Alexandroff, A.D., 78, 79, 114, 115
Arens, R.F., 79, 97, 168, 169
Arkhangel′skiĭ, A.V., 50

B

Badrikian, A., 127, 145, 169
Bentley, H.L., 28
Berezanskiĭ, I.A., 79, 97, 98, 115, 169
Berglund, J.F., 50, 146
Berruyer, J., 169
Billingsley, P., 78, 79
Bogachev, V.I., 78, 79, 116, 127
Bourbaki, N., 28, 98, 145
Bouziad, A., 50
Buchwalter, H., 97, 98, 127, 169, 184
Budak, T., 10, 146

C

Caby, E., 97, 98, 145, 146
Čech, E., 28
Choquet, G., 28, 127, 169
Christensen, J.P.R., 16, 60, 184, 185
Colasante, M.L., 169
Cooper, J.B., 79, 97, 98
Csiszár, I., 97, 145, 146
Császár, Á., 2, 4, 15, 28, 39, 60, 98, 146

D

D'Aristotile, A., 177
Dales, H.G., 51, 147
Davydov, Y., 79, 177
de Vries, J., 50, 146

Deaibes, A., 60, 97, 127, 169
Diaconis, P., 177
Dierolf, S., 50, 146
Dudley, R.M., 15, 79, 127, 177
Dunford, N., 11, 127

E

Edgar, G.A., 185
Eells, Jr., J., 79, 97, 168, 169
Engelking, R., 2, 28, 60

F

Fabian, M., 5, 7
Fedorova, V.P., 60, 79, 97, 98, 115, 126, 169
Ferri, S., 50, 51, 126, 146
Fortet, R., 79
Freedman, D., 177
Fremlin, D.H., 8–12, 14–16, 116, 127, 185
Frolík, Z., 39, 60, 97, 98, 115, 127

G

Gel′fand, I.M., 127
Gibbs, A.L., 177
Gillman, L., 15, 28, 39, 60, 98, 115
Glasner, E., 184
Gnedenko, B.V., 78
Granirer, E.E., 79, 127
Grothendieck, A., 79
Guran, I.I., 50

H

Habala, P., 5, 7
Hager, A.W., 39, 60, 169

J. Pachl, *Uniform Spaces and Measures*, Fields Institute Monographs 30,
DOI 10.1007/978-1-4614-5058-0,
© Springer Science+Business Media New York 2013

Subject Index

A
Alexandroff space, 53
almost all, 2
ambitable semigroup, 45
Ascoli theorem, 7
associated metric space, 3
asymptotic approximation, 174

B
Baire σ-algebra, 100
Baire class 1 function, 15
Baire measure, 100
Baire property, 15
Banach–Steinhaus theorem, 7
band, 9
base of topology, 3
Borel measurable, 4, 15
Borel measure
   τ-additive, 12
   of order 1, 164
   tight, 13
Borel set, σ-algebra, 4
bounded set, 6

C
cardinal reflection, 24
cardinal successor, 1
Cauchy net, 6, 24
CBP measurable function, 16
CBP set, 16
CDE property, 56
characteristic function, 1
coarser uniformity (uniform space),
   22
cofinality, 1

compactification
   Čech–Stone, 3
   $\mathscr{LC}$-, $\mathscr{LUC}$-, 146
   uniform, Samuel, 93, 94
compatible uniformity, 21
complete space, 6, 24
completion, 93, 94
convolution, 135
countably compact space, 4
cozero set, 53
cozero-continuous function, 53
cylindrical measure, 126
   of type 1, 167

D
Daniell representation, 11
Δ-density, 46
diameter, 3
direct product of functionals, 130
distribution, 15, 172
dual space, 5

E
envelope
   open, 104
   reduced, 104
expected value, 15

F
fine uniformity (uniform space), 28
finer uniformity (uniform space), 22
finitely additive function, 16
free uniform measure, 151
   vector-valued, 159

J. Pachl, *Uniform Spaces and Measures*, Fields Institute Monographs 30,
DOI 10.1007/978-1-4614-5058-0,
© Springer Science+Business Media New York 2013